LOGIC, FOUNDATIONS OF MATHEMATICS,
AND COMPUTABILITY THEORY

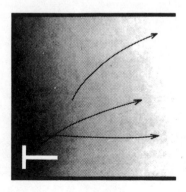

THE UNIVERSITY OF WESTERN ONTARIO
SERIES IN PHILOSOPHY OF SCIENCE

VOLUME 9

LOGIC, FOUNDATIONS OF MATHEMATICS, AND COMPUTABILITY THEORY

PART ONE OF THE PROCEEDINGS
OF THE FIFTH INTERNATIONAL CONGRESS OF
LOGIC, METHODOLOGY AND PHILOSOPHY OF SCIENCE,
LONDON, ONTARIO, CANADA–1975

Edited by

ROBERT E. BUTTS
The University of Western Ontario

and

JAAKKO HINTIKKA
The Academy of Finland and Stanford University

D. REIDEL PUBLISHING COMPANY
DORDRECHT-HOLLAND / BOSTON-U.S.A.

Library of Congress Cataloging in Publication Data

International Congress of Logic, Methodology, and Philosophy of
    Science, 5th, University of Western Ontario, 1975.
    Logic, foundations of mathematics, and computability theory.

    (Proceedings of the Fifth International Congress of Logic,
Methodology, and Philosophy of Science, London, Ontario, Canada,
1975 ; pt. 1) (University of Western Ontario series in philosophy of
science ; v. 9)
    Bibliography: p.
    Includes index.
    1.   Logic, Symbolic and mathematical–Congresses.   2.   Ma-
thematics–Philosophy–Congresses.   3.   Computable functions–Congresses.   I.   Butts, Robert E.   II.   Hintikka, Kaarlo Jaakko Juhani,
1929–   III.   Title.   IV.   Series: University of Western Ontario. The
University of Western Ontario series in philosophy of science ; v. 9.
Q 174.158   1975a pt. 1   [QA9.A1]   501s   [511′.3]   77-22429
ISBN 90-277-0708-1

The set of four volumes (cloth) ISBN 90 277 0706 5

Published by D. Reidel Publishing Company,
P.O. Box 17, Dordrecht, Holland

Sold and distributed in the U.S.A., Canada, and Mexico
by D. Reidel Publishing Company, Inc.,
Lincoln Building, 160 Old Derby Street, Hingham,
Mass. 02043, U.S.A.

Printed in The Netherlands

TABLE OF CONTENTS

PREFACE

The Fifth International Congress of Logic, Methodology and Philosophy of Science was held at the University of Western Ontario, London, Canada, 27 August to 2 September 1975. The Congress was held under the auspices of the International Union of History and Philosophy of Science, Division of Logic, Methodology and Philosophy of Science, and was sponsored by the National Research Council of Canada and the University of Western Ontario. As those associated closely with the work of the Division over the years know well, the work undertaken by its members varies greatly and spans a number of fields not always obviously related. In addition, the volume of work done by first rate scholars and scientists in the various fields of the Division has risen enormously. For these and related reasons it seemed to the editors chosen by the Divisional officers that the usual format of publishing the proceedings of the Congress be abandoned in favour of a somewhat more flexible, and hopefully acceptable, method of presentation.

Accordingly, the work of the invited participants to the Congress has been divided into four volumes appearing in the University of Western Ontario Series in Philosophy of Science. The volumes are entitled, *Logic, Foundations of Mathematics and Computability Theory, Foundational Problems in the Special Sciences, Basic Problems in Methodology and Linguistics,* and *Historical and Philosophical Dimensions of Logic, Methodology and Philosophy of Science.* By means of minor rearrangement of papers in and out of the sections in which they were originally presented the editors hope to have achieved four relatively self-contained volumes.

The papers in this volume consist of all those submitted for publication by invited participants in the fields of mathematical logic, foundations of mathematical theories, computability theory, and the philosophy of logic and mathematics. Contributed papers in these fields appeared in the volume of photo-offset preprints distributed at the Congress. The full programme of the Congress appears in *Historical and Philosophical Dimensions of Logic, Methodology and Philosophy of Science.*

The work of the members of the Division was richly supported by the National Research Council of Canada and the University of Western Ontario. We here thank these two important Canadian institutions. We also thank the Secretary of State Department of the Government of Canada, Canadian Pacific Air, the Bank of Montreal, the *London Free Press,* and I.B.M. Canada for their generous support. Appended to this preface is a list of officers and those responsible for planning the programme and organizing the Congress.

THE EDITORS

February 1977

OFFICERS OF THE DIVISION

A. J. Mostowski	(Poland)	President
Jaakko Hintikka	(Finland)	Vice President
Sir A. J. Ayer	(U.K.)	Vice President
N. Rescher	(U.S.A.)	Secretary
J. F. Staal	(U.S.A.)	Treasurer
S. Körner	(U.K.)	Past President

PROGRAMME COMMITTEE

Jaakko Hintikka (Finland), Chairman
R. E. Butts (Canada)
Brian Ellis (Australia)
Solomon Feferman (U.S.A.)
Adolf Grünbaum (U.S.A.)
M. V. Popovich (U.S.S.R.)
Michael Rabin (Israel)

Evandro Agazzi (Italy)
Bruno de Finetti (Italy)
Wilhelm Essler (B.R.D.)
Dagfinn Føllesdal (Norway)
Rom Harré (U.K.)
Marian Przełęcki (Poland)
Dana Scott (U.K.)

CHAIRMEN OF SECTIONAL COMMITTEES

Y. L. Ershov (U.S.S.R.)	Section I:	Mathematical Logic
Donald A. Martin (U.S.A.)	Section II:	Foundations of Mathematical Theories
Helena Rasiowa (Poland)	Section III:	Computability Theory
Dagfinn Føllesdal (Norway)	Section IV:	Philosophy of Logic and Mathematics
Marian Przełęcki (Poland)	Section V:	General Methodology of Science
J.-E. Fenstad (Norway)	Section VI:	Foundations of Probability and Induction
C. A. Hooker (Canada)	Section VII:	Foundations of Physical Sciences

Lars Walløe   Section VIII: Foundations of Biology
  (Norway)

Brian Farrell   Section IX: Foundations of Psychology
  (U.K.)

J. J. Leach   Section X: Foundations of Social Sciences
  (Canada)

Barbara Hall Partee Section XI: Foundations of Linguistics
  (U.S.A.)

R. E. Butts   Section XII: History of Logic, Methodology
  (Canada)   and Philosophy of Science

LOCAL ORGANIZING COMMITTEE

R. E. Butts (Philosophy, the University of Western Ontario), Chairman

For the University of Western Ontario:

Maxine Abrams (Administrative Assistant)
R. N. Sharvill (Executive Assistant to the President)
G. S. Rose (Assistant Dean of Arts)
R. W. Binkley (Philosophy)
J. J. Leach (Philosophy)
C. A. Hooker (Philosophy)
J. M. Nicholas (Philosophy)
G. A. Pearce (Philosophy)
W. R. Wightman (Geography)
J. D. Talman (Applied Mathematics)
J. M. McArthur (Conference Co-ordinator)

For the City of London:

Betty Hales (Conference Co-ordinator)

For the National Research Council of Canada:

R. Dolan (Executive Secretary)

I

MATHEMATICAL LOGIC

YU. L. ERSHOV

CONSTRUCTIONS 'BY FINITE'

My talk's aim is a description and some grounds for one natural 'good' model $\mathbb{C}$ of finite type functionals over the natural numbers N. In this respect the talk is like D. Scott's report [10] on natural model for type-free λ-calculus at the last Congress.

There are too many distinct models in the mathematics and mathematical logic. So there is a problem: how to distinguish the most important or the most interesting one. For some of them it is not a problem. The natural number set is one example.

The functionals of finite types are an important notion in mathematics and mathematical logic now.

The types are defined by induction as follows:

(i)   0   is type $(0 \in T)$;
(ii)   if σ_0, σ_1 are types $(\sigma_0, \sigma_1 \in T)$ then
            (σ_0, σ_1) is type;
            $(\sigma_0 \mid \sigma_1)$ is type.

The underlying sense of the functionals of the type τ is: the functionals of the type 0 are natural numbers; the functionals of the type (σ_0, σ_1) are the pairs (f, g) of functionals of the types σ_0 and σ_1; the functionals of the type $(\sigma_0 \mid \sigma_1)$ are some functions (maps) from the functionals of the type σ_0 to the functionals of the type σ_1.

The partial functionals are described likely.

One of the first conditions for a model of functionals is to be a λ-model. This condition implies, in essence, the 'identification' of the functionals of types $((\sigma_0, \sigma_1) \mid \sigma_2)$ and $(\sigma_0 \mid (\sigma_1 \mid \sigma_2))$. A general scheme of definitions for such models can be described in terms of category theory (the useful language of modern mathematics) as follows.

One must:

(a) Define a useful Cartesian closed category $\mathbb{K}$;

(b) Define an object $K_0 \in Ob\,\mathbb{K}$–that is a 'model' for the natural numbers;

Butts and Hintikka (eds.), Logic, Foundations of Mathematics and Computability Theory, 3–9.

(c) Define by induction for every finite type τ an object K_τ such that

$$K_{(\sigma_0,\sigma_1)} = K_{\sigma_0} \times K_{\sigma_1}$$

and

$$K_{(\sigma_0 \mid \sigma_1)} = K_{\sigma_1}^{K_{\sigma_0}},$$

where $\times$ is the operation of the direct product and c^f is an object $\mathbb{K}$ such that there is a natural equivalence between the sets $Mor_\mathbb{K}(a \times b, c)$ and $Mor_\mathbb{K}(a, c^b)$ for every $a \in Ob\mathbb{K}$, where $b, c \in Ob\mathbb{K}$. The existence of the operation $\times$ and objects c^b for every $c, b \in Ob\mathbb{K}$ is the essence of the notion of Cartesian closed category.

The 'simplest' (in some sense) Cartesian closed category is the category **Set** of all sets. The corresponding model of the functionals – the 'complete' model is, of course, an important and interesting one. But this model is too big and non-constructive. Given below are the two approaches to the definitions of the other models. These approaches give the same class $\mathbb{C}$ of the functionals. The approaches are distinct and the sources of them are very general ideas. The coincidence of the corresponding models, a clear structure of the model and some other reasons (some of them are indicated below) serve as a foundation for an interest and importance of the model $\mathbb{C}$.

1. The first approach is using (and extending) one of the leading ideas in the development (of foundations) of mathematics; the construction of all 'ideal' structures of mathematics by 'finite' objects. The approach is a source of the title of my talk. For more creative use of the idea the notion of 'finite' must be taken as a primitive undefinable notion. This idea of 'generalized' finite can be clarified by the following philosophical analogy. One of the elementary fragments of the human knowledge consists of the two consecutive steps: an analysis and (following) synthesis. In more details: an object under research is analyzed by some definite set of the analysis tools. The obtained data are then generalized and synthesized in some 'ideal' reconstruction of the object under research. So, these very data obtained by analysis are just 'finite' and used for constructing the 'ideal' (infinite?!) synthetic objects.

One of the simplest mathematical realizations of the approach is the theory of f_0-spaces [1] (or more general A_0-spaces [2]).

Now I give some exact definitions.

Let $\langle X, X_0, \leqslant \rangle$ be a partially ordered set $\langle X, \leqslant \rangle$ (one of the senses of the relation $x \leqslant y$ is 'x is appoximation (subset, part) of y'); $X_0 \subseteq X$ is the set of the 'finite' objects (elements) of X. These are the data of some (fixed) analysis. Require for any $x \in X$ to have rather many finite approximations:

(1)   if $x \not\leqslant y$, then there is $x_0 \in X_0$ such that $x_0 \leqslant x$ but $x_0 \not\leqslant y$.

A natural 'completeness' of the analysis may be formulated as:

(2)   if $x_0, x_1 \in X_0$, $x \in X$ and x_0 and x_1 are the approximations of x, then

$$\exists x_2 \in X_0 \, \forall y \in X (x_0 \leqslant y \ \& \ x_1 \leqslant y \Leftrightarrow x_2 \leqslant y)$$

The following condition is a simple technical one on the existence of a trivial ('finite', of course) approximation:

(3) $\exists x_0 \in X_0 \, \forall y \in X (x_0 \leqslant y)$

Using another mathematical language – the language of topology – we can state the following:

If $\langle X, X_0, \leqslant \rangle$ satisfies the conditions (1)–(3), then the set of a subset of the form

$$\check{x}_0 = \{x \mid x \in X, x_0 \leqslant x\}, \qquad x_0 \in X_0$$

is a basis for some topology on X. On the contrary, from this topology we can find order $\leqslant$ and the set X_0. Below I shall speak only of such topologies. Such topological spaces are called f_0-spaces.

If X, Y are f_0-spaces, then the set $C(X, Y)$ of all continuous maps from X to Y has a natural topology. And this topological space $C(X, Y)$ is also f_0-space.

PROPOSITION. *The category* $\mathbb{F}_0$ *of all* f_0-*spaces is the Cartesian closed one.*

In particular, there is a natural homeomorphism of the spaces $C(X \times Y, Z)$ and $C(X, C(Y, Z))$ for any f_0-spaces X, Y, Z.

Now we define the class $\mathbb{C} = \{C_\sigma \mid \sigma \in T\}$ as above, putting for $\sigma = 0$, $C_0 = N^*$, where N^* is the set $\{S \mid S \subseteq N, |S| \leq 1\}$ i.e. $N^* = \{\varnothing, \{0\}, \{1\}, \ldots\}$; $N_0^* = N^*$, and the order $\leq$ on N^* is inclusion.

In the category $\mathbb{F}_0$ a notion of a complete f_0-space is defined and the subcategory $C\mathbb{F}_0$ of complete f_0-spaces is a reflective subcategory of $\mathbb{F}_0$.

Technically speaking: the functor of including $I: C\mathbb{F}_0 \to \mathbb{F}_0$ has a left adjoint functor, the functor of completing $C^*: \mathbb{F}_0 \to C\mathbb{F}_0$. All the spaces from $\mathbb{C}$ are complete.

2. The second approach to definition of $\mathbb{C}$ is based on a quite different idea of the definition of a partial calculable functionals of the finite types.

The main idea of the definition of the partial calculable functionals is an inductive definition of the classes of functionals together with the 'right' (principal, Gödel's) enumeration of them. So, the objects under consideration are now the enumerated sets.

Give some exact definitions of the enumeration theory.

An enumeration of the set S is a map $\nu: N \xrightarrow{\text{on}} S$. The pair $\gamma = (S, \nu)$, where ν is the enumeration S, is called an enumerated set. Morphism from the enumerated set $\gamma_0 = (S_0, \nu_0)$ to enumerated set $\gamma_1 = (S_1, \nu_1)$ is any map $\mu: S_0 \to S_1$ for which there is a general recursive function f such that $\mu\nu_0 = \nu_1 f$. The set of all morphisms from γ_0 to γ_1 is designated as $Mor(\gamma_0, \gamma_1)$.

The question is: how to find the 'right' enumeration of the set $Mor(\gamma_0, \gamma_1)$?

First we define when a map $\varphi: N \to Mor(\gamma_0, \gamma_1)$ is calculable. φ is calculable if a map $\varphi^*: N \times S_0 \to S_1$ naturally defined by $\varphi(\varphi^*(n, s_0) = [\varphi(n)](s_0))$ is a morphism from $N \gamma_0$ to γ_1, where N is (N, id_N). The direct product operation exists in the category $\mathfrak{N}$ of enumerated sets. In other words, φ is calculable if there is a general recursive function g such that for any $m, n \in N$ $[\varphi(n)](\nu_0 m) = \nu_1 g(n, m)$. If the set of the calculable enumerations of the set $Mor(\gamma_0, \gamma_1)$ is not empty we want to find a principal enumeration φ among them. It means that any calculable enumeration ψ is reducible to φ, i.e. there exists a general recursive function h such that $\psi = \varphi h$. If the principal enumeration φ of $Mor(\gamma_0, \gamma_1)$ exists, then it is a decision of a problem of the 'right' enumeration of $Mor(\gamma_0, \gamma_1)$. In this case the enumerated set $(Mor(\gamma_0, \gamma_1), \varphi)$ is denoted as $\mathfrak{Mor}(\gamma_0, \gamma_1)$.

Define now $\mathbb{F} = \{F_\sigma \mid \sigma \in T\}$ as follows: $F_0 = (N^*, \nu^*)$, where ν^* is a principal calculable enumeration of the set N^*; $F_{(\sigma_0, \sigma_1)} = F_{\sigma_0} \times F_{\sigma_1}$; $F_{(\sigma_0 \mid \sigma_1)} = \mathfrak{Mor}(F_{\sigma_0}, F_{\sigma_1})$. There are theorems on the existence of $\mathfrak{Mor}$ in this case.

Without exact definitions (they can be found in my book [1]) I formulate some exact proposition.

PROPOSITION. *The subcategory* C_{20}^* *of the category* $\mathfrak{N}$ *is a Cartesian closed category*; $F_\sigma \in ObC_{20}^*$, $\sigma \in T$.

So, $\mathbb{F}$ is a model of the partial calculable functionals of finite type. Now we find a connection with topology.

On every enumerated set one can define naturally some topology.

If $\gamma = (S, \nu)$ is an enumerated set, then $S' \subseteq S$ is called ν-enumerable iff $\nu^{-1}(S')$ is a recursively enumerable set. The topology T_ν on S is defined as the topology with the basis of the ν-enumerable subsets of S.

So, the following theorem is true.

THEOREM. *The topological spaces* F_σ, $\sigma \in T$ *are* f_0-*spaces and the completion* $C^*(F_\sigma)$ *is naturally coincident with* C_σ.

The second approach is a definition of $\mathbb{C}$ as the biggest 'natural' domain of an action of the calculable partial functionals $\mathbb{F}$.

So, the approaches to the definition of the class of partial continuous functionals of finite types give the same class $\mathbb{C}$.

3. Another reason for 'rightness' of the class $\mathbb{C}$ is a good definition of a class of everywhere defined functionals:

Define the subsets of everywhere defined functionals $\bar{G}_\sigma$ of C_σ as follows:

$$\bar{G}_0 = \{\{0\}, \{1\}, ...\} \subseteq C_0;$$
$$\bar{G}_{(\sigma_0, \sigma_1)} = \bar{G}_{\sigma_0} \times \bar{G}_{\sigma_1};$$
$$\bar{G}_{(\sigma_0 \mid \sigma_1)} = \{F \mid F \in C_{(\sigma_0 \mid \sigma_1)}, \forall g \in \bar{G}_{\sigma_0}(F(g) \in \bar{G}_{\sigma_1})\}.$$

The 'model' $\bar{\mathbb{G}}$ is not an extensional one. Now we define the equivalence relations $\sim_\sigma$ on $\bar{G}_\sigma$ as usual:

$\sim_0$ is equality;

if $\sim_{\sigma_0}$ and $\sim_{\sigma_1}$ are defined: for $f, f' \in \bar{G}_{\sigma_0}$, $g, g' \in \bar{G}_{\sigma_1}$
$(f, g) \sim_{(\sigma_0, \sigma_1)} (f', g') \Leftrightarrow f \sim_{\sigma_0} f'$ and $g \sim_{\sigma_1} g'$

and for $F, F' \in \bar{G}_{(\sigma_0 \mid \sigma_1)}$

$$F \sim_{(\sigma_0 \mid \sigma_1)} F' \Leftrightarrow \forall g \in \bar{G}_{\sigma_0} (F(g) \sim_{\sigma_1} F'(g)).$$

Let $G_\sigma = \bar{G}_\sigma / \sim_\sigma$ so, we have an extensional model $\mathbb{G} = \{ G_\sigma \mid \sigma \in T \}$.

PROPOSITION [4]. *The model $\mathbb{G}$ is the same as the countable functionals of Kleene-Kreisel.*

In his thesis M. Hyland gave good reasoning for distinguishing the class of countable functionals. In particular, he also proved the above proposition. From his research it can be seen that so far two models of functionals are known that are the models for Spector's theory **BR**. There are the model $\mathbb{G}$ and the termal model [6, 8]. Scarpellini's model [7] is the same as $\mathbb{G}$.

The proof that $\mathbb{G}$ is a model **BR** can be done by the same means as Luckhardt [6]. It was proved by me [3] and independently by Troelstra [9].

In conclusion I would like to say that, in my opinion, dealing with the model $\mathbb{G}$ is simplified by working with the model $\mathbb{C}$. So, the proof of the existence of bar-recursive functionals in $\mathbb{C}$ is obtained by simply using the fixed-point theorem.

BIBLIOGRAPHY

[1] Ershov, Yu. L., *Theory of Enumeration 2*, Novosibirsk, 1973, 1–170.
[2] Ershov, Yu. L., 'Theory of A-Spaces', *Algebra and Logic* **12,** No. 9 (1973), 369–416.
[3] Ershov, Yu. L., 'On Model G of the Theory BR', *Dokl. Akad. Nauk* **217,** No. 5 (1974), 1004–1006.
[4] Ershov, Yu. L., 'Maximal and Everywhere Defined Functionals', *Algebra and Logic* **13,** No. 4 (1974), 374–397.

[5] Hyland, J. M. E., 'Recursion Theory on the Countable Functionals', Thesis, 1975.
[6] Luckhardt, H., 'Extensional Gödel Functional Interpretation: A Consistency Proof of Classical Analysis', *Springer Lecture Notes* No. 306 (1973).
[7] Scarpellini, B., 'A Model for Bar-recursion of Higher Types', *Coms. Math.* **23** (1971), 123–153.
[8] Tait, W. W., 'Normal Form Theorem for Bar-recursive Functions of Finite Type', in *Proceed. Second. Scand. Log. Symp.*, Amsterdam, 1971, 353–367.
[9] Troelstra, A. S., 'Metamathematical Investigation of Intuitionistic Arithmetic and Analysis', *Springer Lecture Notes*, No. 344 (1973).
[10] Scott, D., 'Models for Various Type-free Calculi', in *Proceed. IVth Congr. LMPHS*', Amsterdam, 1973, 157–187.

JON BARWISE[1]

SOME EASTERN TWO CARDINAL THEOREMS*

1. INTRODUCTION

A more descriptive title of this paper might be 'Elaborations on Morley's proof of Vaught's gap ω two cardinal theorem'. The title given above comes from the main result of Section 2, possibly the most interesting result of the paper. Let us begin by introducing some terminology and stating Vaught's Theorem.

Throughout this paper L is a finite or countable first order language with a distinguished unary predicate symbol U. We use μ, κ, λ to denote infinite cardinals and always assume $\lambda \leqslant \kappa \leqslant \mu$. A structure $\mathfrak{M} = \langle M, U, ... \rangle$ is of *type* (κ, λ) if Card $(M) = \kappa$, Card $(U) = \lambda$. Given λ, let $\exp_0 (\lambda) = \lambda$ and $\exp_{n+1} (\lambda) = 2^{\exp_n (\lambda)}$. A structure $\mathfrak{M}$ of type (κ, λ) is a *gap n model* if $\kappa \geqslant \exp_n (\lambda)$ and is a *gap $> n$ model* if $\kappa > \exp_n (\lambda)$. A theory T *admits* (κ, λ) if T has a model of type (κ, λ) and *admits gap n* if T has a gap n model.

1.1. THEOREM [Vaught (1965)]. *Let T be a theory of L which admits gap n for each $n < \omega$. Then T admits all types (κ, λ).*

In Section 2 we show how, for certain theories, the hypothesis of 1.1 can be weakened to bring the results more in line with mathematical experience. The discovery of these results came in our first attempts to solve the problem solved in Section 3. For this reason, the rest of this introduction is devoted to a discussion of the original problem. We could have written this paper so as to derive the results of Section 2 from the methods of Section 3, the way we discovered them, but we feel that the direct proof given in Section 2 is much easier to understand. This also allows us to have more freedom in Section 3. In particular, in Section 3 we use the ordinary Erdös-Rado Theorem, while in Section 2 we need a new 'simultaneous' version of the Erdös-Rado Theorem.

Butts and Hintikka (eds.), Logic, Foundations of Mathematics and Computability Theory, 11–31.
Copyright © 1977 by Reidel Publishing Company, Dordrecht-Holland. All Rights Reserved.

A model $\mathfrak{M}$ is a gap ω model if it is a gap n model, for all $n < \omega$; i.e. if it is of type (κ, λ) where $\kappa \geqslant \sup_n \exp_n (\lambda)$. It follows from Vaught's proof that the class of gap ω models is compact and that there is a recursive set CFA of axioms of L for the sentences true in all gap ω models. This set CFA is unique, up to logical equivalence, and has the property that a theory T of L admits all types (κ, λ) iff $T \cup \text{CFA}$ is consistent. The aim of Section 3 is to find such a set of axioms.

While Vaught's proof shows that there is such a recursive set CFA of axioms, it gives no hint as to what they might be. On the other hand, Morley's proof of Vaught's Theorem (see, e.g. Chang-Keisler, 1973, pp. 436–438) does give a hint. In fact, his proof shows that if T is a theory *with built-in Skolem functions*, then T has a gap ω model iff T has a model $\mathfrak{M} = \langle M, U, ... \rangle$ with an infinite set $\langle Z, < \rangle$ of U-indiscernibles. (This means that $Z \subseteq M - U$, $<$ is a linear ordering of Z and for any $z_1 < \cdots < z_n$, $z_1' < \cdots < z_n'$ and $u_1 \cdots u_m \in U$, $(\mathfrak{M}, z_1 \cdots z_n, u_1 \cdots u_m) \equiv (\mathfrak{M}, z_1' \cdots z_n', u_1 \cdots u_m)$.) It is quite easy to write a theory Ind such that for any T, T has a model with an infinite set of U-indiscernible iff $T \cup \text{Ind}$ is consistent. One simply writes down for each finite set Φ of formulas $\varphi(x_1 \cdots x_n, u_1 \cdots u_m)$ and each $k \geqslant n$,

$$\exists z_1 \cdots z_k [\{z_1 < \cdots < z_k\} \text{ is a set of } U\text{-indiscernibles for those}$$
$$\varphi \in \Phi].$$

Morley's proof comes in two parts. He shows that if T has gap n-models, for each $n < \omega$, then T is consistent with Ind, by means of the Erdös-Rado Theorem. The second step shows that if T has built in Skolem functions and if $\mathfrak{M} = \langle M, U, ... \rangle$ is a model of T of type (μ, μ) with a set $(Z, <)$ of μ U-indiscernibles, then, for any (κ, λ), $\mu \geqslant \kappa \geqslant \lambda \geqslant \omega$, then there is an $\mathfrak{M}_0 < \mathfrak{M}$, $\mathfrak{M}_0$ of type (κ, λ). Our proofs will be elaborations on this proof. Note that since the first part of Morley's proof does not use built in Skolem functions, our eventual set CFA of axioms must imply all of Ind. On the other hand, Example 3.1 gives an example of a theory T consistent with Ind which has no gap >1 model, so CFA must be definitely stronger than Ind. What CFA says, roughly, is that for each k and n and each finite set Φ of formulas of L,

$$\exists z_1 \cdots z_k [z_1 \cdots z_k \text{ act as } U\text{-indiscernibles for a sequence}$$
of n choices of partial Skolem functions for the formulas in $\Phi]$.

To see a more precise statement,[2] the reader will have to read Section 3.

The two sections of this paper can be read independently. They both use the following notational conventions:

v, x, y, z arbitrary variables,

u, w range over U,

$\mathbf{x}, \mathbf{y}$ denote finite sequences $x_1, ..., x_n, y_1, ..., y_m$,

$\mathbf{u}$ denotes a finite sequence $u_1 \cdots u_n$ of u's.

Thus $\exists u(\cdots)$ really means $\exists u(U(u) \wedge (\cdots))$. Sometimes, just for emphasis, we write this as $\exists u \in U(\cdots)$. In Section 2 we assume L has no function symbols.

2. An Eastern Two Cardinal Theorem

In this section we examine Vaught's gap ω two cardinal theorem from an eastern (or at least midwestern) point of view. The sobriquets 'western model theory' and 'eastern model theory' have been applied[3] to emphasize differences in the Tarski-Berkeley and the Robinson-Yale attitudes toward model theory. One difference, the one we have in mind, is that western model theory looks for elegant results which apply to arbitrary first order theories whereas eastern model theory is more concerned with accurately reflecting and explaining observed mathematical phenomena. It is in this sense that the result of this section can be called an eastern two-cardinal theorem.

There are reasonably simple examples (see 2.13 e.g.) which show that the hypothesis of Vaught's Theorem cannot be weakened in general. However, the unmovable gaps that seem to occur in real mathematics are seldom very large. It is this phenomenon which we wish to examine. To do so we need some terminology.

2.1. DEFINITION. Let L be a countable language (without function symbols) with a distinguished unary predicate symbol U. A formula φ is *universal mod* U if, once φ is put in prenex form, all existential quantifiers of φ are bounded by U.

A result of Feferman (1974) shows that a sentence φ is logically equivalent to a universal mod U sentence iff for all $\mathfrak{M}$, $\mathfrak{N}$, $\mathfrak{M} \subseteq \mathfrak{N}$, $U^{\mathfrak{M}} = U^{\mathfrak{N}}$ and $\mathfrak{N} \vDash \varphi$ implies $\mathfrak{M} \vDash \varphi$. This result shows us that we are dealing with a natural class of formulas. We will use only the trivial half ($\Rightarrow$) of the result in our proof.

There is a natural hierarchy on the universal mod U formulas, determined by the actual number of universal quantifiers which occur in front of a bounded existential quantifier. More precisely, given a universal mod U formula φ, put φ in prenex normal form

$$\boxed{} \forall x_1 \boxed{} \forall x_2 \boxed{} \cdots \forall x_n \boxed{} \forall y_1 \cdots \forall y_m \text{ [quantifier free]},$$

where the rectangles are either empty or enclose only bounded existential quantifiers, and where n is as small as possible. For this purpose bounded universal quantifiers are treated just like unbounded universals. The integer n is called the *width* of φ. Since we allow the rectangles to be empty, we are counting $\forall$'s, not numbers of alternations, in determining width. (In the counterexamples below, all the rectangles but the last are indeed empty.) If φ has width less than or equal to n, we call φ a $\forall^{(n)}$-*mod* U *formula*. Note that the $\forall^{(0)}$ mod U formulas are simply the ordinary universal formulas of L. A formula obtained from a $\forall^{(n)}$-*mod* U formula by means of existential quantification (any number of $\exists$'s, bounded or not) is an $\exists\forall^{(n)}$-*mod* U formula. Most naturally occurring examples of sentences not admitting all types (κ, λ) are $\exists\forall^{(n)}$-mod U, for some n. We begin with some examples which suggest a theorem and then prove the theorem.

2.2. EXAMPLE.   The sentence

$$\exists z_1 \cdots z_n \forall x [x = z_1 \vee \cdots \vee x = z_n \vee U(x)]$$

is $\exists\forall^{(0)}$-mod U. It has models $\mathfrak{M} = \langle M, U, ... \rangle$, but none where $M - U$ is infinite.

The next result is trivial but illustrates a method.

2.3. PROPOSITION.   *Let T be a $\exists\forall^{(0)}$-mod U theory which has a model $\mathfrak{M} = \langle M, U, ... \rangle$ with U and $M - U$ both infinite. Then T admits all types (κ, λ), $\kappa \geqslant \lambda$.*

Proof. By the Compactness Theorem, T has a model $\mathfrak{M}$ with Card $(U) \geqslant \lambda$ and Card $(M) = \kappa$. Choose a submodel $\mathfrak{M}_0 = \langle M_0, U_0, \ldots \rangle$ of $\mathfrak{M}$ with witnesses for all the existential quantifiers in T, with Card $(U_0) = \lambda$ and Card $(M_0) = \kappa$. Then, because $\mathfrak{M}_0 \subseteq \mathfrak{M}$, all the universal formulas true in $\mathfrak{M}$ are true in $\mathfrak{M}_0$, so $\mathfrak{M}_0 \vDash T$. $\mathfrak{M}_0$ is of type (κ, λ). ∎

2.4. EXAMPLE.   Consider the sentence

$$\forall x \exists u \in U[R(u, x) \wedge \forall y(R(u, y) \rightarrow y = x)].$$

This is a $\forall^{(1)}$-mod U sentence and asserts that R maps some subset of U onto M. It can have models $\mathfrak{M} = \langle M, U, R \rangle$ with $M - U$ infinite, but none with Card $(M) >$ Card (U).

2.5. EXAMPLE.   Consider a vector space V over a field U as a structure $\mathfrak{M} = \langle U \cup V, U, V, \ldots \rangle$. The sentence '$V$ is of dimension $\leqslant k$' can be written as $\exists \forall^{(1)}$-mod U sentence which, of course, has no two cardinal model. Namely,

$$\exists v_1, \ldots, v_k \forall w[w \text{ can be written, uniquely, as a linear combination of } v_1, \ldots, v_k].$$

The part in brackets asks for unique $r_1, \ldots, r_k \in U$ such that $w = \sum r_i v_i$. Since everything is in terms of relations, not functions, $w = \sum r_i v_i$ must be written by a universal formula.

2.6. EXAMPLE.   Let $\mathfrak{M} = \langle [0, 1], \mathbb{Q} \cap [0, 1], < \rangle$ where $\mathbb{Q}$ is the set of rationals. This is a gap-1 model of type $(2^{\aleph_0}, \aleph_0)$. Let T be a theory asserting that $<$ is a linear ordering with first and last elements and that

$$\forall x \forall y \exists q \in U[x < y \rightarrow x < q < y].$$

It is easily seen that T has no model of type $(\kappa, \aleph_0)$ for $\kappa > \aleph_0$. T is written out as a $\exists \forall^{(2)}$-mod U theory.

2.7. EXAMPLE.   It is not easy to find really practical examples of theories which admit $(2^{2^{\aleph_0}}, \aleph_0)$ but not larger gaps. One such comes from the known hieroglyphics

$$(2^{2^{\aleph_0}})^+ \to (\aleph_1)^3_{\aleph_0} \quad \text{but} \quad (2^{2^{\aleph_0}}) \not\to (4)^3_{\aleph_0}.$$

The first is an instance of the Erdös-Rado Theorem. The second is a special case of $\exp_n(\lambda) \not\to (n+2)^{n+1}_\lambda$; see 6.5(b) in Kunen (1976). Using these, one can write a sentence of the form $\forall^{(3)}$-mod U which expresses the first $\not\to$, and so has models of type $(2^{2^{\aleph_0}}, \aleph_0)$ but none of type $(\kappa, \aleph_0)$ with $\kappa > 2^{2^{\aleph_0}}$. (Use U for the indices of the partition, $M - U = 2^{2^\omega}$ and $R(x_1, x_2, x_3, u)$ for

$$\{x_1, x_2, x_3\} \text{ is in the } u\text{th element of the partition.})$$

The second $\not\to$ gives rise to $\forall^{(n+1)}$-mod U theories with gap n models but no gap $>n$ models.

This sequence of examples suggests that for $\forall^{(n)}$-mod U theories T, if T admits a gap $>(n-1)$ model then T admits all types. This turns out to be correct. It is convenient to state the basic construction in terms of resplendent models: see Barwise-Schlipf (1976).

2.8. THEOREM. *Let $n \geq 1$. Let $\mathfrak{M} = \langle M, U, ... \rangle$ be a resplendent model of power μ which is elementarily equivalent to some gap $>(n-1)$ model $\mathfrak{M}' = \langle M', U', ... \rangle$. Then, for $\lambda \leq \kappa \leq \mu$, there is a submodel $\mathfrak{M}_1$ of $\mathfrak{M}$ which is a model of all $\exists \forall^{(n)}$-mod U sentences true in $\mathfrak{M}$.*

Proof. By throwing in constant symbols, it suffices to find $\mathfrak{M}_1$ satisfying all $\forall^{(n)}$-mod U sentences true in $\mathfrak{M}$. We add new symbols to L to obtain a language L^+ as follows: Z, V(unary), $<$(binary) and for each $\forall^{(n)}$-mod U sentence σ, say $\sigma = \forall x_1 \exists u_1 \forall x_2 \exists u_2 \cdots \forall x_n \exists u_n$ $\varphi(x_1, ..., u_n)$, φ universal, we add unary F_1^σ, binary F_2^σ, ..., and n-ary F_n^σ. (Actually F_1^σ needs to be a sequence of unary function symbols as long as the sequence u_1, etc., so we ought to write $\mathbf{F}_1^\sigma$, but we won't.) Note that all function symbols of L^+ have 'arity' $\leq n$. Consider the following recursive theory T^+ of L^+:

(i) $\sigma \to \forall x_1, ..., x_n \varphi(x_1, F_1^\sigma(x_1), ..., x_n, F_n^\sigma(x_1, ..., x_n))$, (for all σ as above)

(ii) "Z is infinite and linearly ordered by $<$"

(iii) "$U \subseteq V$" $\wedge$ "$V \cap Z = 0$"

(iv) V(c) (for all constants c of L).

There is one instance of (v)$_F$ and (vi)$_F$ for each m-ary function symbol F.

(v)$_F$ $\forall x_1, ..., x_m, y_1, ..., y_m$ [if $x_1, ..., x_m, y_1, ..., y_m \in V \cup Z$, if for all $i \leq m$, $x_i \in Z \leftrightarrow y_i \in Z$, if whenever $x_i, x_j \in Z$ then $(x_i < x_j \leftrightarrow y_i < y_j)$, and if whenever $x_i \in V$ then $x_i = y_i$, then $F(x_1, ..., x_m) = F(y_1, ..., y_m)$]

(vi)$_F$ $\forall x_1, ..., x_m [F(x_1, ..., x_m) \in U]$.

There are two distinct parts of the proof. One part is to show:

(1)   Th$(\mathfrak{M}) \cup T^+$ is consistent.

From this it follows, by the definition of resplendency, that $\mathfrak{M}$ can be expanded to a model of T^+. In fact, using resplendency again, it follows that

(2) $\mathfrak{M}$ can be expanded to a model $\mathfrak{M}^+ = (\mathfrak{M}, Z, <, V, ...)$ of T^+ with Z and V of power μ.

The other part of the proof consists of using (2) to construct the desired submodel $\mathfrak{M}_1$. Let us carry out this part of the proof and then return to prove (1). Since $\mathfrak{M}$ is resplendent and U is infinite, U has power μ. Let $\mathfrak{M}_0^+ = (M_0, U_0, ..., V_0, ...)$ be a submodel of $\mathfrak{M}^+$ of type (λ, λ) (closed under all the functions), with universe contained in V. This is possible by (iii), (iv), and (vi). Also require of $\mathfrak{M}_0^+$ that if $x_1, ..., x_n \in Z \cup V_0$ then $F(x_1, ..., x_n) \in U_0$. This is possible by (v) and (vi). Let $Z_0 \subseteq Z$ be of power κ and let $\mathfrak{M}_1^+ = (M_1, U_0, ..., Z_0, <, V_0, ...)$ be the smallest submodel of $\mathfrak{M}^+$ containing M_0 and Z_0. By the construction of $\mathfrak{M}_0^+$ (the 'also require' part), $M_1 = Z_0 \cup V_0$, and hence $\mathfrak{M}_1^+$ is of type (κ, λ). We claim that $\mathfrak{M}_1$, the reduct of $\mathfrak{M}_1^+$ to the language L, satisfies every $\forall^{(n)}$-mod U sentence true in $\mathfrak{M}$. To see this, let $\sigma = \forall x_1 \exists u_1, ..., \forall x_n \exists u_n \varphi$ be such a true sentence. By (i)$_\sigma$, $\mathfrak{M}^+$ is a model of $\forall x_1, ..., x_n \varphi(x_1, F_1^\sigma(x_1), ..., x_n, F_n^\sigma(x_1, ..., x_n))$. Since this is universal and $\mathfrak{M}_1^+ \subseteq \mathfrak{M}^+$, $\mathfrak{M}_1^+$ is a model of this sentence. But then $\mathfrak{M}_1 \models \forall x_1 \exists u_1 \cdots \forall x_n \exists u_n \varphi$, as desired.

To finish the proof, it remains only to prove (1). As might be expected, this uses the Erdös-Rado Theorem, but, because we don't have any room for play, it needs a stronger version than is usually applied in model theory. We state the general result, even though we only need it for the case $k = 1$. ∎

2.9. LEMMA. *Let* $\text{Card}(X) > \exp_{n+k-1}(\lambda) = \exp_{n-1}(\exp_k(\lambda))$. *Let the* $n + k$ *element subsets be partitioned into* $\leq \lambda$ *sets,*

$$[X]^{n+k} = \bigcup_{i < \lambda} C_i,$$

and let the n *element subsets be partitioned into* $\leq \exp_k(\lambda)$ *sets*

$$[X]^n = \bigcup_{j < \exp_k(\lambda)} D_j.$$

There is a set $Z \subset X$ *of power* $> \lambda$ *such that:*

$$\text{for some } i, \quad [Z]^{n+k} \subset C_i$$
$$\text{for some } j, \quad [Z]^n \subset D_j.$$

This lemma is easily proved by induction on n using the main lemma (on the existence of prehomogeneous sets) that goes into the proof of the Erdös-Rado Theorem; see Lemma 6.8b in Kunen (to appear). One can also give a simple model theoretic proof as in Chang-Keisler (1973) or derive it from Theorem 1 of Baumgartner (to appear).

We now return to the proof of (1), that $\text{Th}(\mathfrak{M}) \cup T^+$ is consistent. Recall, from the statement of the theorem, that $\mathfrak{M} \equiv \mathfrak{M}'$, where $\mathfrak{M}' = \langle M', U', ... \rangle$ is a gap $> (n-1)$ L-structure, say of type (κ, λ). To prove (1) it suffices, by the Compactness Theorem, to show that for any finite subset $T_0 \subseteq T^+$, $\mathfrak{M}'$ can be expanded to a model of T_0. Actually, we show how to expand $\mathfrak{M}'$ to a model for all $(i)_\sigma$, (ii), (iii), (iv), and $(vi)_F$, and of any finite number of instances of $(v)_F$. For any $\forall^{(n)}$-mod U sentence $\sigma = \forall x_1 \exists u_1 \cdots \forall x_i \exists u_i \varphi$, $i \leq n$, φ universal, choose functions $F_1 \cdots F_i$ mapping M' into U' making $(i)_\sigma$ true. Let $V' = U'$ union the set of interpretations of the constants of L. Thus $\text{Card}(V') = \text{Card}(U') = \lambda$.

Let $<$ be any linear ordering of M^+. Let $F_1 \cdots F_{113}$ be a typical finite number of function symbols of L'. We need only find an infinite set Z disjoint from V' such that for all $x_1 \cdots x_m, y_1 \cdots y_m \in V' \cup Z$, if the sequences $x_1 \cdots x_m$ and $y_1 \cdots y_m$ are ordered the same way on the elements of Z and have the other terms (those from V') equal, then $F_1(x_1 \cdots x_m) = F_1(y_1 \cdots y_m), ..., F_{113}(x_1 \cdots x_m) = F_{113}(y_1 \cdots y_m)$.   Since Card $(V') = \lambda$ we can ignore the condition $V' \cap Z = 0$ as long as we get Card $(Z) > \lambda$, for we can then throw away anything in $V' \cap Z$. The case where n (the n of the theorem) $= 1$ is easy, so we treat only the case $n > 1$. Recall that all of $F_1, ..., F_{113}$ have arity $\leq n$.

Partition $[M]^n$ into equivalence classes by defining, for $a, b \in [M]^n$, $a \sim b$ iff:

> for any listings $a = \{x_1, ..., x_n\}$, $b = \{y_1, ..., y_n\}$ of a, b which satisfies $x_i < x_j$ iff $y_i < y_j$, one has $F(x_1, ..., x_n) = F(y_1, ..., y_n)$ for all symbols F in the list $F_1, ..., F_{113}$ which happen to be n-ary.

The equivalence class of $a = \{x_1 < \cdots < x_n\}$ is determined by a function which assigns to each permutation π of $\{1, ..., n\}$ and each F as above, an element $u \in U$ to act as the value of $F(x_{\pi(1)}, ..., x_{\pi(n)})$, so there are $\leq \lambda$ equivalence classes.

We now partition $[M]^{n-1}$ into equivalence classes by defining, for $a, b \in [M]^{n-1}$, $a \approx b$ iff

> for any F in the list $F_1, ..., F_{113}$, F m-ary $(m \leq n)$, if $x_1, ..., x_m, y_1, ..., y_m$ are sequences from $a \cup V'$, $b \cup V'$ respectively, such that $x_i \in a$ iff $y_i \in b$, $x_i < x_j$ iff $y_i < y_j$ and $x_i = y_i$ if $x_i \in V'$, then $F(x_1, ..., x_m) = F(y_1, ..., y_m)$.

Given $a \in [M]^{n-1}$, there are at most λ ways its elements can appear in a sequence $x_1 \cdots x_n$ from $a \cup V'$. The equivalence class of a is determined by assigning, to each such sequence, and each F, a value from U to act as $F(x_1, ..., x_n)$. Thus, since Card $(U) = \lambda$ there are at most $\lambda^\lambda = 2^\lambda = \exp_1(\lambda)$ equivalence classes of $\approx$.

Applying the lemma (with the n of that lemma set at $n - 1$ and $k = 1$), we get a set Z of power $> \lambda$ such that $[Z]^n$ lies in one equivalence class of $\sim$ and $[Z]^{n-1}$ lies in one equivalence class of $\approx$. Thus, for this Z, $(v)_F$ is true for all F in the list $F_1, ..., F_{113}$. This completes the proof of (1) and hence of the theorem. ■

2.10. COROLLARY. *Let* $n \geq 1$. *If* T *is a* $\exists \forall^{(n)}$*-mod* U *theory of* L, *then* T *admits all types iff* T *has a gap* $> (n-1)$ *model.*

Proof. This is an immediate consequence of the theorem and the fact that resplendent models exist of every cardinality.

On the other hand, our examples show that, for each n, there is a $\forall^{(n)}$-mod U sentence φ with gap $n-1$ models but with no gap $> (n-1)$ model. Also, our examples combined with 2.10 show that, for each $n \geq 1$, there is a sentence $\forall x_1, ..., \forall x_n \exists u \in U$ [quantifier free] not equivalent to any $\forall^{(n-1)}$-mod U sentence.

We conclude this section with an open problem, a relevant example, and two stray remarks. ∎

2.11. PROBLEM. Is there a simple function f such that:

  (i)   for all sentences φ, if φ has a gap $> f(\varphi)$ model, then φ admits all types (κ, λ);
  (ii) if φ, ψ have the same syntactic form (in some sense) then $f(\varphi) = f(\psi)$; and
  (iii) $f(\varphi)$ is as small as possible?

Our theorem answers this for $\varphi \in \bigcup_n \exists \forall^{(n)}$-mod U, by letting $f(\varphi) =$ width $(\varphi) - 1$. Our final example shows that no such trivial quantifier counting procedure works in general.

2.12. EXAMPLE. For any $n > 0$ there is a sentence φ of the form

$$\forall x \forall y \exists z \ [\text{quantifier free}]$$

which has a gap n model but not a gap $> n$ model. For example, consider $n = 3$. Let $U = M_0$ be a countable infinite set of urelements, $M_1 = \text{Power}(U)$, $M_2 = \text{Power}(M_1)$, $M_3 = \text{Power}(M_2)$ and $M = M_3 \cup M_2 \cup M_1 \cup U$. Let $\mathfrak{M} = \langle M, U, M_1, M_2, M_3, \varepsilon \rangle$. Thus $\mathfrak{M}$ is a gap 3 model. Let φ say

$$\forall x \forall y \exists z [x \neq y \rightarrow \text{``}z \in (x \cup y - x \cap y)\text{''} \wedge$$
$$\bigwedge_{0 \leq i < 3} (M_{i+1}(x) \wedge M_{i+1}(y) \rightarrow M_i(z))].$$

Then φ has no model (κ, λ) for $\kappa > \exp_3(\lambda)$. This example shows that if we want to get an *a priori* bound on the unspreadable gaps that can occur as models of a sentence φ, even an $\forall\forall\exists$ sentence φ, we would have to consider not just the quantifier prefix of φ but some measure of the complexity of the matrix. (It is easy to see that at least two $\forall$'s are needed for example 2.12.)

2.13. REMARK. Most results in eastern model theory hold for some prefix class determined entirely by the number of *alternations* of quantifiers. The Linear Prefix Theorem of Walkoe (1970) tells us that there ought to be theorems of model theory which deal with the actual quantifier prefix configuration, theorems where there is a real difference between, say, $\forall\forall\forall\exists$ and $\forall\forall\exists\exists$. The main result of this section is one such result.

2.14. REMARK. Our first proof of Theorem 2.8 gave an explicit set S_n of axioms such that a $\forall^{(n)}$-mod U theory T has a gap $> (n-1)$ model iff $T \cup S_n$ is consistent. The reader of the next section should be able to recapture S_n. We have eliminated it to make the proof of 2.8 more transparent.

3. THE FIRST ORDER PROPERTIES OF PAIRS OF CARDINALS REASONABLY FAR APART

In this section we describe the set CFA of axioms[3] of L such that, for any theory T of L, $T \cup$ CFA is consistent iff T admits all types (κ, λ). Vaught's Theorem 1.1 will fall out of our proof. The main result of this section is Theorem 3.11. Our proof uses the method of straightening out Henkin quantifiers, as discussed in Barwise (to appear). We assume the reader is faimilar with Section 2 of that paper.

We begin with an example to show that the theory Ind discussed in Section 1, though strong enough for theories with built in Skolem functions, is not strong enough for other theories.

3.1. EXAMPLE. A model $\mathfrak{M} = \langle M, U, \ldots \rangle$ with an infinite set $\langle Z, < \rangle$ of U-indiscernibles such that its theory Th $(\mathfrak{M})$ does not admit gap > 1 models. Let $\langle \mathbb{Q}, < \rangle$ be the rationals, let U be a countable set disjoint from $\mathbb{Q}$ and let $U^{\mathbb{Q}}$ be the set of all functions mapping $\mathbb{Q}$ into U. Let App (f, q, u) iff $f \in U^{\mathbb{Q}}$, $q \in \mathbb{Q}$ and $f(q) = u$. Let $M = U \cup \mathbb{Q} \cup U^{\mathbb{Q}}$ and $\mathfrak{M} = \langle M, U, \mathbb{Q}, <, \text{App} \rangle$. It is easy to see that $\langle \mathbb{Q}, < \rangle$ is a set of U-indiscernibles in $\mathfrak{M}$, since any automorphism σ of $\langle \mathbb{Q}, < \rangle$ can be extended to an automorphism $\bar{\sigma}$ of $\mathfrak{M}$ with $\bar{\sigma}(u) = u$, for all $u \in U$. (Simply set $\bar{\sigma}(f) =$ the unique g such that $g(q) = f(\sigma^{-1}(q))$, for all q.) On the other hand the sentences

(1) $\forall f, g \in U^{\mathbb{Q}} \exists q \in \mathbb{Q}[f \neq g \rightarrow {}'f(q) \neq g(q)']$

(2) $\exists f \in U^{\mathbb{Q}}['f$ is one-one$']$

in conjunction obviously prevent Th $(\mathfrak{M})$ from having a gap > 1 model.

The trouble in 3.1 is obvious. Though $\langle \mathbb{Q}, < \rangle$ is a set of U-indiscernibles, once one chooses an f satisfying (2), $\langle \mathbb{Q}, < \rangle$ is no longer a set of U-indiscernibles. Our set CFA of axioms must assert (the first order part of) the existence of U-indiscernibles which are indiscernible enough that we can add Skolem functions and have them stay indiscernible. The way to express all this by Henkin quantifiers, and hence by first order logic, is by using 'Skolem producers'.

DEFINITION.   A *Skolem producer* for a structure $\mathfrak{M}$ is a collection $\mathcal{G} = \{G_i \mid i \in I\}$ of functions such that, for any linearly order set $\langle X, < \rangle$, $X \subseteq M$, the set

$$M_0 = \{G_i(x_1 \cdots x_n) \mid i \in I, x_1 < \cdots < x_n, x_1, ..., x_n \in X\}$$

produced by $\mathcal{G}$ forms the universe of an elementary submodel $\mathfrak{M}_0 < \mathfrak{M}$.

Let us look at a special case which illustrates the most important part of this definition. Consider the problem of constructing a submodel $\mathfrak{M}_0$ with some specified property of a given model $\mathfrak{M}$ of some $\forall \exists$-sentence $\sigma = \forall x \exists y \varphi(x, y)$. We usually choose a Skolem function F for σ,

$$\mathfrak{M} \vDash \forall x \varphi(x, F(x)).$$

Then, given any X, the *closure* of X under F is a model of σ. Note, however, that $X \cup \{F(x) \mid x \in X\}$ is not usually big enough. One also needs $F^2(x)$, $F^3(x)$, ... Suppose, however, that we replace $\forall x \exists y \varphi(x, y)$ by the equivalent

$$\forall x \exists y_1 \exists y_2 \cdots [\varphi(x, y_1) \wedge \varphi(y_1, y_2) \wedge \cdots].$$

Now choose Skolem functions G_1, G_2, ... such that, for all x,

$$\varphi(x, G_1(x)) \wedge \varphi(G_1(x), G_2(x)) \wedge \cdots .$$

Then, given X, $M_0 = X \cup \{G_n(x) \mid n = 1, 2, ..., x \in X\}$ is the universe of a model $\mathfrak{M}_0$ of σ, even though $\mathfrak{M}_0$ may not be closed under any of G_1, G_2, ... The set $\{G_1, G_2, ...\}$ might be called a *Skolem producer for* σ.

Our plan of attack is as follows. We introduce an expanded language L^+ of L and a theory CFA^+ of L^+. This theory asserts that a certain family is a Skolem producer and that an infinite linearly ordered set $\langle Z, < \rangle$ is U-indiscernible with respect to this family. We then transform CFA^+ into a Henkin theory CFA^H in the original language L and take CFA to the set of first order approximations of CFA^H. Thus the intuitive content of CFA is quite clear, even though it might be a mess to write out a typical axiom. A more accurate English translation of a typical $\varphi \in CFA$ than that given in Section 1 is:

$\exists z_1 \cdots z_k$ [The ordered set $\{z_1 < \cdots < z_k\}$ is U-indiscernible for a sequence of n steps in the construction of a Skolem producer for formulas in Φ],

where Φ is a finite set of formulas and k is at least as large as the number of free variables in Φ.

Now down to work.

The Auxiliary Language L^*

To each formula $\sigma(\mathbf{x})$ of L of the form $\exists y \varphi(\mathbf{x}, y)$ add a new function symbol F_σ with as many places as there are elements in the sequence $\mathbf{x}$. Call the resulting language L^*. Our only interest in L^* is in the terms, and our only use for these terms is as convenient indices for the elements of a Skolem producer. Define var (t) to be the set of variables occurring in the term t and $\#(t) = \text{Card (var}(t))$. Define an equivalence relation $\sim$ on terms of L^* by:

> $t_1 \sim t_2$ iff there is a map π of var (t_1) onto var (t_2) such that π preserves the order of subscripts and t_2 is the result of replacing each x_i in t_1 by $\pi(x_i)$.

Examples:

$$F(x_1, x_3) \sim F(x_3, x_4), \qquad F(x_1, x_3) \not\sim F(x_3, x_1),$$

$$F(x_1, x_3) \not\sim F(x_2, x_2).$$

3.2.   *The Auxiliary Language* L^+

To form L^+, add to the original L a unary Z, binary $<$, and, for each term t of L^* a new $\#(t)$-ary function symbol G_t. This makes sense even if $\#(t) = 0$ for then G_t is simply a new constant symbol. We often write G_t in formulas without displaying its arguments. Whenever we do this, we intend G_t to apply to the variables in var (t), in increasing order. For example, if t is $F(x_1, x_3)$ then $\varphi(G_t)$ means $\varphi(G_t(x_1, x_3))$. If t' is $F(x_3, x_1)$ then $\varphi(G_{t'})$ means $\varphi(G_{t'}(x_1, x_3))$, but G_t and $G_{t'}$ are different function symbols.

Our first objective is to write out a set of axioms SP^+ of L^+ about the G_t to insure that they form a Skolem producer for L, and to do this in a way that can be easily transcribed into Henkin sentences. For all terms t_1, t_2 of L^* such that $t_1 \sim t_2$, let $B_{t_1 t_2}$ be the axioms

$$\forall x_1, ..., x_n[G_{t_1}(x_1, ..., x_n) = G_{t_2}(x_1, ..., x_n)]$$

where $n = \#(t_1) = \#(t_2)$. For each term t of L^* let A_t be defined as follows. If t is a constant symbol or variable of L, let A_t be the universal closure of

$$(G_t = t).$$

If t is $F(t_1, ..., t_n)$ where F is a symbol of the original L, let A_t be the universal closure of

$$G_t = F(G_{t_1}, ..., G_{t_2}).$$

The important case is when t is of the form $F_\sigma(t_1, ..., t_n)$ where $\sigma = \exists y \varphi(x_1, ..., x_n, y)$, in which case A_t is the universal closure of

$$\exists y \varphi(G_{t_1}, ..., G_{t_n}, y) \to \varphi(G_{t_1}, ..., G_{t_n}, G_t).$$

The set SP^+ consists, by definition, of all the A_t and all the $B_{t_1 t_2}$ where $t_1 \sim t_2$.

3.3. SKOLEM PRODUCER LEMMA.   *Let $\mathfrak{M}$ be a structure and let $\mathcal{G} = \{G_t \mid t \text{ a term of } L^*\}$ be a set of functions such that $(\mathfrak{M}, G_t)_t$ is a model of SP^+. Then $\mathcal{G}$ is a Skolem producer.*

Proof.   Let $\langle X, < \rangle$ be a linearly ordered subset of M and let $M_0 = \{G_t(x_1, ..., x_n) \mid t \text{ is a term of } L^*, n = \#(t) \text{ and } x_1 < \cdots < x_n \text{ in } \langle X, < \rangle\}$. We claim that if $a_1, ..., a_m \in M_0$ and $\mathfrak{M} \vDash \exists y \varphi(a_1, ..., a_m, y)$ then there is a $b \in M_0$ such that $\mathfrak{M} \vDash \varphi(a_1, ..., a_m, b)$. To simplify matters, assume $m = 2$. We illustrate the proof by a special case. Suppose $a_1 = G_{t_1}(z_1, z_2)$, $a_2 = G_{t_2}(z_3, z_4, z_5)$ where $z_1 < z_2$ in $\langle X, < \rangle$ and $z_3 < z_4 < z_5$ in $\langle X, < \rangle$. Arrange $z_1, ..., z_5$ in an increasing sequence, for example, if $z_1 < z_3 < z_4 = z_2 < z_5$, then call these elements $\bar{x}_1, \bar{x}_2, \bar{x}_3, \bar{x}_4$ respectively. By appropriately invoking the B axioms, we may assume that t_1 has variables x_1, x_3 while t_2 has x_2, x_3, x_4. Let t be $F_{\exists y \varphi}(t_1, t_2)$. Then A_t assures us that

$$\mathfrak{M} \vDash \varphi[a_1, a_2, G_t(\bar{x}_1, \bar{x}_2, \bar{x}_3, \bar{x}_4)]$$

since $a_1 = G_{t_1}(\bar{x}_1, \bar{x}_3)$ and $a_2 = G_{t_2}(\bar{x}_2, \bar{x}_3, \bar{x}_4)$. The general case presents only notational complexities over this case. ∎

We obtain CFA^+ by adding certain axioms C_t to SP^+. Given two sequences $x_1, ..., x_n, y_1, ..., y_n$ of distinct variables we write $x_1, ..., x_n \approx y_1, ..., y_n$ for the first order formula:

$$\bigwedge_{\substack{1 \leq i \leq n \\ 1 \leq j \leq n}} [Z(x_i) \leftrightarrow Z(y_i) \wedge (x_i < x_j \leftrightarrow y_i < y_j) \wedge U(x_i) \to x_i = y_i].$$

Then, for any term t let $n = \#(t)$ and let C_t be:

$$\forall x_1, ..., x_n \forall y_1, ..., y_n [x_1, ..., x_n \approx y_1, ..., y_n \wedge U(G_t(x_1, ..., x_n))$$
$$\to G_t(x_1, ..., x_n) = G_t(y_1, ..., y_n)].$$

3.4. DEFINITION.   Let CFA^+ consist of all the axioms in SP^+, all the C_t and axioms asserting:

$\langle Z, < \rangle$ is an infinite linearly ordered set,

$Z \cap U = 0$.

Let CFA_k^+ be the above except with C_t only for those t with $\#(t) \leqslant k$.

3.5. LEMMA.   Let $\mathfrak{M} = \langle M, U, ... \rangle$ be a gap $> n$ model. There is an expansion $\mathfrak{M}^+$ of $\mathfrak{M}$ to an L^+-structure such that $\mathfrak{M}^+ \vDash CFA_n^+$.

Proof.   The proof is a routine application of the Erdös-Rado Theorem. First, for any function symbol $F_{\exists x\varphi}$ of L^* (not L^+) choose a function $F_{\exists x\varphi}$ to act as a Skolem function for $\exists x\varphi$. Then each term t of L^* has a natural interpretation G_t. Use these G_t to interpret the symbols G_t. This will make the A and B axioms true automatically. Now let $<$ be any linear ordering of M. We define an equivalence relation on $[M]^n$, the set of n element subsets of M as follows. Given $a, b \in [M]^n$, call a, b equivalent if

for each t, $\#(t) = k \leqslant n$, and all sequences $x_1, ..., x_k$, $y_1, ..., y_k$ from $a \cup U$, $b \cup U$ respectively, such that $x_i \in a$ iff $y_i \in b$, $x_i < x_j$ iff $y_i < y_j$ and $x_i \in U \to x_i = y_i$, if $G_t(x_1, ..., x_k) \in U$ then $G_t(y_1, ..., y_k) = G_t(x_1, ..., x_k)$.

This partitions $[M]^n$ into $\leqslant 2^{Card(U)}$ sets. Since $Card(M) > \exp_n(Card(U)) = \exp_{n-1}(2^{Card(U)})$, the Erdös-Rado Theorem gives a set Z of power $> Card(U)$ such that all elements of $[Z]^n$ lie in one class. Since $Card(Z) > Card(U)$ we can throw away some points and get $Z \cap U = 0$. This gives all of CFA_n^+. ∎

Let us say that an L-structure $\mathfrak{M} = \langle M, U, ... \rangle$ satisfies condition (+) if there is an expansion $\mathfrak{M}^+ = (\mathfrak{M}, Z, <, G_t)_{t \in L^*}$ which is a model of CFA^+ and such that $Card(Z) = Card(M)$.

3.6. LEMMA.   Let $\mathfrak{M} = \langle M, U^{\mathfrak{M}}, ... \rangle$ be an infinite L-structure of type (μ, μ) which satisfies condition (+). Then for any type (κ, λ) with $\mu \geqslant \kappa \geqslant \lambda \geqslant \omega$ there is an elementary submodel $\mathfrak{M}_0 < \mathfrak{M}$ of type (κ, λ).

Moreover, $U^{\mathfrak{M}_0}$ *can be taken to contain any preassigned set* $X \subseteq U^{\mathfrak{M}}$ *of power* $\leq \lambda$.

Proof. Assume Card $(X) = \lambda$. Expand $\mathfrak{M}$ to an $\mathfrak{M}^+$ as in condition (+). Let U^0 contain X and contain, for any n-tuple $x_1, ..., x_n$ from $Z \cup X$, all values of $G_t(x_1, ..., x_n)$ which lie in U, and have Card $(U^0) = \lambda$. This is possible by axiom (C). Given U^n, let $U^{n+1} \supseteq U^n$ have power λ and, for each $x_1, ..., x_n \in Z \cup U^n$, if $G_t(x_1, ..., x_n) \in U$ then $G_t(x_1, ..., x_n) \in U^{n+1}$. Let $U_0 = \cup_n U^n$. Then U_0 has power λ and if $x_1, ..., x_n \in Z \cup U_0$ then $G_t(x_1, ..., x_n)$ is in U_0 if it is in U. Extend $<$ on Z to a linear ordering of $Z \cup U_0$ and let

$$M_0 = \{G_t(x_1, ..., x_n) \mid x_1 < \cdots < x_n, x_1 \cdots x_n \in Z \cup U_0, t \in L^*\}.$$

By the Skolem producer lemma, M_0 is the universe of a submodel $\mathfrak{M}_0 < \mathfrak{M}$ and, by the construction of U_0, $\mathfrak{M}_0$ is of the form $(M_0, U_0, ...)$. Thus $\mathfrak{M}_0$ is of type (μ, λ). An application of the ordinary downward Lowenheim-Skolem-Tarski Theorem to $\mathfrak{M}_0$ gives the desired conclusion. ∎

Note that the gap ω two cardinal theorem is an immediate consequence of Lemmas 3.5 and 3.6.

3.7. *The Henkin Version of the Axioms*

Our next task is to rewrite CFA^+ of L^+ as a Henkin theory CFA^H of the original L. First some notation.

For any term t of L^* let $\mathbf{x}_t = x_{i_1}, ..., x_{i_n}$ be the sequence of variables in var (t) listed in order of increasing subscripts. Let $\mathbf{x}'_t$ be a sequence of new variables $x'_{i_1}, ..., x'_{i_n}$. Let v_t, v'_t be other new variables. If $\mathbf{x} = x_1, ..., x_n$, $\mathbf{y} = y_1, ..., y_n$ then write Eqv $(\mathbf{x}, \mathbf{y}, z_1, ..., z_k)$ for the first order formula which is the conjunction, over $1 \leq i \leq n$, $1 \leq j \leq n$, of

$$(U(x_i) \vee x_i = z_1 \vee \cdots \vee x_i = z_k)$$

$$(U(y_j) \vee y_j = z_1 \vee \cdots \vee y_j = z_k)$$

$$U(x_i) \leftrightarrow U(y_i)$$

$$U(x_i) \to x_i = y_i$$

$$\wedge \{\neg(x_i = z_{i_1} \wedge x_j = z_{j_1} \wedge y_i = z_{i_2} \wedge y_j = z_{j_2}) \mid$$

$$\neg(i_1 < i_2 \leftrightarrow j_1 < j_2), i_1, i_2, j_1, j_2 \in \{1, ..., k\}\}.$$

Thus, Eqv $(\mathbf{x}, \mathbf{y}, z_1, ..., z_k)$ simply asserts that the sequences $\mathbf{x}$, $\mathbf{y}$ are each sequences from $U \cup \{z_1, ..., z_k\}$, that they are equal in those places where they are from U, and that the subsequences from $\{z_1, ..., z_k\}$ are ordered in the same way with respect to the ordering $z_1 < \cdots < z_k$.

For any finite set s of terms of L* and any $k < \omega$ let $H_s^k(z_1, ..., z_k)$ be the Henkin formula whose quantifier prefix is:

$$\left. \begin{array}{l} \forall \mathbf{x}_t \exists v_t \\ \forall \mathbf{x}_t' \exists v_t' \\ (\text{all } t \in s) \end{array} \right\},$$

and whose matrix is the conjunction of:

(1)　　　"$z_1, ..., z_k$ are all distinct and not in U"

(2)　　　$\mathbf{x}_t = \mathbf{x}_t' \rightarrow v_t = v_t'$ 　　　　　　　　$(t \in s)$

(3)　　　$\mathbf{x}_{t_1} = \mathbf{x}_{t_2} \rightarrow v_{t_1} = v_{t_2}$ 　　　　　$(t_1, t_2 \in s, t_1 \sim t_2)$

(4)　　　Eqv $(\mathbf{x}_t, \mathbf{x}_t', z_1, ..., z_k) \wedge U(v_t) \rightarrow v_t = v_t'$ 　　$(t \in s)$

(5)　　　$v_t = c$

　　　　　　　　　　　　　　　　　　$(t \in s, t \text{ the constant } c)$

(6)　　　$v_t = x_n$

　　　　　　　　　　　　　　　　　$(t \in s, t \text{ the variable } x_n)$

(7)　　　$v_t = F(v_{t_1}, ..., v_{t_n})$

　　　　　　　　　　　　$(t \in s, t = F(t_1, ..., t_n), F \in L)$

(8)　　　$\exists y \varphi(v_{t_1}, ..., v_{t_n}, y) \rightarrow \varphi(v_{t_1}, ..., v_{t_n}, v_t)$

　　　　　　　　　　　　　　$(\text{if } t \in s, t = F_{\exists y \varphi}(t_1 \cdots t_n)).$

These conjunctions perform the following functions. Read v_t as $G_t(\mathbf{x}_t)$. Then (2) insures that v_t' is also $G_t(\mathbf{x}_t')$; (3) takes care of $B_{t_1 t_2}$, (4) insures C_t for $Z = \{z_1 < \cdots < z_k\}$, and (5)–(8) insure $(A)_t$. Thus in total, $H_s^k(z_1, ..., z_k)$ asserts, for $t \in s$, the existence of functions G_t such that

$$(\mathfrak{M}, Z, <, G_t)_{t \in s} \models A_t, B_{t_1 t_2}, C_t,$$

where $Z = \{z_1, ..., z_k\}$ is ordered by $z_i < z_j$ iff $i < j$.

DEFINITION. The theory CFA^H is, by definition, the set of all Henkin sentences of the form $\exists z_1 \cdots z_k H_s^k(z_1, ..., z_k)$. The theory CFA_n^H consists of all $\exists z_1 \cdots z_k H_s^k(z_1, ..., z_k)$ such that $\#(t) \leq n$ for all $t \in s$.

3.8. LEMMA. *If $\mathfrak{M}$ is a gap $> n$ model then $\mathfrak{M} \models \text{CFA}_n^H$.*

Proof. Immediate, from 3.5. ■

3.9. LEMMA. *Let $\mathfrak{M} = \langle M, U, ... \rangle$ be a resplendent model of CFA^H. Then $\mathfrak{M}$ satisfies condition $(+)$.*

Proof. By Theorem 2.8 of Barwise (1976) and the resplendency of $\mathfrak{M}$, $\mathfrak{M}$ is a model of

$$\exists z_1 \cdots z_k \cdots \left. \begin{cases} \forall \mathbf{x}_t \exists v_t \\ \forall \mathbf{x}_t' \exists v_t' \\ (\text{all } t) \end{cases} \right\} \begin{cases} z_1, ..., z_k, ... \text{ are all distinct, not in } U \\ \text{and all instances of (2)–(8) hold.} \end{cases}$$

Let $G_t(\mathbf{x}_t) = v_t$, let $Z = \{z_1, z_2, ...\}$, ordered by $z_i < z_j$ iff $i < j$. Then $(\mathfrak{M}, Z, <, G_t)_{t \in L^*}$ is a model of CFA^+. The only trouble is that Z is countable. But, by resplendency again, since CFA^+ can be made true in an expansion of $\mathfrak{M}$, it can be made true in an expansion where $\text{Card}(Z) = \text{Card}(M)$. ■

3.10. *The Final First Order Theory CFA of L*

The theory CFA simply consists of all first order approximations of the theory CFA^H. Similarly, CFA_n is the set of first order approximations to sentences in CFA_n^H. Thus $\text{CFA} = \bigcup_n \text{CFA}_n$.

3.11. THEOREM. (a) *Every gap $> n$ model is a model of CFA_n. Hence every gap ω model is a model of CFA.*

(b) (Downward Lowenheim-Skolem for two cardinals.) *Let $\mathfrak{M} = \langle M, U, ... \rangle$ be a resplendent model of CFA of power μ. Then, for each type (κ, λ), $\lambda \leq \kappa \leq \mu$, there is an elementary submodel $\mathfrak{M}_0 < \mathfrak{M}$ of type (κ, λ). Moreover, $\mathfrak{M}_0$ can be chosen to contain any preassigned set $X \subseteq U$ of power $\leq \lambda$.*

Proof. Part (a) is immediate from 3.8. Assume the hypothesis of (b). Since resplendent models are uniform, $\text{Card}(U) = \text{Card}(M)$ so $\mathfrak{M}$

is of type (μ, μ). Since $\mathfrak{M}$ is resplendent, CFA implies CFA^H, by Theorem 2.4 of Barwise (1976). Thus, by 3.9, $\mathfrak{M}$ satisfies condition (+). The conclusion follows from Lemma 3.6. ∎

3.12. COROLLARY. CFA *is a set of axioms for the first order properties of gap ω models.*

Proof. Immediate from 3.11. ∎

3.13. COROLLARY. (Upward Lowenheim-Skolem for two cardinals.) *Let $\mathfrak{M} = \langle M, U, ... \rangle$ be a resplendent model of CFA of power λ. For any $\kappa \geqslant \lambda$ there is an elementary extension $\mathfrak{M}' = \langle M', U, ... \rangle$ with the same U of type (κ, λ).*

Proof. Let T' be the recursive theory of an expansion $L' \supseteq L$ used by Vaught (1965) in the original proof of his theorem. By 3.12, CFA is a set of axioms for the L-consequences of T'. Thus, by resplendency, $\mathfrak{M}$ can be expanded to a model $\mathfrak{M}'$ of T'. But every model $\mathfrak{M}'$ of T' has a proper elementary extension with the same U. Iterate this κ times. ∎

Thus we see that Morley's proof leads to a downward Lowenheim-Skolem Theorem while Vaught's leads to an upward result. Since downward results are always much more constructive than upward results, it is not surprising that Morley's proof is the one that yields an explicit set CFA of axioms for two cardinal models.

University of Wisconsin, Madison
University of California, Los Angeles

NOTES

* The preparation of this paper was supported by NSF Grant No. MPS74-06355-1.
[1] The author is an Alfred P. Sloan Fellow.
[2] This axiomatization provides a solution to Open problem 11 in Chang-Keisler (1973). A different solution was obtained independently by J. Schmerl (to appear).
[3] They were first applied by a recursion-theorist at work on a book on model theory.

BIBLIOGRAPHY

Barwise, J.: to appear, 'Some Applications of Henkin Quantifiers, *Israel J. Math.*
Barwise, J. and Schlipf, J.: (1976), 'An Introduction to Recursively Saturated and Resplendent Models', *J. Symbolic Logic.*

Baumgartner, J. E.: to appear, 'Canonical Partition Relations', *J. Symbolic Logic*.

Chang, C. C. and Keisler, H. J.: 1973, *Model Theory*, North Holland, Amsterdam.

Feferman, S.: 1974, 'Applications of Many-sorted Interpolation Theorems', in the *Proceedings of the Tarski Symposium*, Amer. Math. Soc., Providence, pp. 205–224.

Kunen, K.: to appear, 'Combinatorics', *Handbook of mathematical Logic*, North-Holland, Amsterdam.

Schmerl, J.: to appear, 'An Axiomatization for a Class of Two-cardinal Models'.

Vaught, R.: 1965, 'A Lowenheim-Skolem Theorem for Cardinals far Apart', *The Theory of Models*, North Holland, Amsterdam, pp. 390–401.

Walkoe, W. J., Jr.: 1970, 'Finite partially-Ordered Quantification', *J. Symbolic Logic* **35**, 535–555.

Note Added in Proof. June 10, 1976: Schmerl, in a paper to appear in the *Proceedings of the A.M.S.*, has given a strong negative answer to Problem 2.11. A similar negative answer was obtained by Shelah. Shelah also pointed out that every first-order sentence σ is equivalent to a universal mod U theory T in an expanded language. He further points out that Theorem 2.8 remains true if one allows unary function symbols in L.

JEAN-YVES GIRARD

FUNCTIONAL INTERPRETATION
AND KRIPKE MODELS

In this paper, we shall investigate the relationship between Gödel's functional interpretation and validity in a Kripke structure built up from the concept of functional system.

– a functional system is defined by a set of equations $t = u$, with *free variables*. Then a formula $A^* = \exists x^\sigma \forall y^\tau A'[x, y]$ is said to be *valid* in functional system F if there is a term t of F of type σ such that $A[t, y]$ is derivable in $F \cdot (A[t, y]$ may be brought to the form $u = o)$. It turns out that one may restrict to equations $t = t'$, where the terms t and t' can only take the values 0 and 1. Then it is simpler to add a specific type fm for such terms, with two distinguished elements $\mathbf{T}$ and $\mathbf{\bot}$.

Among functional systems, two are especially interesting

$+ T$, which is the smallest one, and has the nice property of having a recursive set of provable equations $t = u$ (t, u of type fm)

$+ IT$, which is adequate for the functional interpretation of HA. In the paper, functional systems are treated as logical systems; this is a way of proving more direct (and more clear) results.

– the Kripke structure $\mathbb{K}$ is obtained by taking as indexing set the set of functional systems, since a functional system defines clearly a structure for the language of HA (with domain the closed terms of type o); the (pre)-ordering is the relation 'is an extension'.

One proves that A is valid at level F in $\mathbb{K}$ iff the functional interpretation A^* is valid in the functional system F.

So, in the restricted Kripke structure $\mathbb{K}(IT)$ every provable formula of HA is valid.

– As a corollary, we get the following theorem:

Let A be a purely universal formula $\forall x p(x)$ of HA, with p primitive recursive predicate, interpreted by the term p^* in system T; then $\forall x p(x)$ is provable in HA if and only if there are closed q, r of T such

Butts and Hintikka (eds.), Logic, Foundations of Mathematics and Computability Theory, 33–57.
Copyright © 1977 *by Reidel Publishing Company, Dordrecht-Holland. All Rights Reserved.*

that

$$T \vdash q(\bar{o}) = \mathbf{T}$$
$$T \vdash q(x) = \mathbf{T} \to q(Sx) = \mathbf{T}$$
$$T \vdash q(r(x)) = \mathbf{T} \to p^*(x) = \mathbf{T}.$$

(Recall that the set of quantifier-free B such that $T \vdash B$ is recursive).

A part of this paper is a new version of part IV of the author's doctoral dissertation, namely the properties of T(VF in [1]), while IT is an improvement of the former VI in [1]. (In [1], more general functional systems are considered; it follows that results similar to Theorem IV.3.2 hold for provability of purely universal formulas in the full theory of finite types).

We have assumed that the reader is familiar with the functional interpretation. All basic knowledge may be found in [7]. Similarly for natural deduction: see [4].

We have avoided all normalization and Church-Rosser properties proofs, since this is only routine. (However, complete proofs of the properties needed in the paper may be found in [1]).

I am grateful to Joël Combase for the stimulating discussions he had with me on the subject of this paper.

I. FUNCTIONAL SYSTEMS

In the sequel, we shall consider a very general definition of 'functional systems'. However, there is an important restriction on the systems described here: the set of all types is fixed in advance.

1. Types

DEFINITION 1.1. The *types* are given by the following inductive definition:
- o is a type.
- fm is a type.
- if σ and τ are types, then $\sigma \to \tau$ is a type.
- if σ and τ are types, then $\sigma \times \tau$ is a type.

REMARKS 1.2. o is the type of integers; fm is the type of truth-values; $\sigma \to \tau$ is the type of functions (for instance constructive) from σ to τ; $\sigma \times \tau$ is the Cartesian product of σ and τ.

We shall denote by $(\sigma_1, \ldots, \sigma_n \to \tau)$ the type $\sigma_1 \to (\cdots \to (\sigma_n \to \tau) \cdots)$.

We shall denote by $(\sigma \times \tau \to \rho)$ the type $(\sigma \times \tau) \to \rho$, and by $(\rho \to \sigma \times \tau)$ the type $\rho \to (\sigma \times \tau) \cdots$.

2. Functional Languages

A functional language is a language with several sorts of objects (one sort for each type described in 1.)

2.1. Objects common to all languages

—variables x^σ, y^σ, z^σ, ... for each type σ.
—the binary predicate letter $=$ (the two places of this predicate are of type fm)
—the connectives $\wedge$, $\vee$, $\to$, $\neg$.
—the typed quantifiers $\forall x^\sigma$ and $\exists x^\sigma$ for each type σ.
—the symbol AP:
—the constants

(i) $\bar{o}$ of type o; S of type $o \to o$.
(ii) R^σ of type $(\sigma, (\sigma, o \to \sigma), o \to \sigma)$ for each type σ.
(iii) $\mathbf{T}$ and $\bot$ of type fm.
(iv) D^σ of type $(\sigma, \sigma, fm \to \sigma)$ for each type σ.
(v) $\Pi^{\sigma,\tau}$ of type $(\sigma, \tau \to \sigma)$ for all types σ and τ.
(vi) $\Sigma^{\rho,\sigma,\tau}$ of type $((\rho, \sigma \to \tau), (\rho \to \sigma), \rho \to \tau)$ for all types ρ, σ and τ.
(vii) $\otimes^{\sigma,\tau}$ of type $(\sigma, \tau \to \sigma \times \tau)$ for all types σ and τ.
(viii) $\Pi_1^{\sigma,\tau}$ and $\Pi_2^{\sigma,\tau}$ of respective types $(\sigma \times \tau \to \sigma)$ and $(\sigma \times \tau \to \tau)$ for all types σ and τ.

2.2. Proper symbols

For each type σ, a set (at most denumerable) of objects of type σ. These objects are the *proper constants* of the language.

2.3. *Terms*

Let L be a functional language; the terms of L are given inductively:

  − the variables x^σ, y^σ, z^σ are terms of type σ.

  − a constant (proper or not) of type σ is a term of type σ.

  − if t and u are terms of respective types $\sigma \to \tau$ and τ, then $APtu$ is a term of type τ.

2.4. *Notational conventions*

$APtu$ will be abbreviated as $t(u)$; $t(u_1) \cdots (u_n)$ will be abbreviated as $t(u_1, ..., u_n)$. $a \otimes b$ will abbreviate $\otimes(a, b)$.

Unless absolutely necessary, all types will be dropped from terms and formulas; it will not be difficult to find the possible types of an expression (and its subexpressions) provided we know that all written expressions are correct.

If the variable x occurs (or not occurs!) in t, we may denote t by $t[x]$; if u has the same type as x, we shall denote by $t_x[u]$, or even $t[u]$ the term obtained by replacing in t every occurrence of x by u. Notice the difference between ( ) and [ ].

2.5. *λ-abstractor*

Let $t = t[x]$ a term of type τ, and let x be of type σ. We recall that $\lambda x t = \lambda x t[x]$ is defined to be the term of type $\sigma \to \tau$ given by:

$$\lambda xx = \Sigma(\Pi, \Pi)$$

$$\lambda xu = \Pi(u) \text{ if } x \text{ does not occur in } u$$

$$\lambda x \cdot (u(v)) = \Sigma(\lambda xu, \lambda xv) \text{ otherwise}$$

2.6. *Formulas* (inductive definition)

  − if t and u are terms of type fm, then $t = u$ is an (atomic) *formula*.

  − if A and B are formulas, $A \to B$, $A \wedge B$, $A \vee B$, $\neg A$, $\forall x^\sigma A$, $\exists x^\sigma B$ are formulas (x^σ variable of any type σ).

2.7. Free and bound variables are defined as usual. The technical problem of free and bound variables is supposed to be solved by an arbitrary method.

If σ is a type distinct from fm, we define a binary predicate $\simeq$ by $x \simeq y$: $\forall z^{\sigma \to fm}(z(x) = z(y))$. We agree that, if x and y are of type fm, $x \simeq y$ is simply $x = y$; the same for terms.

3. Functional Theories

3.1. Axioms (and rules) common to all functional theories

 – the axioms and rules of *classical* predicate calculus (with infinitely many types)

 – the equality axioms: $x = x$; $x = y \to y = x$; $x = y \wedge y = z \to x = z$; $x = y \to Z(x) = Z(y) \cdot (x, y, z$: type fm; Z of type $fm \to fm)$.

 – The axioms

(T1) $R(x, y, \bar{o}) \simeq x$

(T2) $R(x, y, S(z)) \simeq y(R(x, y, z), z)$

(T3) $D(x, y, \mathbf{T}) \simeq x$

(T4) $D(x, y, \mathbf{\perp}) \simeq y$

(T5) $\Pi(x, y) \simeq x$

(T6) $\Sigma(x, y, z) \simeq x(z, y(z))$

(T7) $\Pi_1(x \otimes y) \simeq x$

(T8) $\Pi_2(x \otimes y) \simeq y$

(T9) $(\lambda xx(z))(x) \simeq x(z)$; $(\lambda xy(x(z)))(x) \simeq y(x(z))$

$\quad\quad\quad (\lambda xy(z(x)))(x) \simeq y(z(x))$

 – the axioms

$$x = \mathbf{T} \vee x = \mathbf{\perp} \quad\quad (x \text{ of type } fm)$$

$$\mathbf{T} \neq \mathbf{\perp}.$$

3.2. Proper Axioms

Every set of axioms of the form

$\quad\quad\quad\quad t = u$, where t and u are (non necessarily closed) terms of type fm.

So, the non-logical axioms (proper or not) of a functional theory are quantifier-free (or purely universal).

3.3. *Properties of* $\simeq$

PROPOSITION. (i) *for each type* σ, $\simeq$ *is provably an equivalence relation between terms of type* σ.

(ii) *In all functional theories, one can prove* $x \simeq x' \wedge y \simeq y' \rightarrow x(y) \simeq x'(y')$ (y, y': *type* σ; x, x': *type* $\sigma \rightarrow \tau$).

(iii) $(\lambda x t)(u) \simeq t_x[u]$.

Proof. (i) is trivial if σ is fm; the other case is also trivial.

(ii) We show that, if x is of type $\sigma \rightarrow \tau$ and y of type σ,

$$x \simeq x' \rightarrow x(y) \simeq x'(y) \qquad \text{and} \qquad y \simeq y' \rightarrow x(y) \simeq x(y').$$

First case: $\tau = fm$; then, if $y \simeq y'$ then trivially $x(y) \simeq x(y')$ (consider the two subcases $\sigma = fm$ and $\sigma \neq fm$). If $x \simeq x'$, then, by (T9) $x(y) \simeq (\lambda x x(y))(x)$ and $x'(y) \simeq (\lambda x x(y))(x')$; so $x(y) \simeq x'(y)$.

Second case: $\tau \neq fm$; let $y \simeq y'$ and let z of type $\tau \rightarrow fm$; then by (T9) $z(x(y)) \simeq (\lambda yz(x(y)))(y)$ and $z(x(y')) \simeq (\lambda yz(x(y)))(y')$, so $z(x(y)) \simeq z(x(y'))$. Suppose now $x \simeq x'$; since $z(x(y)) \simeq (\lambda xz(x(y)))(x)$ and $z(x'(y)) \simeq (\lambda xz(x(y)))(x)$ (again T9), we get $z(x(y)) \simeq z(x'(y))$.

(iii) By induction over the length of t, using (ii):

$- (\lambda x x)(u) \simeq \Sigma(\Pi, \Pi, u) \simeq \Pi(u, \Pi(u)) \simeq u$.

$- (\lambda x t)(u) \simeq \Pi(t, u) \simeq t$ (x does not occur in t).

$- (\lambda x t(t'))(u) \simeq \Sigma(\lambda x t, \lambda x t', u) \simeq (\lambda x t)(u)((\lambda x t')(u)) \simeq (t_x[u])(t'_x[u])$.

COROLLARY. *The predicates* $\simeq$ *are equality predicates in functional theories.* (*That is* $x^\sigma \simeq y^\sigma \rightarrow (A[x] \leftrightarrow A[y])$ *is provable*).

4. *Functional Systems*

4.1. *Definition*

A functional theory F is called a *functional system* if
(i) F is consistent
(ii) for all closed terms t of type fm, $F \vdash t = \mathbf{T}$ or $F \vdash t = \mathbf{\perp}$.

4.2. *Example: the system T*

Let T be the functional theory without proper constants and proper axioms.

THEOREM (Hanatani, Tait).   (i) *T is a functional system.*

(ii) *for all closed terms t of type o, there exists an unique integer n such that $T \vdash t \simeq \bar{n}$ (as usual $\bar{n}$ is $S(\cdots S(\bar{o}) \cdots)$).*

Proof.   Easy modification of the proof in [7] or in [3].

4.3. *Remarks*

Question: It is clear that there is no elementary proof of 4.2(ii) since (ii) implies the consistency of arithmetic. However, the weaker statement 4.2(i) does not seem to have a great strength. So one may ask for an elementary proof of (i), that is, since the consistency of T is a trivial consequence of the Church-Rosser property, that every closed term of type *fm* is provably equal to **T** or **⊥**. Moreover, do we have an elementary *weak* normalization theorem for closed terms of type *fm*? (strong normalization implies normalization for closed terms of type *o*).

Let us remark that in a recursively axiomatizable functional system, the set $\{(t, u); t, u$ closed of type *fm*, and $F \vdash t = u\}$ is recursive.

4.4. *Extensions*

Let F and F' be functional systems; F' is said to be an extension of F if F' is an extension of F when considered as logical systems, namely, every constant of F is in F', and every axiom of F is provable in F'.

5. *Useful Terms*

$\bar{n}$: $\overline{n+1} = S(\bar{n})$ ($\bar{o}$ is the constant $\bar{o}$)

O: $O^o = \bar{o}$; $O^{fm} = \mathbf{T}$; $O^{\sigma \to \tau} = \Pi(O^\tau)$; $O^{\sigma \times \tau} = O^\sigma \otimes O^\tau$.

$\hat{+}$, $\hat{\cdot}$, $\doteq$ of type $(o, o \to o)$ and P of type $(o \to o)$ such that

$\hat{+}$: $x \hat{+} \bar{o} \simeq x$; $x \hat{+} s(y) \simeq S(x \hat{+} y)$

$\hat{\cdot}$: $x \hat{\cdot} \bar{o} \simeq \bar{o}$; $x \hat{\cdot} S(y) \simeq (x \hat{\cdot} y) \hat{+} x$

P: $P(\bar{o}) \simeq \bar{o}$; $P(S(x)) \simeq x$.

$\doteq$: $x \doteq \bar{o} \simeq x$; $x \doteq S(y) \simeq P(x \doteq y)$.

E of type $(o, o \rightarrow fm)$: $E(\bar{o}, \bar{o}) = \mathbf{T}$; $E(S(x), \bar{o}) = E(\bar{o}, S(x)) = \mathbf{\bot}$; $E(S(x), S(y)) = E(x, y)$.

$\hat{\wedge}$, $\hat{\vee}$, $\hat{\rightarrow}$, $\hat{\leftrightarrow}$ of type $(fm, fm \rightarrow fm)$ and $\hat{\neg}$ of type $(fm \rightarrow fm)$, corresponding to the usual propositional connectives; for instance $\hat{\neg}$ is $D^{fm}(\mathbf{\bot}, \mathbf{T})$, while $\hat{\vee}$ is $D^{fm}(\mathbf{T})$.

II. Properties of functional systems

1. Proposition

Let A be a prenex formula of a functional theory F. Then A is provably equivalent to a formula $Q_1 x_1^{\sigma_1} \cdots Q_n x_m^{\sigma_n} t(x_1, ..., x_n) = \mathbf{T}$, where the quantifiers Q_i are such that $Q_i \neq Q_{i+1}$ $(i < n)$, and t is a term of type $(\sigma_1, ..., \sigma_n \rightarrow fm)$.

Proof. (i) Every quantifier-free formula is equivalent to a formula $u = \mathbf{T}$:

$$- t = t' \leftrightarrow (t \hat{\leftrightarrow} t') = \mathbf{T}$$
$$- \neg (t = \mathbf{T}) \leftrightarrow (\hat{\neg} t) = \mathbf{T}$$
$$- (t = \mathbf{T} \vee t' = \mathbf{T}) \leftrightarrow (t \hat{\vee} t') = \mathbf{T}$$

(ii) Two adjacent quantifiers of the same kind may be contracted; for instance $\exists x^{\sigma} \exists y^{\tau} A[x, y] \leftrightarrow \exists z^{\sigma \times \tau} A[\Pi_1 z, \Pi_2 z]$: $\rightarrow$ holds, because, if $A[x, y]$, then $A[\Pi_1 z, \Pi_2 z]$, with $z = x \otimes y$, using T7, T8 and corollary to Proposition I.3.3. Conversely, if $A[\Pi_1 z, \Pi_2 z]$, then take x to be $\Pi_1 z$ and y to be $\Pi_2 z$.

(iii) from (i) and (ii), one gets a formula of the form $Q_1 x_1 \cdots Q_n x_n u[x_1, ..., x_n] = \mathbf{T}$, with alternate Q_i's. ·Let t be $\lambda x_1 \cdots \lambda x_n u[x_1, ..., x_n]$. ∎

2. Theorem

Let F be a functional system, $A[x^{\sigma}, y^{\tau}]$ a formula of F with x and y as its only two free variables. Suppose condition (C) is satisfied:

(C) for any functional system F' which is an extension of F, and for any closed term a of type σ in F', there is a closed term b of type τ in F' such that $F' \vdash A[a, b]$.

Then there is a closed term c of type $\sigma \to \tau$ in F such that

$$F \vdash \forall x A[x, c(x)].$$

COROLLARY. *Suppose the formula $A[x^\sigma]$ of F has x as its only free variable. Then $F \vdash \forall x A$ iff for any functional system F' which is an extension of F, and for all closed a of type σ in F' $F' \vdash A[a]$.*

Remark. It will be clear from the proof below that the theorem and its corollary still hold if we restrict to systems F' which are obtained from F by adding only finitely many new constants.

Proof. Suppose (C) is true.

Let $(u_n)_{n \in \mathbb{N}}$ an enumeration of the closed terms of type $\sigma \to fm$ of F, and let a be a new constant of type σ.

Let f be a partial function from $\mathbb{N}$ into $2 = \{0, 1\}$ (when defined); we put: F_f = functional theory obtained from F by adding a, and the proper axioms

$$u_n(a) = \mathbf{T} \text{ if } f(n) \text{ defined and } f(n) = 0$$
$$u_n(a) = \mathbf{\bot} \text{ if } f(n) \text{ defined and } f(n) = 1.$$

LEMMA 1. *For all f in $2^{\mathbb{N}}$ (that is for all total f), there exists a closed term c_f of F such that $F_f \vdash A[a, c_f(a)]$.*

Proof. Either F_f is inconsistent, so we may take c_f to be $O^{\sigma \to \tau}$, or F_f is consistent. Let t be a closed term of type fm of F_f; by λ-abstraction, is provably equal to $u_n(a)$, for some n. Since f is total, $t = \mathbf{T}$ or $t = \mathbf{\bot}$ is provable. It follows that F_f is a functional system, and trivially an extension of F. From (C), one can find b_f such that $F_f \vdash A[a, b_f]$; let c_f a closed term of F with $c_f(a) \simeq b_f$ (c_f exists by λ-abstraction). ∎ Now, let U_A be the set of functions f from a segment $\{0, ..., n\}$ of $\mathbb{N}$ into 2, such that $\exists c_f F_f \vdash A[a, c_f(a)]$. For $f \in U_A$, let $O_f = \{g; g \in 2^{\mathbb{N}} \text{ and } g \text{ extends } f\}$.

LEMMA 2.

$$2^{\mathbb{N}} = \bigcup_{f \in U_A} O_f.$$

Proof. Let g be in 2^N, and let $c_g: F_g \vdash A[a, c_g(a)]$. Since finitely many axioms are needed to prove this theorem, we have $F_f \vdash A[a, c_g(a)]$ where f is the restriction of g to some initial segment; clearly $g \in O_f$ and $f \in U_A$. ∎

Since 2^N is compact, 2^N is the union of finitely many (open) O_f's:

$$2^N = O_{f_1} \cup \cdots \cup O_{f_n} \qquad (f_i \in U_A).$$

Since F_{f_i} is obtained from $F_\varnothing$ ($\varnothing$ is to the totally undefined function) by adding finitely many closed quantifier-free axioms, it is clear (see Proposition 1) that F_{f_i} is equivalent to the functional theory $F_\varnothing + p_i = \mathbf{T}$, where p_i is a closed term of type fm of $F_\varnothing$.

It is a routine exercise (familiar from propositional calculus) to show that $F_\varnothing \vdash p_1 = \mathbf{T} \vee \cdots p_n = \mathbf{T}$.

Since $f_i \in U_A$, we have terms c_i of F, such that

$$F_\varnothing \vdash p_i = \mathbf{T} \rightarrow A[a, c_i(a)]$$

Let p_i be equal to $q_i(a)$, where q_i is a closed term of F; then

$$F \vdash q_1(x) = \mathbf{T} \vee \cdots \vee q_n(x) = \mathbf{T}$$

and

$$F \vdash q_i(x) = \mathbf{T} \rightarrow A[x, c_i(x)].$$

Let $d_1, ..., d_n$ be the terms of type τ (with free variable x) given by $d_1 \simeq c_1(x)$; $d_{i+1} \simeq D(c_{i+1}(x), d_i, q_{i+1}(x))$.

One shows, by metamathematical induction over i that $q_i(x) = \mathbf{T} \wedge q_{i+1}(x) = \perp \cdots \wedge q_i(x) = \perp \rightarrow d_i \simeq c_i(x)$ is a theorem of F. It follows that $F \vdash q_1(x) = \mathbf{T} \vee \cdots \vee q_n(x) = \mathbf{T} \rightarrow A[x, d_n]$; so, $F \vdash A[x, d_n]$. The theorem follows now by taking c to be $\lambda x d_n$. ∎

3.   Theorem

Let a, b be closed terms of the functional system F of respective types $(\sigma, \tau \rightarrow fm)$ and $(\sigma \rightarrow fm)$. Suppose that

$$F \vdash \forall x^\sigma ((\forall y^\tau a(x, y) = \mathbf{T}) \rightarrow b(x) = \mathbf{T}).$$

Then there exists a closed term t of F such that (*t of type* $\sigma \to \tau$)

$$F \vdash \forall x (a(x, t(x)) = \mathbf{T} \to b(x) = \mathbf{T}).$$

Proof. Let e be a new constant of type σ, and let F_e be the functional theory obtained from F by adding e. Then

$$F_e \vdash (\forall y a(e, y) = \mathbf{T} \to b(e) = \mathbf{T}).$$

Let F' be $F_e + \forall y a(e, y) = \mathbf{T}$; then F' is a functional theory, since its axioms are purely universal. $F' \vdash b(e) = \mathbf{T}$.

As a corollary to the cut-elimination theorem applied to theories with purely universal axioms, we get that $b(e) = \mathbf{T}$ is provable from closed quantifier-free instanciations of the proper axioms of F'. (Sketch of the proof: using $\neg\neg$-translation, we consider the case of a closed quantifier-free formula A derived intuitionistically from purely universal closed formulas $B_1 \cdots B_n$ in the natural deduction system *NJ*; from the subformula property, every formula occurring in a cut-free proof must be at most purely universal. From this remark, the only rule applicable to a purely universal formula in the cut-free deduction is $\forall$-introd or $\forall$-elim. Since a $\forall$-elim cannot follow a $\forall$-introd in a cut-free proof, it is clear that each formula B_i is followed by the number of $\forall$-elim necessary to get a quantifier-free formula. If we drop in the deduction tree all formulas containing quantifiers, and put closed terms at the place of eventual free variables we get the property).[1]

So, there are $t_1, ..., t_n$ closed terms of F_e such that

$$F_e \vdash a(e, t_1) = \mathbf{T} \wedge \cdots \wedge a(e, t_n) = \mathbf{T} \to b(e) = \mathbf{T}.$$

Let $d_1, ..., d_n$ be defined by $d_1 \simeq t_1$; $d_{i+1} \simeq D(d_i, t_{i+1}, a(e, t_{i+1}))$; clearly, d_n is defined to be t_k, where k is the greatest integer such that $a(e, t_k) = \perp$, if there is one, 1 otherwise.

From this it follows clearly that $F_e \vdash a(e, d_n) = \mathbf{T} \to a(e, t_i) = \mathbf{T}$. So, $F_e \vdash a(e, d_n) = \mathbf{T} \to b(e) = \mathbf{T}$. Using λ-abstraction, and replacing e by x, one gets a term t such that $F \vdash a(x, t(x)) = \mathbf{T} \to b(x) = \mathbf{T}$. ∎

Remark.   The theorem is stated for a functional system F; however, inspection of the proof shows that it holds for a functional theory.

4.   *Intuitionistic Functional Theories*

PROPOSITION.   *If F is a functional theory, let F^* be the intuitionistic system with the same set of axioms. Then*

(i) *If $F^* \vdash \exists y A[y]$, there is a term t of F such that $F^* \vdash A[t]$*

(ii) *If F is a functional system, A and B closed and $F^* \vdash A \vee B$, then $F^* \vdash A$ or $F^* \vdash B$.*

Proof.   (i) Observe that the axioms of F are of the form $t = u$, except $x = \mathbf{T} \vee x = \mathbf{\bot}$ and $\mathbf{T} \neq \mathbf{\bot}$. From a cut-free proof in the natural deduction system NJ of $\exists y A[y]$ under the universal closure of the axioms of F, one extracts t, by induction on the length of the proof: let the last rule be a $\exists$-introd; then the immediate formula before $\exists y A[y]$ is $A[t]$ for some t. Otherwise, the last rule must be a $\vee$-elim applied to $u = \mathbf{T} \vee u = \mathbf{\bot}$ for some term u of type fm. This means that $\exists y A[y]$ is provable in F_1^* and F_2^* with strictly shorter proofs, with $F_1 = F + u = \mathbf{T}$, $F_2 = F + u = \mathbf{\bot}$; let t_1 and t_2 be the terms obtained from these proofs; then we may take t to be $D(t_1, t_2, u)$.

(ii) $A \vee B$ is provably equivalent in F^* to $\exists x^{fm}(x = \mathbf{T} \wedge A) \vee (x = \mathbf{\bot} \wedge B)$. By (i) there is a term t such that $F^* \vdash (t = \mathbf{T} \wedge A) \vee (t = \mathbf{\bot} \wedge B)$; since A and B are closed, t may be chosen closed; since F is a functional system, $F^* \vdash t = \mathbf{T}$ or $F^* \vdash t = \mathbf{\bot}$; from this, it follows that $F^* \vdash A$ or $F^* \vdash B$. ∎

III.   The Functional Systems T and IT

1.   *System T*

We shall prove that the class of purely universal theorems of T is recursive.

1.1.   *Valuations*

In the sequel, we shall consider the terms of T built up from a given denumerable sequence of variables.

Call a term *weakly normal* if none of its subterms are of the form:

$$R(a, b, \bar{o}), \quad R(a, b, S(t)), \quad \Pi(a, b), \quad \Sigma(a, b, c), \quad \Pi_1(a \otimes b),$$
$$\Pi_2(a \otimes b), \quad D(a, b, \mathbf{T}), \quad D(a, b, \mathbf{\bot}).$$

Call a term *strongly normal* if it is weakly normal and none of its subterms is a term of type *fm* distinct from $\mathbf{T}$ and $\mathbf{\bot}$. (So, for closed terms, weakly normal terms are strongly normal).

Call a term a *key-term* if it is of type *fm*, it is weakly normal, not strongly normal, and all its strict subterms are strongly normal. (Example: a variable of type *fm* is a key-term. Clearly every weakly normal not strongly normal term has a key-term among its subterms).

Let $(u_n)_{n \in \mathbb{N}}$ an enumeration of key-terms.

A *valuation* is a member of $2^{\mathbb{N}}$

1.2. Reductions

Let f be a valuation, that is $f \in 2^{\mathbb{N}}$. We give here an inductive definition of the relation $a = /_f b$ '*a f-reduces to b*': $a = /_f a$; if $a = /_f b$ and $b = /_f c$, then $a = /_f c$; if $a = /_f a'$ and $b = /_f b'$, then $a(b) = /_f a'(b')$.

$$R(a, b, \bar{o}) = /_f a; \qquad R(a, b, S(c)) = /_f b(R(a, b, c), c);$$
$$D(a, b, \mathbf{T}) = /_f a; \qquad D(a, b, \mathbf{\bot}) = /_f b.$$
$$\Pi_1(a \otimes b) = /_f a; \qquad \Pi_2(a \otimes b) = /_f b$$
$$\Pi(a, b) = /_f a; \qquad \Sigma(a, b, c) = /_f a(c, b(c))$$
$$u_n = /_f \mathbf{T} \quad \text{if} \quad f(n) = 0; \qquad u_n = /_f \mathbf{\bot} \quad \text{if} \quad f(n) = 1.$$

Similarly, one defines $a = /b$ by dropping the last two clauses which are the only ones which make use of f.

1.3. Theorem

Let f be a valuation; then for any term t, there is a unique strongly normal term u such that $t = /_f u$.

Proof. Existence: trivial variant of known proofs of the literature (for instance [6]). Unicity: from the Church-Rosser property: if $a = /_f a'$ and $a = /_f a''$, there is an a''' such that $a' = /_f a'''$ and $a'' = /_f a'''$.

From a given proof of the Church-Rosser property, it is easy to get a proof of this property for $=/_f$ (using the fact that if t is a key-term, then t has the property). ■

1.4. Weak validity

Let t be a term of type fm; we shall say that $t = \mathbf{T}$ is *weakly valid* if for all $f \in 2^{\mathbb{N}}$ $t = /_f \mathbf{T}$.

THEOREM. *The set $\{t; t = \mathbf{T}$ weakly valid$\}$ is recursive.*

Proof. Let $\varphi, \psi \cdots$ range over functions with domain a finite segment $[0, n]$, and values into $2 = \{0, 1\}$, and define $=/_\varphi$ as $=/_f$; then $a = /_f b$ iff there is φ, f extends φ and $a = /_\varphi b$.

If $t = \mathbf{T}$ is not weakly valid, one can find a φ such that $t = /_\varphi \mathbf{\bot}$ (and conversely).

If $t = \mathbf{T}$ is weakly valid, there exists $\varphi_1, ..., \varphi_n$ such that every f extends at least one φ_i, and $a = /\varphi_i \mathbf{T}$ $(i = 1 \cdots n)$ (and conversely). So weak validity and its negation are Σ_1^0. ■

1.5. Weak validity and system T

THEOREM. $t = \mathbf{T}$ *is weakly valid iff* $T \vdash t = \mathbf{T}$.

Proof. (i) Let F be a functional system which extends T, $c_1, ..., c_n, ...$ of F (corresponding to the free variables $x_1, ..., x_n, ...$).
Let f be defined by

$$f(k) = 0 \quad \text{if} \quad F \vdash u_k[c_1, ..., c_n, ...] = \mathbf{T}$$
$$f(k) = 1 \quad \text{if} \quad F \vdash u_k[c_1, ..., c_n, ...] = \mathbf{\bot}.$$

Then, since F is a functional system, and $u_k[c_1, ..., c_n, ...]$ is closed, it follows that f is a total function.
Trivially, we have:

$$a = /_f b \rightarrow F \vdash a[c_1, ..., c_n, ...] \simeq b[c_1, ..., c_n, ...].$$

Suppose $t = \mathbf{T}$ is weakly valid; then for any closed $d_1, ..., d_n$ of F corresponding to the free variables of t, we have $F \vdash t[d_1, ..., d_n] = \mathbf{T}$; from this and corollary to Theorem II.2, we conclude that $T \vdash t = \mathbf{T}$.

(ii) Conversely, observe that a valuation defines a model of T: objects of type σ: terms of type σ with free variables; equality of type fm: to have the same $=/_f$ normal form. So, if $T \vdash t = \mathbf{T}$, then $t =/_f \mathbf{T}$ for all valuations f.

COROLLARY. *The set of purely universal theorems of T is recursive.*

Proof. Trivial consequence of the two last theorems. ∎

The similarity between the quantifier-free part of T and the propositional calculus is emphasized by the

1.6. *Theorem* (universal interpolation)

Let A be a quantifier-free formula of T, y a variable. Then there exists a quantifier-free B with $\mathrm{Var}\,(B) \subset \mathrm{Var}\,(A) - \{y\}$ *such that*

$$T \vdash B \rightarrow A$$

and for all C quantifier-free not depending on y such that

$$T \vdash C \rightarrow A, \quad \text{then} \quad T \vdash C \rightarrow B$$

Proof. Remark that in T, we have the conjunctive normal form theorem: every quantifier-free formula is equivalent to a formula $\bigwedge_i \bigvee_j a_{ij}$ with a_{ij} of the form $t_{ij} = \mathbf{T}$ or $t_{ij} = \mathbf{\bot}$ for key-terms t_{ij}. (This is trivial from the calculus of valuations.)

Now, if A is in conjunctive normal form, let B be obtained by dropping all a_{ij} depending on y (with the convention· that a void disjunction is $\mathbf{\bot}$ and a void conjunction is $\mathbf{T}$). B will clearly be an universal left interpolant for the variable y and A.

2. *System IT*

Let u be the term of type $(o \rightarrow fm) \rightarrow (o \rightarrow o)$ given by

$$u(x, \bar{o}) \simeq o$$

$$u(x, S(y)) \simeq \begin{cases} u(x, y) & \text{if} \quad x(y) \overset{\cdot}{\rightarrow} x(S(y)) = \mathbf{T}. \\ y & \text{otherwise} \end{cases}$$

2.1. *Definition of IT*

DEFINITION 2.1.1. *IT* is the functional theory obtained from T by adding the induction axiom: $(x(\bar{o}) \wedge (x(u(x, z)) \stackrel{\rightarrow}{\rightarrow} x(S(u(x, z))))) \stackrel{\rightarrow}{\rightarrow} x(z)) = \mathbf{T}$.

PROPOSITION 2.1.2. $IT \vdash p = \mathbf{T}$ *iff there exist q of type $o \to fm$, t of type o, such that*

$$T \vdash q(\bar{o}) = \mathbf{T}$$
$$T \vdash (q(x) \stackrel{\rightarrow}{\rightarrow} q(S(x))) = \mathbf{T}$$
$$T \vdash (q(t) \stackrel{\rightarrow}{\rightarrow} p) = \mathbf{T}$$

and $\mathrm{var}\,(q) \subset \mathrm{var}\,(p)$, $\mathrm{var}\,(t) \subset \mathrm{var}\,(p)$, $x \notin \mathrm{var}\,(p)$.

Proof. $\to$: from Theorem II.3 (which holds for functional theories), if $IT \vdash p = \mathbf{T}$, there are terms r and t of respective types $o \to fm$ and o, such that $(r(\bar{o}) \wedge (r(u(r, t)) \stackrel{\rightarrow}{\rightarrow} r(S(u(r, t)))) \stackrel{\rightarrow}{\rightarrow} r(t)) = \mathbf{T} \to p = \mathbf{T}$ is provable in T.

Let x be a new variable of type o, and let $Q[x]$ be the quantifier-free formula: $r(\bar{o}) = \mathbf{T} \wedge (r(u(r, x)) \stackrel{\rightarrow}{\rightarrow} r(S(u(r, x))) = \mathbf{T}) \to r(x) = \mathbf{T}$; Clearly $Q[\bar{o}]$ and $Q[t] \to p = \mathbf{T}$ are provable in T; we show that $Q[x] \to Q[S(x)]$ is provable in T:

  — either $r(x) \stackrel{\rightarrow}{\rightarrow} r(S(x)) = \mathbf{T}$; then $u(r, S(x)) \simeq u(r, x)$; in this case, it amounts to verify that $(P[x] \to r(x) = \mathbf{T}) \to (P[x] \to r(S(x)) = \mathbf{T})$, for a formula $P[x]$: $r(\bar{o}) = \mathbf{T} \wedge (r(u(r, x) \stackrel{\rightarrow}{\rightarrow} r(S(u(r, x))) = \mathbf{T})$

  — otherwise $r(x) \stackrel{\rightarrow}{\rightarrow} r(S(x)) = \bot$; then $u(r, S(x)) \simeq x$; so $Q(S(x))$ is equivalent to $r(\bar{o}) = \mathbf{T} \wedge (r(x) \stackrel{\rightarrow}{\rightarrow} r(s(x)) = \mathbf{T}) \to r(S(x)) = \mathbf{T}$ which is trivially true.

One may clearly replace $Q[x]$ by $q(x) = \mathbf{T}$ for a q of type $o \to fm$. $\leftarrow$: if $q(\bar{o}) = \mathbf{T}$, $q(x) \stackrel{\rightarrow}{\rightarrow} q(S(x)) = \mathbf{T}$ are provable, then $q(u(q, t)) \stackrel{\rightarrow}{\rightarrow} q(S(u(q, t))) = \mathbf{T}$ is provable; the induction axiom gives us $q(t) = \mathbf{T}$; if $(q(t) \stackrel{\rightarrow}{\rightarrow} p) = \mathbf{T}$ is provable, then $p = \mathbf{T}$ is provable. ■

COROLLARY 2.1.3. *IT is a functional system.*

Proof. It suffices to show that *IT* is consistent; suppose $IT \vdash \mathbf{T} = \bot$. Then, by Prop. 2.1.2, there are closed q, t, such that $T \vdash q(\bar{o}) = \mathbf{T}$; $T \vdash q(x) \stackrel{\rightarrow}{\rightarrow} q(S(x)) = \mathbf{T}$; $T \vdash q(t) = \bot$. By Theorem I.4.2 (ii), one may suppose that t is $\bar{n}$ for some n; then we get a contradiction in T, which is consistent. ∎

COROLLARY 2.1.4. *Let F be a functional system which is an extension of IT; then we have the derived rule: if $F \vdash A[\bar{o}]$ and $F \vdash A[x] \to A[S(x)]$ then $F \vdash A[x]$ for any quantifier-free formula A.*

Proof. Similar to the proof of 2.1.2, replacing A by $q = \mathbf{T}$. ∎

2.2. *Induction over purely universal formulas*

TECHNICAL LEMMAS 2.2.1. *Recall the terms E, P, $\dot{-}$ of I.5; then the following are theorems of IT,*

(i) $E(x, x) = \mathbf{T}$

(ii) $E(x, \bar{o}) = \mathbf{T} \vee E(x, S(P(x))) = \mathbf{T}$

(iii) $E(x, \bar{o}) = \mathbf{T} \wedge p(x) = \mathbf{T} \to p(\bar{o}) = \mathbf{T}$

(iv) $E(x, \bar{1}) = \mathbf{T} \wedge p(x) = \mathbf{T} \to p(\bar{1}) = \mathbf{T}$

(v) $E(x, S(P(x))) = \mathbf{T} \wedge p(x) = \mathbf{T} \to p(S(P(x))) = \mathbf{T}$

(vi) $E(x, y) = \mathbf{T} \to E(P(x), P(y)) = \mathbf{T}$

(vii) $E(S(x) \dot{-} S(y), x \dot{-} y) = \mathbf{T}$

(viii) $E(S(x) \dot{-} x, \bar{1}) = \mathbf{T}$

(ix) $E(S(x) \dot{-} S(x), \bar{o}) = \mathbf{T}$

(x) $E(x \dot{-} P(y), \bar{o}) = \mathbf{T} \to E(x \dot{-} y, \bar{o}) = \mathbf{T}$

Proof. (i) Because $E(\bar{o}, \bar{o}) = \mathbf{T}$ and $E(S(x), S(x)) = E(x, x)$.

(ii) Using (i), one gets $E(\bar{o}, \bar{o}) = \mathbf{T}$ and $E(S(x), S(P(S(x)))) = \mathbf{T}$.

(iii) using $E(\bar{o}, \bar{o}) = \mathbf{T}$ and $E(S(x), \bar{o}) = \mathbf{\bot}$.

(iv) basis: trivial; induction step: (iii)

(v) trivial

(vi) using (ii), (iii), (v), there are 4 cases to consider, for instance, $E(x, S(P(x))) = \mathbf{T}$ and $E(y, S(P(y))) = \mathbf{T}$; by (v), $E(x, y) = E(S(P(x)), S(P(y)))$; from this, $E(x, y) = E(P(x), P(y))$.

(vii) induction on y: $E(Sx \dot{-} S\bar{o}, x \dot{-} \bar{o})$ comes from $Sx \dot{-} S(\bar{o}) \simeq P(S(x)) \simeq x$ and $x \dot{-} \bar{o} \simeq x$, together with (i). If $E(S(x) \dot{-} S(y), x \dot{-} y) = \mathbf{T}$, then, by (vi) $E(P(S(x) \dot{-} S(y)), P(x \dot{-} y)) = \mathbf{T}$ that is

$$E(S(x) \dot{-} S(S(y)), x \dot{-} S(y)) = \mathbf{T}.$$

(viii)   Because $E(S(\bar{o}) \dot{-} \bar{o}, \bar{1}) = \mathbf{T}$ and $E(S(x) \dot{-} x, \bar{1}) = \mathbf{T} \wedge E(S(S(x)) \dot{-} S(x), S(x) \dot{-} x) = \mathbf{T} \to E(S(S(x)) \dot{-} S(x), \bar{1}) = \mathbf{T}$ by (iv); also $E(S(S(x)) \dot{-} S(x), S(x) \dot{-} x) = \mathbf{T}$ by (vii).

(ix) trivial from (vi) and (viii).

(x) induction on y: basis trivial; induction step by (vi). ∎

THEOREM 2.2.2.　*Let F be a functional system which is an extension of IT, and let $A[x, z]$ be a quantifier-free formula, x and z variables of type o.*

Suppose $F \vdash A[\bar{o}, z]$ and $F \vdash A[x, S(z)] \to A[S(x), z]$ then $F \vdash A[x, \bar{o}]$.

Proof.　One may suppose that A is $p(x, z) = \mathbf{T}$. Let $B[y]$ be the formula $E(x \dotminus P(y), \bar{o}) = \mathbf{T} \vee p(y, x \dotminus y) = \mathbf{T}$.

Then $F \vdash B[\bar{o}]$ because $F \vdash p(\bar{o}, z) = \mathbf{T}$. Also $B[y] \to B[S(y)]$; it suffices to show that

$$(E(x \dotminus P(y), \bar{o}) = \mathbf{T} \vee p(y, x \dotminus y) = \mathbf{T})$$
$$\to (E(x \dotminus y, \bar{o}) = \mathbf{T} \vee p(S(y), x \dotminus S(y)) = \mathbf{T})$$

if $E(x \dotminus y, \bar{o}) = \mathbf{\perp}$, then by 2.2.1 (ii) $E(x \dotminus y, S(P(x \dotminus y)a = \mathbf{T}$ that is $E(x \dotminus y, S(x \dotminus S(y))) = \mathbf{T}$;　then　by　(v)　$p(y, x \dotminus y) = \mathbf{T} \to$ $p(y, S(x \dotminus S(y))) = \mathbf{T}$, and, using the hypothesis of the theorem, $p(y, x \dotminus y) = \mathbf{T} \to p(S(y), x \dotminus S(y)) = \mathbf{T}$. Now, observe that, by 2.2.1 (x) we have $E(x \dotminus P(t), \bar{o}) = \mathbf{T}$; if $E(x \dotminus y, \bar{o}) = \mathbf{T}$ then the formula to prove is trivially true. From Corollary 2.1.4, we get $F \vdash B[y]$, and it follows that

$$F \vdash E(S(x) \dotminus x, \bar{o}) = \mathbf{T} \vee p(S(x), S(x) \dotminus S(x)) = \mathbf{T}.$$

Since by 2.2.1 (viii) and (ix), one gets $E(S(x) \dotminus x, \bar{1}) = \mathbf{T}$ and $E(S(x) \dotminus S(x), \bar{o}) = \mathbf{T}$, one concludes, using (iii) and (iv) that $p(S(x), \bar{o}) = \mathbf{T}$. From $p(\bar{o}, \bar{o}) = \mathbf{T}$ and $p(S(x), \bar{o}) = \mathbf{T}$ one gets $p(x, \bar{o}) = \mathbf{T}$. ■

COROLLARY 2.2.3.　*Let F be functional system, extension of IT; let $A[x, z]$ be a quantifier-free formula, x of type o, z of type σ, and let t be a term of type $(o, \sigma \to \sigma)$ such that:*

$$F \vdash A[\bar{o}, z]$$
$$F \vdash A[x, t(x, z)] \to A[S(x), z]$$

then $F \vdash A[x, z]$.

Proof.　Let $B[x, y]$ $(x, y$ of type $o)$ be defined by $A[x, f(x, z, y)]$ with f such that: $f(x, z, \bar{o}) \simeq z$ and $f(x, z, S(y)) \simeq t(x, f(x, z, y))$. Then

$F \vdash B[\bar{o}, y]$ and $F \vdash B[x, S(y)] \to B[S(x), y]$; from Theorem 2.2.2, one gets $F \vdash B[x, \bar{o}]$ so $F \vdash A[x, z]$.

THEOREM 2.2.4. *Let $A[x]$ (x of type o) be a purely universal formula, and let F be a functional system which extends IT; then, if $F \vdash A[\bar{o}]$ and $F \vdash A[x] \to A[S(x)]$ we have $F \vdash A[x]$.*

Proof. By II.1, let A be $\forall y p(y, x) = \mathbf{T}$; since $F \vdash (\forall y (p(y, x) = \mathbf{T}) \to p(y, S(x)) = \mathbf{T}$ we get, using II.3, a term t such that $F \vdash p(t(x, y), x) = \mathbf{T} \to p(y, S(x)) = \mathbf{T}$. Now, it remains to apply Corollary 2.2.3.

COROLLARY 2.2.5. *Let F be a functional system, extension of IT; let $C[x]$ be a formula of the form $B \to A[x]$, with A, B purely universal, x not free in B. Then, if $F \vdash C[\bar{o}]$ and $F \vdash C[x] \to C[S(x)]$, we have $F \vdash C[x]$.*

Proof. If B is closed, the result is trivial, since $F + B$ is a functional system extension of IT (if consistent). Otherwise, let F be an extension of F, and A', B' be obtained from A, B, by substituting closed terms for the variables distinct from x; then, by the previous remark, $A'[x]$ is provable; now, Theorem II.2 shows that $A[x]$ is provable.

2.3. *Equality theorem*

PROPOSITION 2.3.1. *The following are theorems of IT*

$$E(x, x) = \mathbf{T}; \qquad E(x, y) = \mathbf{T} \wedge p(x) = \mathbf{T} \to p(y) = \mathbf{T}.$$

Proof. $E(x, x) = \mathbf{T}$ 2.2.1(i).

Let $C[x, y, p]$ be the formula $E(x, y) = \mathbf{T} \wedge p(x) = \mathbf{T} \to p(y) = \mathbf{T}$ and let $B[x]$ be the formula $\forall p \forall y C[x, y, p]$.

Then $\vdash B[\bar{o}]$ by a proof analogous to 2.2.1(iii). Also $B[x] \to B[S(x)]$: one may prove $B[S(x)]$ by induction on y, under the hypothesis $B[x]$ (2.2.5) $C(S(x), \bar{o}, p)$ is trivial from $E(S(x), \bar{o}) = \mathbf{\perp}$, and $B[x] \to C(S(x), S(y), p)$ follows from $E(S(x), S(y)) = E(x, y)$. ∎

THEOREM 2.3.2.

$$IT \vdash x^0 \simeq y^0 \leftrightarrow E(x, y) = \mathbf{T}.$$

Proof. From $E(x, x) = \mathbf{T}$, one gets, using the fact that $\simeq$ is an equality, that $x \simeq y \to E(x, y) = \mathbf{T}$. Conversely, since

$$E(x, y) = \mathbf{T} \to \forall p(p(x) = p(y))$$

by 2.3.1, we get $E(x, y) = \mathbf{T} \to x \simeq y$. ∎

IV. FUNCTIONAL INTERPRETATION AND THE KRIPKE MODEL $\mathbb{K}$

1. *Basic Definitions*

1.1. *Functional interpretation of HA*

1.1.1. Functional interpretation of terms

Suppose we have identified the variables of *HA* with the variables of type o of a functional system F; by induction on terms t, we define t^*, which is a term of type o of F:

$$x^* = x$$
$$(S(t))^* = S(t^*)$$
$$\bar{o}^* = \bar{o}$$
$$(t + u)^* = t^* \dotplus u^*$$
$$(t \cdot u)^* = t^* \mathbin{\dot{\cdot}} u^*$$

1.1.2. Functional interpretation of formulas; to each formula A of *HA*, we associate a formula A^* of the functional system F; A^* is to be of the form $\exists x \forall y t(x, y) = \mathbf{T}$, where x and y are variables of F of respective type σ and τ, and t is of type $(\sigma, \tau \to fm)$ (σ and τ depend on A). In the inductive definition one may encounter expressions of the form $\exists x_1 \cdots x_n \forall y_1 \cdots y_m A'[x_1, ..., x_n, y_1, ..., y_n]$. By II.1, these expressions may be brought into the form given above, adding, if necessary superfluous quantifiers. For the inductive step, suppose B^* is $\exists x_1 \forall y_1 t_1(x_1, y_1) = \mathbf{T}$ *and* C^* is $\exists x_2 \forall y_2 t_2(x_2, y_2) = \mathbf{T}$.

$$(t = u)^* = (E(t^*, u^*) = \mathbf{T})$$
$$(B \wedge C)^* = \exists x_1 \exists x_2 \forall y_1 \forall y_2 (t_1(x_1, y_1) = \mathbf{T} \wedge t_2(x_2, y_2) = \mathbf{T})$$
$$(B \vee C)^* = \exists w^{fm} \exists x_1 \exists x_2 \forall y_1 \forall y_2 ((w = \mathbf{T} \wedge t_1(x_1, y_1) = \mathbf{T})$$
$$\vee (w = \bot \wedge t_2(x_2, y_2) = \mathbf{T}))$$

$$(B \to C)^* = \exists X_2 \exists Y_1 \forall x_1 \forall y_1 (t_1(x_1, X_2(x_1, y_2)) = \mathbf{T}$$
$$\to t_2(X_2(x_1), y_2) = \mathbf{T})$$

$$(\mathbf{1})^* = (\mathbf{1} = \mathbf{T})$$

$$(\forall z B)^* = \exists X_1 \forall y_1 \forall z (t_1(X_1(z), y_1)[z] = \mathbf{T})$$

$$(\exists z B)^* = \exists z \exists x_1 \forall y_1 (t_1(x_1, y_1)[z] = \mathbf{T}).$$

1.1.3. If F is a functional system, let F^* be the intuitionistic system with the same set of axioms. If A is a formula of HA, we say that A^* is *valid* in P if A^* is (intuitionistically) derivable in F, that is if A^* is $\exists x^\sigma \forall y t(x, y) = \mathbf{T}$, there is a term u of type σ of F, not depending on y, such that $F \vdash \forall y t(u, y) = \mathbf{T}$. (Since F and F^* have clearly the same universal theorems).

Explicit realizability of existential theorems of F^* is proved in II.4(i).

Remark. It is also possible to extend the functional interpretation of HA to a functional interpretation of a theory with infinitely many types (namely the types common to all functional systems), and containing all constants of system F. We mention this only as a possibility.

1.2. *The Kripke structure* $\mathbb{K}$

We assume that the reader is familiar with Kripke semantics.

Let (c_n) be a denumerable set of proper constants containing infinitely many constants of each type.

Indexing set of $\mathbb{K}$: we put $I = \{F; F$ is a functional system with as set of proper constants a finite subset of the sequence $(c_n)\}$. I is preordered by the relation; $F < F'$ iff F' is an extension of F

$$\mathbb{K} = \{\mathbb{K}_F; F \in I\};$$

$\mathbb{K}_F$ is defined to be the structure for the language of HA defined by: *domain:* the set of closed terms of type o of F; the interpretation of the function letters $S, +, \cdot$ is given by the functions $t \mapsto S(t)$, $t, u \mapsto t \dotplus u$, $t, u \mapsto t \hat{\ } u$. We put

$$\mathbb{K}_F \Vdash t = u \quad \text{iff} \quad F \vdash E(t, u) = \mathbf{T}.$$

Remark. It is clear that it is possible to extend $\mathbb{K}$ to a Kripke structure for a language with infinitely many types; again, we mention this only as a possibility.

In the sequel, we shall denote by $\mathbb{K}(IT)$ the Kripke structure $\{\mathbb{K}_F; F \in I; F > IT\}$.

1.3. Language of $\mathbb{K}_F$

The language of $\mathbb{K}_F$ is defined as usual, that is we add new constants **t, u** $\cdots$ for the closed terms of type o of F. If A is a formula of the language of $\mathbb{K}_F$, its functional interpretation is obtained by adding the clause $(\mathbf{t})^* = t$; if A is a formula of the language of $\mathbb{K}_F$ and A^* is its interpretation, the validity of A^* in F is defined as in 1.1.3.

2. Main Result

THEOREM.   *Let $F \in I$, and let A be a closed formula of the language of $\mathbb{K}_F$, then*

$$\mathbb{K}_F \Vdash A \quad iff \quad A^* \text{ is valid in } F \text{ (equivalently } F^* \vdash A^*).$$

Remark. If one extends the functional interpretation to higher order objects, the theorem may be stated as $\mathbb{K} \Vdash (A \leftrightarrow A^*)$ for all formulas A.

Proof. Induction on the logical complexity of A

$- A$ is atomic $t = u$; by the definition of $\mathbb{K}_F \Vdash A$ and of A^*.

$- A$ is $B \wedge C$; then $\mathbb{K}_F \Vdash C$ iff $\mathbb{K}_F \Vdash C$ and $\mathbb{K}_F \Vdash B$; also A^* is provably equivalent to $B^* \wedge C^*$; so the property holds, using induction hypothesis.

$- A$ is $B \vee C$; then $\mathbb{K}_F \Vdash B \vee C$ iff $\mathbb{K}_F \Vdash B$ or $\mathbb{K}_F \Vdash C$; also A^* is provably equivalent (in F^*) to $B^* \vee C^*$ (trivial) $\cdots$ then apply II.4(ii).

$- A$ is $\exists z B$; then $\mathbb{K}_F \Vdash A$ iff there is a closed t, $\mathbb{K}_F \Vdash B[t]$; also, $F^* \vdash A$ iff there is a term t (which may be taken closed, since A is closed) such that $F^* \vdash B[t]$.

$- A$ is $\forall z B$. (1°) suppose $\mathbb{K}_F \Vdash A$; then, for all $F' > F$, and for all closed t in F', $\mathbb{K}_{F'} \Vdash B[t]$, so, by induction hypothesis $F'^* \vdash B^*[t]$. Suppose the interpretation of $B[z]$ is $\exists x^\sigma \forall y t(x, y, z) = \mathbf{T}$. Then, for all $F' > F$ and all closed t of type o in F', there is an a of type σ such that $F' \vdash \forall y u(a, y, t) = \mathbf{T}$. Applying II.2, we get a term a' of type $o \to \sigma$ such that $F \vdash \forall y u(a'(z), y, z) = \mathbf{T}$.

(2°) Conversely, suppose $F^* \vdash A^*$; with the same notations as in (1°), this means that there an a' of F such that $F \vdash u(a'(z), y, z) = \mathbf{T}$, so in any $F' > F$, we have $u(a'(t), y, t) = \mathbf{T}$, and by induction hypothesis $\mathbb{K}_F \Vdash B[t]$ for every closed t. This means that $\mathbb{K}_F \Vdash A$.

$- A$ is $B \to C$; suppose B^*, C^*, A^* are given by 1.1.2.

(1°) Suppose $\mathbb{K}_F \Vdash B \to C$. Let $F' > F$, and let a be a closed term of F'; we claim that $F' \vdash (\forall y_1 t_1(a, y_1) = \mathbf{T}) \to (\forall y_2 t_2(b, y_2) = \mathbf{T})$ for some closed term b: if $F' + \forall y_1 t_1(a, y_1) = \mathbf{T}$ is inconsistent, this is trivial; otherwise, $F' + \forall y_1 t_1(a, y_1) = \mathbf{T}$ is a functional system F'' s.t., by induction hypothesis, $\mathbb{K}_{F''} \Vdash B^*$; so $\mathbb{K}_{F''} \Vdash C^*$, and, applying the induction hypothesis and the deduction theorem, we prove the claim. So, applying II.2, we conclude that there is a term b such that

$$F \vdash (\forall y_1 t_1(x_1, y_1) = \mathbf{T}) \to t_2(b(x_1), y_2) = \mathbf{T}.$$

Now, an application of II.3 gives a term c such that

$$F \vdash t_1(x_1, c(x_1, y_2)) = \mathbf{T} \to t_2(b(x_1), y_2) = \mathbf{T}.$$

From this, it follows that A^* is valid in F.

(2°) Suppose A^* is valid in F; then A^* is valid in any extension F' of F. Suppose $\mathbb{K}_{F'} \vdash B$; then by induction hypothesis, $F' \vdash t_1(a, y_1) = \mathbf{T}$ for some closed a; if $F' \vdash t_1(x_1, c(x_1, y_2)) = \mathbf{T} \to t_2(b(x_1), y_2) = \mathbf{T}$ then $F' \vdash t_2(b(a), y_2) = \mathbf{T}$ and, by induction hypothesis, $\mathbb{K}_{F'} \vdash C$. ∎

3. *Corollaries*

THEOREM 3.1. (i) $\mathbb{K}(IT)$ *is a Kripke model of HA.*

(ii) *If A is provable in HA, then $IT^* \vdash A^*$.*

Proof. (i) Clearly, all axioms of HA, except induction are valid in $\mathbb{K}(IT)$: (equality axioms are provable in extensions of IT by 2.3.1). The other axioms are consequences of $E(x, x) = \mathbf{T}$ and the explicit definition of under. Now, we show that the set of formulas valid in $\mathbb{K}(IT)$ is closed under the rule of induction: let A be a formula of HA such that $A[\bar{o}]$ and $A[x] \to A[S(x)]$ are valid in $\mathbb{K}(IT)$, then, by Theorem 2, the functional interpretation of these formulas is valid in IT: there are terms a and b of IT such that $IT \vdash t(a, y, \bar{o}) = \mathbf{T}$ and

$$IT \vdash \forall y'(t(z, y', x) = \mathbf{T}) \to t(b(z, x), y, S(x)) = \mathbf{T}.$$

By Theorem III.2.2.4 applied to the formula $\forall y t(c(x), y, x) = \mathbf{T}$, with $c = R(a, b)$, we get $A^*[x]$ valid in IT, that is $\mathbb{K}_{IT} \Vdash A$.

(ii) Trivial consequence of (i) and Theorem 2. ∎

THEOREM 3.2. *Let p be a predicate letter for a primitive recursive predicate in an extension by definition of HA. Then $HA \vdash \forall x p(x)$ if there exist closed terms q and r of respective types $o \to fm$ and $o \to o$ such that*

$$T \vdash q(\bar{o}) = \mathbf{T}$$
$$T \vdash q(x) = \mathbf{T} \to q(S(x)) = \mathbf{T}$$
$$T \vdash q(r(x)) = \mathbf{T} \to p^*(x) = \mathbf{T},$$

where p^* is the term of T canonically associated with p.

Proof. One easily proves that all results remain unchanged if one works in an extension by definition of HA containing defining equations for *p.r.* functions and predicates.

The 'if' part of the theorem follows from the fact that every finitely axiomatizable part of IT, involving only finitely many types has an ω-model in HA. The 'only if' part is a trivial consequence of theorem 3.1(ii) and Proposition III.2.1.1. ∎

COROLLARY. *The set of purely universal theorems of IT is not recursive.*

NOTE

[1] A more direct proof can be given with the help of Herbrand's theorem.

BIBLIOGRAPHY

[1] Girard, J. Y.: 'Interpretation fonctionnelle et élimination des coupures dans l'arithmétique d'ordre supérieur'; thèse de doctorat d'état, université Paris VII, 1972.

[2] Gödel, K.: 'Über eine bisher noch nicht Erweiterung des finiten Standpunktes', *Dialectica* **12** (1958).
[3] Hanatani, Y.: 'Calculabilité des fonctionnelles récursives primitives de type fini sur les nombres naturels', *Ann. Jap. Ass. Phil. Soc.* **3,** No. 1, 1966. (Revised english version in: 'Proof theory symposium, Kiel 1974. Springer Lecture Notes, No. 500.
[4] Prawitz, D.: *Natural Deduction*, Almqvitz & Wiksell, Stockholm 1965.
[5] Spector, C.: 'Provably Recursive Functionals of Analysis: A Consistency Proof by an Extension of Principles Formulated in Current Intuitionistic Mathematics. Recursive Function Theory', *Proc. Symp. Pure Math.* **V,** AMS Providence RI, 1962.
[6] Tait, W. W.: 'Intensional Interpretation of Functionals of Finite Type', *JSL* **32** (1967).
[7] Troelstra, A. S. (ed.): 'Metamathematical Investigation of Intuitionistic Arithmetic and Analysis', *Springer Lecture Notes*, No. 344.

AXIOMS FOR INTUITIONISTIC MATHEMATICS
INCOMPATIBLE WITH CLASSICAL LOGIC

1. INTRODUCTION

1.1. Standard formalizations of constructive mathematics ('constructive' here in the narrow sense of Bishop (1967): choice sequences are regarded as inacceptable, and Church's thesis is not assumed) can be carried out in formal systems based on intuitionistic logic which become classical formal systems on addition of the principle of the excluded third. The fact that in such systems for constructive mathematics the logical operations permit an interpretation different from the classical truth-functional one is then solely expressed by the fact that *less* axioms are assumed. As is well-known, this results in formal properties[1] such as

$$\vdash A \lor B \;\Rightarrow\; \vdash A \quad or \quad \vdash B,$$

$$\vdash \exists x A x \;\Rightarrow\; \vdash A t \quad \text{for some term } t \quad (\exists x A x \text{ closed}).$$

But given an interpretation of the logical operations different from the classical one, there is also the possibility of axioms which are valid on this interpretation, but not on the classical one. It goes without saying that, granted the fact that our formal system was intended to code (part of) mathematical practice based on this interpretation, special interest attaches to such axioms; and as we shall see from the discussion below, this is not the only reason for being interested in them.

As to the justification of such axioms, we can apply various standards of informal rigour. The strictest requirement is a precise description of the mathematical objects of a theory ('conceptual analysis') followed by an informal, but rigorous derivation of the axioms proposed for these objects. We are rarely in a position to present such an analysis; an example taken from intuitionism is the theory of lawless sequences (see e.g. Kreisel, 1968; Troelstra, 1969 § 9, 1976 § 2).

In most cases, we can at best produce plausibility arguments, not unlike the way in which many new axioms for set theory (e.g. those on

Butts and Hintikka (eds.), Logic, Foundations of Mathematics and Computability Theory, 59–84.
Copyright © 1977 by Reidel Publishing Company, Dordrecht-Holland. All Rights Reserved.

large cardinals) are proposed; the plausibility arguments lead us to expect consistency of the new axioms with the already accepted ones. Investigation of formal properties of systems including the proposed axioms, and exploration of the mathematical consequences may then help us to obtain a better insight into what is involved in the justification of such axioms.

For the examples of possible axioms conflicting with classical logic presented in this paper, we shall be content with plausibility arguments and formal analogies, without hiding (to ourselves) the weak epistemological basis for our axioms. It is worth mentioning however, that in one respect the status of our examples is different from large-cardinal axioms in set theory: for at least three of our five examples, the axioms conflicting with classical logic do not increase proof-theoretic strength, i.e. relative consistency proofs are available; for the other two examples, relative consistency is at least a plausible conjecture.

Our aim in this paper is not the presentation of new results, but to draw attention to what is at present a neglected area.

1.2. *Contents of the Paper*

In Sections 3–8 examples of axioms for intuitionistic mathematics incompatible with classical logic are presented. For some of these, such as Church's thesis and the continuity axioms, mathematical consequences and metamathematical properties have been studied extensively; others have scarcely been investigated. As we go along, we mention various open problems connected with these examples, and give reasons for being interested in axioms of this type.

Section 9 discusses admissible (or derived) rules corresponding to the axioms discussed in our examples; Section 10 connects the problem of the completeness of intuitionistic predicate logic with the addition of new axioms.

2. PRELIMINARIES

2.1. *Variables*

We shall use x, y, z, u, v, w to indicate numerical variables. We distinguish two disjoint classes of variables for number theoretic functions: 'constructive-function' variables (a, b, c, d) and 'choice-sequence' variables $(\alpha, \beta, \gamma, \delta)$. X, Y, Z are used as meta-variables for variables for sets of natural numbers; similarly X', Y', Z' for sets of (choice) sequences.

2.2. Intuitionistic first-order arithmetic **HA**, intuitionistic second order arithmetic **HAS**, and elementary analysis **EL** are described in Troelstra (1973). **EL** is essentially **HA** with function variables and quantifiers, ranging over sequences closed under 'recursive in'.

EL* is actually the same as **EL** but with choice-sequence variables instead of constructive function variables. **HAS**^{1} is a combination of **HAS** and **EL***, with variables for sets of sequences instead of sets of natural numbers.

2.3. *Some Notations*

Let j be a primitive recursive pairing from N^2 onto N, with primitive recursive inverses j_1, j_2.

$$j(\alpha, \beta) \equiv_{\text{def}} \lambda x.j(\alpha x, \beta x),$$

$$X(x, y) \equiv_{\text{def}} Xj(x, y), \qquad (X)_x \equiv_{\text{def}} \lambda y.X(x, y),$$

$$X'(\alpha, \beta) \equiv_{\text{def}} Xj(\alpha, \beta), \qquad (X')_\alpha \equiv_{\text{def}} \lambda\beta.X(\alpha, \beta),$$

For any formal system **H** let $L(\mathbf{H})$ denote its language. For other notations, not explained in this paper, see Troelstra (1973).

3. EXAMPLE I: CHURCH'S THESIS

3.1. In $L(\mathbf{EL})$ we can express Church's thesis formally as

CT $\qquad \forall a \exists z \forall x \exists y[Tzxy \ \& \ ax = Uy],$

where T, U are Kleene's T-predicate and the result-extracting function respectively. In the presence of

$\text{AC}_{00} \qquad \forall x \exists y A(x, y) \rightarrow \exists a \forall x A(x, ax)$

CT is equivalent to

$$CT_0 \qquad \forall x \exists y A(x, y) \rightarrow \exists z \forall x \exists y [Tzxy \wedge A(x, Uy)].$$

Note that in **EL**, obtained by restricting AC_{00} of EL_1 to quantifier free A (call the restricted schema $QF - AC_{00}$), CT is unproblematic on the interpretation of the range of the function variables as ranging over total recursive functions and hence **EL** + CT is conservative over **HA**. CT_0 can be used to express Church's thesis within $L(\mathbf{HA})$.

The intuitive justification for CT_0 (or $CT + AC_{00}$) is as follows: we think of the function variables as ranging over lawlike sequences (sequences given by a law for computing the value to each argument). The standard 'constructive' reading of '$\forall x \exists y A(x, y)$' is: we have a method for which we can show that it yields a y to each x such that $A(x, y)$. If we leave non-lawlike, 'indefinite' or 'incomplete' objects such as choice sequences out of consideration, this means obviously the same as $\exists b \forall x A(x, bx)$; and this justifies AC_{00}. CT then expresses that those laws determining sequences must be identified with mechanical (recursive) laws. From an intuitionistic point of view, this amounts to the identification 'humanly computable' = 'mechanically computable' (cf. Kreisel, 1972).

3.2. CT_0 obviously conflicts with classical logic. By classical logic,

$$\forall x \exists z [(z = 0 \ \& \ \neg \exists y Txxy) \vee (z \neq 0 \ \& \ Txx(z \dot- 1))],$$

but there is no recursive b giving z for each x, since then $bx = 0 \leftrightarrow \neg \exists y Txxy$, which would make $\{x : \exists y Txxy\}$ recursive. Thus CT_0 refutes an instance of $\forall x [Ax \vee \neg Ax]$, namely for $Ax \equiv \exists y Txxy$.

Via the existence of r.e., recursively inseparable sets, which follows from CT_0, we can refute e.g. an instance of $(A \rightarrow B) \vee (B \rightarrow A)$. Many counterexamples to classical theorems on assumption of CT_0 are in fact already consequences of the existence of an Enumerable but Not Decidable set

END $\exists a \neg \forall x (\exists y (ay = x) \vee \neg \exists y (ay = x))$

or the more sophisticated 'existence of enumerable but not effectively separable sets':

INSEP $\quad \exists a \exists b(\forall x(\neg \exists y(ay = x) \wedge \exists z(bz = x)) \wedge$

$\qquad \neg \exists c \forall x[(\exists y(ay = x) \to cx = 0) \wedge (\exists y(by = x) \to cx \neq 0)].$

The mathematical consequences of e.g. END+INSEP, END+ ¬INSEP have not been systematically investigated so far.

3.3. It is certainly not quite obvious that CT_0 can be consistently added to **HA**. There is no difficulty in interpreting $\forall x \exists y$ by recursive dependence of y from x, but can this consistently be extended to formulae of arbitrary complexity such that all schemata of intuitionistic predicate logic are validated?

The positive answer is provided by Kleene's realizability, which associates to each formula A an interpretation $\exists x(xrA)$ ('A is realizable') such that we have

THEOREM. (i) $\mathbf{HA} + ECT_0 \vdash A \leftrightarrow \exists x(xrA)$

$\qquad$ (ii) $\mathbf{HA} + ECT_0 \vdash A \Leftrightarrow \mathbf{HA} \vdash \exists x(xrA),$

where ECT_0 is a schema which generalizes CT_0:

ECT₀ $\quad \forall x[Ax \to \exists yB(x, y)] \to \exists u \forall x[Ax \to \exists z(Tuxz \wedge B(x, Uz))],$

where A does not contain $\vee$, and $\exists$ only in front of prime formulae (i.e. $\exists x(t = s)$).

Beeson (1972; see Troelstra 1973, 3.4.14) has shown that ECT_0 is essentially stronger: $\mathbf{HA} + CT_0 \nvdash ECT_0$.

The mapping $\varphi: A \mapsto \exists x(xrA)$ can be said to make explicit the recursive dependence of y on x in $\forall x \exists y$, *hereditarily*. The crucial problem is here that we also have to define $\exists x(xr(A \to B))$ for given $\exists x(xrA)$, $\exists x(xrB)$. This is done by

$$xr(A \to B) \equiv_{\text{def}} \forall y(yrA \to !\{x\}(y) \wedge \{x\}(y)rB).$$

It is an *open problem* whether there also exists a mapping φ such that

$$\mathbf{HA} + CT_0 \vdash A \leftrightarrow \varphi A, \qquad \mathbf{HA} + CT_0 \vdash A \Leftrightarrow \mathbf{HA} \vdash \varphi A.$$

3.4. *Markov's Schema*

M $\qquad \forall x[Ax \vee \neg Ax] \wedge \neg\neg \exists x Ax \rightarrow \exists x Ax$

is modulo CT_0 equivalent to

M_{PR} $\qquad \neg\neg \exists x Ax \rightarrow \exists x Ax$ (*A* primitive recursive) or equivalently

$\qquad \forall xy (\neg\neg \exists z Txyz \rightarrow \exists z Txyz).$

The mathematical consequences of $\mathbf{HA} + ECT_0 + M$ have been extensively explored – this system may be regarded as a codification of mathematical practice of the Russian constructivist school. Some examples

  (i)   Mappings from complete separable metric spaces into separable metric spaces are continuous. (Čeitin, 1959; Y. N. Moschovakis, 1964)
  (ii) Real-valued functions on [0, 1] are not always uniformly continuous. (Beeson A)
  (iii) There exists a continuous mapping φ of the square $I^2 \equiv \{(x, y): x \in [0, 1], y \in [0, 1]\}$ into I^2 such that $\rho((x, y), \varphi(x, y)) \geqslant 8^{-1}$ for all $(x, y) \in I^2$, ρ an Euclidean metric; and there is a uniformly continuous mapping ψ of I^2 into I^2 such that $\forall z \in I^2$ $(\psi z \neq z)$. (Refutation of Brouwer's fixed point theorem, due to Orevkov (1963, 1964).)

3.5. We note in passing that instead of ECT_0 one sometimes considers (e.g. Dragalin, 1973)

ECT' $\qquad \forall x[\neg Ax \rightarrow \exists y B(x, y)] \rightarrow \exists u[\neg Ax \rightarrow \exists z (Tuxz \wedge B(x, Uz))].$

However, this is equivalent to ECT_0 relative to $\mathbf{HA} + M$, since a formula A' with $\exists$ only in front of prime formulae is via M equivalent to a negative formula A (i.e. a formula without $\vee$, $\exists$) and for such A, $A \leftrightarrow \neg\neg A$.

4. EXAMPLE II: CONTINUITY AXIOMS

4.1. In the formulation of the continuity schemata we shall assume all predicates to be extensional w.r.t. numerical and function variables, i.e.

$$x = x' \wedge \alpha = \alpha' \wedge A(\alpha, x) \to (\alpha', x').$$

This is automatically ensured if for example we think of the language $L(\mathbf{EL}^*)$. The following continuity schemata are of special interest here.

CONT$_0$! $\quad \forall \alpha \exists! x A(\alpha, x) \to \exists \gamma \in K_0 \forall \alpha A(\alpha, \gamma(\alpha))$,

CONT$_0$ $\quad \forall \alpha \exists x A(\alpha, x) \to \exists \gamma \in K_0 \forall \alpha A(\alpha, \gamma(\alpha))$,

CONT$_1$ $\quad \forall \alpha \exists \beta A(\alpha, \beta) \to \exists \gamma \in K_0 \forall \alpha A(\alpha, \gamma \mid \alpha)$.

Here

$$K_0 \gamma \equiv \forall \alpha \exists x (\gamma(\bar{\alpha}x) \neq 0) \wedge \forall nm(\gamma n \neq 0 \to \gamma n = \gamma(n * m))$$

and

$$\gamma(\alpha) = x \equiv \exists y(\gamma(\bar{\alpha}y) = x + 1)$$

$$\gamma \mid \alpha = \beta \equiv \forall x \exists y(\gamma(\langle x \rangle * \bar{\alpha}y) = \beta x + 1).$$

$K_0 \gamma$ expresses: γ is a neighbourhood function; i.e. γ codes a continuous functional $\Gamma: N^N \to N$ or $\Gamma: N^N \to N^N$. (If we permit functionals we could e.g. express CONT$_0$ as $\forall \alpha \exists x A(\alpha, x) \to \exists \Gamma \in \text{Cont} \; \forall \alpha A(\alpha, \Gamma\alpha)$.)

4.2. There is a strong analogy between CONT$_0$, CONT$_1$, and CT$_0$; CONT$_0$ and CONT$_1$ can be used in the same way as CT$_0$ in refuting schemata of classical predicate logic; so is e.g. $\forall \alpha(A\alpha \vee \neg A\alpha)$ refuted by CONT$_0$ for $A\alpha \equiv \exists x(\alpha x = 0)$. On the other hand there is a conflict between CONT$_0$ and CT; in

CT $\quad \forall \alpha \exists x \forall y \exists z(Txyz \; \& \; \alpha y = Uz)$

it is required by $CONT_0$ that the x depends continuously on α (i.e. can be found from an initial segment) which is obviously impossible. $CONT_0! + CT$ on the other hand is consistent (see, e.g. Troelstra, 1973, 2.6.15, 3.2).

4.3. Consistency of $EL + CONT_1$ presents similar problems as the consistency of $HA + CT_0$ did before, which can be resolved by similar methods: in this case Kleene's realizability of functions (for a short account see Troelstra, 1973, 3.3). However, in this case there is another method available which opens up different perspectives: the elimination of choice sequences. It is well known that the usual intuitive justifications and plausibility arguments for the continuity schemata depend on the interpretation of the function variables as ranging over choice sequences. But the so called elimination theorems provide a quite different solution: they permit us to regard quantification over choice sequences[2] as a 'figure of speech'. A bit more precisely, we consider systems H (in the language of $EL +$ constant K for the class of neighbourhood functions); we think of H as containing, besides number quantifiers, only ordinary function quantifiers $\forall a, \exists a$; CS_H contains in addition special 'choice' quantifiers $\forall \alpha, \exists \alpha$ for which it is required that in combinations $\forall \alpha \exists x, \forall \alpha \exists \beta$ the x, resp. β depend continuously on α.

The elimination theorems now take the form:

THEOREM. (Kreisel and Troelstra, 1970; Troelstra, 1974). *There is a mapping σ defined for formulae not containing choice parameters free such that for suitable H, CS_H with $H \subseteq CS_H$*

$$CS_H \vdash \sigma(A) \leftrightarrow A$$
$$CS_H \vdash A \Leftrightarrow H \vdash \sigma(A)$$
$$\sigma(A) \equiv A \quad \text{for} \quad A \in L[H].$$

(For certain choices of H, A must be restricted to formulae not containing quantifiers $\exists a$.)

The study of these elimination mappings has led to interesting metamathematical results, such as: $EL +$ Fan theorem ($=$ intuitionistic form of König's lemma) is conservative over intuitionistic arithmetic (Troelstra, 1974).

4.4. Already from traditional intuitionistic literature various striking mathematical consequences of the continuity postulates (primarily $CONT_0$) are known. The mathematical applications can be roughly divided into two groups

(a) theorems which are actually classically false when we interpret the choice quantifiers as ranging over classical number theoretic functions and apply classical logic; example: all real-valued functions on $\mathbb{R}$ are continuous;

and

(b) theorems which also hold classically, but where the classical proof depends on classical logic, and the constructive proof uses the continuity axioms. Example: one direction of Riemann's permutation theorem: if all permutations of terms of a convergent series $\sum x_n$ yield convergent series, then $\sum x_n$ is absolutely convergent.

(For a detailed exposition of the examples see e.g. Troelstra, 1976, § 6.)

Combination of theorems of type (b) with the elimination theorem opens the following, *largely unexplored* possibility: after establishing a (classical) theorem A in $\mathbf{CS_H}$, we obtain a constructive version (in Bishop's sense) $\sigma(A)$ which can be established in $\mathbf{H}$. Another method for converting a classical theorem with a classical proof, into a constructive version is provided by the Dialectica interpretation – the preceding method will be applicable in fewer cases, but may yield better results when applicable. This is clearly a subject for further investigation, and requires a hunt for suitable examples from mathematical practice.

4.5. There are many open problems regarding logical relationships between various forms of continuity schemata. For example, I do not know of an interpretation or model validating $CONT_0$ but not $CONT_1$ – although there are various models validating $CONT_0$! but not $CONT_0$ (see e.g. J. R. Moschovakis, 1973; in combination with van Dalen, 1974).

Some attention has been given to 'Kripke's schema'

KS $\exists \alpha [A \leftrightarrow \exists x(\alpha x = 0)]$ (α not free in A).

KS is consistent with $\mathrm{CONT_0}$! (J. R. Moschovakis, 1973) but on the other hand inconsistent with $\mathrm{CONT_1}$, *at least if we permit A in KS to contain choice parameters*: in

$$\forall \beta \exists \alpha [\forall x(\beta x = 0) \leftrightarrow \exists x(\alpha x = 0)]$$

α cannot possibly depend continuously on β.

5. EXAMPLE III: THE UNIFORMITY PRINCIPLE

5.1.   This was first introduced, together with a discussion of its intuitive meaning, by Troelstra (1973 A)[3].

We are primarily interested in the following two forms of the uniformity principle for numbers:

UP! $\forall X \exists ! x A(X, x) \rightarrow \exists x \forall X A(X, x)$

UP $\forall X \exists ! x A(X, x) \rightarrow \exists x \forall X A(X, x)$

and the uniformity principle for functions:

$\mathrm{UP_1}$ $\forall X \exists \alpha A(X, \alpha) \rightarrow \exists \alpha \forall X A(X, \alpha)$.

5.2.   We briefly recapitulate the remarks on the plausibility of these principles from Troelstra (1973 A).

The argument is similar to plausibility arguments for continuity schemata after a rough description of choice sequences. '$\forall X \exists x$' requires a method producing an x for each X. But sets are much more 'diffuse' than sequences: with sequences it is possible even if they are not given by a law, to approximate them to any desired degree of accuracy by constructing initial segments, whereas for sets we do not need to know even a single element belonging to the sets. Hence, if we have to assign something so definite (so 'discrete') as a natural number to each set, it can only be done by assigning a number which serves the purpose for all sets.

There is a weakness in this rough argumentation: sets are 'diffuse' only when we think of their *extension*, but there is also the possibility of constructing an x from the *definition* of x, in the same way as a natural number assigned to each *recursive* function may depend on the gödel number of the function. Now three remarks are relevant here.

(1) In the language of **HAS** we cannot refer to definitions of sets, only to their extensions. Hence we still expect, as is indeed the case, that **HAS** + UP is consistent.

(2) Quite apart from the restriction to the language of **HAS**, the possible effect of referring to definitions instead of extensions, becomes considerably less in the case of an $\forall X \exists ! x$-assertion, (which implies $\forall X \exists x \forall z (A(X, x) \ \& \ \forall Y \forall y (Xy \leftrightarrow Yy) \ \& \ A(Y, z) \rightarrow x = z))$ hence this makes the preceding plausibility argument more cogent. Compare this to the compatibility of CT_0 and $CONT_0!$.

(3) Similarly, in **HAS** nothing precludes the sets to be 'choice' sets (i.e. depending on non-lawlike parameters); and for a choice-set we cannot expect in general to be able to refer to its definition.

5.3. An illustration of remarks 2^0 and 3^0 is found in the following theorem and its proof:

THEOREM. $CONT_0!$ *implies* UP! (*in* **HAS**[1]).

Proof. Assume $\forall X \exists ! x A(X, x)$; let X be a set and define

$$Z_\alpha = \{x : Xx \ \& \ \exists x (\alpha x = 0)\}.$$

Then

$$\forall \alpha \exists ! x A(Z_\alpha, x)$$

and hence for some x, y

$$\forall \beta \in (\overline{\lambda z.1}) y A(Z_\beta, x).$$

Apply this to $\beta = \lambda x.1$ and $\beta' \in (\overline{\lambda z.1}) y * \langle 0 \rangle$, then $A(X, x)$, $A(\emptyset, x)$ and thus $\forall X A(X, x)$. ∎

5.4.   Now we wish to make a few remarks concerning parametrization principles. The prime example is Kripke's schema KS; in $L(\mathbf{HAS}^1)$ this may be expressed as

$$\forall X \exists \alpha [X \leftrightarrow \exists x (\alpha x = 0)]$$

where X ranges over zero-place relations. Or, admitting a choice axiom AC_{01}:

$$\forall X \exists \alpha \forall x [Xx \leftrightarrow \exists y (\alpha(x, y) = 0)].$$

All sets are parametrized by functions with the help of the predicate $A(\alpha, x) \equiv \exists y (\alpha(x, y) = 0)$.

As we shall demonstrate, $CONT_0 +$ a parametrization principle implies UP relative to $\mathbf{HAS}^1$:

THEOREM[4].   *Assume* $CONT_0$ *and a parametrization principle of the form*

$$PF_A \qquad \forall X \exists \alpha \forall x [A(\alpha, x) \leftrightarrow Xx].$$

(PF_A: *parametrization by functions via A.*) *Then* UP *holds.*

Proof.   (i) We first show that PF_A implies a corresponding principle PF_B where B satisfies ($*$ also for concatenation of a finite sequence and an infinite one)

$$(1) \qquad \forall \alpha \forall n \exists \beta \forall x [B(\alpha, x) \leftrightarrow B(n * \beta, x)].$$

To see this, we define $C(\alpha, \beta)$ ('β is a code for α') by

$$C(\alpha, \beta) \equiv_{def} \forall k [\forall x \exists y > x ((\beta)_k y = \alpha k)$$
$$\wedge \forall z (z \neq \alpha k \rightarrow \neg \forall x \exists y > x ((\beta)_k y = z)].$$

We note that

$$(2) \qquad \forall \alpha \exists \beta C(\alpha, \beta)$$
$$(3) \qquad C(\alpha, \beta) \wedge C(\alpha', \beta) \rightarrow \alpha = \alpha'$$
$$(4) \qquad C(\alpha, \beta) \rightarrow \exists \beta' C(\alpha, n * \beta').$$

(2) is immediate if, for any α, we take for β the sequence given by $(\beta)_k = \lambda x.\alpha k$ for all k. (3) is immediate; (4) is easily verified by choosing β'' such that $(\beta'')_k x = (\beta)_k x$ for all $x > 1\text{th}(n)$, and $\beta'' \in n$; then $\beta'' = n * \beta'$.

Now if we put

$$B(\beta, x) \equiv \exists\alpha(C(\alpha, \beta) \wedge A(\alpha, x))$$

then PF_A obviously implies

PF_B $\qquad \forall X \exists \beta \forall x [B(\beta, x) \leftrightarrow Xx]$

and also (1) is satisfied because of (4).

(ii) For a B satisfying (1), PF_B can be shown to imply UP. Assume

$$\forall X \exists x D(X, x);$$

by PF_B

$$\forall \beta \exists x D(\lambda y.B(\beta, y), x)$$

and thus with $CONT_0$, for some $\gamma \in K$

$$\forall \beta D(\lambda y.B(\beta, y), \gamma(\beta)).$$

Assume $\gamma n \neq 0$ for say $\overline{(\lambda x.0)}y = n$. Since $\lambda y.B(n * \beta, y)$ ranges over all X (by (1)), it follows that since $\forall \beta D(\lambda y.B(n * \beta, y), \gamma n \doteq 1)$,

$$\exists x \forall X D(X, x) \quad \blacksquare$$

We contrast this with

THEOREM. *Any parametrization principle* PF_A *implies the negation of a suitable instance of* UP_1

UP_1 $\qquad \forall X \exists \alpha B(X, \alpha) \rightarrow \exists \alpha \forall X B(X, \alpha).$

Proof. Apply UP_1 to $B(X, \alpha) \equiv_{\text{def}} \forall x[A(\alpha, x) \leftrightarrow Xx]$, then $\exists \alpha \forall X \forall x[A(\alpha, x) \leftrightarrow Xx]$; and this implies $\forall XY(X = Y)$ which is obviously false.

As to the plausibility of UP_1: this schema is analogous to UP, and the argument is once again that species are so 'diffuse' when compared with functions that $\forall X \exists \alpha A(x, \alpha)$ can only be satisfied by finding an α which does not depend on x such that $\forall X A(X, \alpha)$.

5.5. *Survey of Some Results*

This section may be skipped by a reader who is only interested in the main ideas. Information on the consistency of UP!, UP, UP^c (restriction of UP to closed instances) is contained in Troelstra (1973A), de Jongh-Smorynksi (1974), van Dalen (1974). Let us introduce some schemata

IP $(\neg A \to \exists x B) \to \exists x(\neg A \to B)$ (x not free in A)

IP^1 $(\neg A \to \exists \alpha B) \to \exists \alpha(\neg A \to B)$ (α not free in A)

IP_0 $\forall x(A \vee \neg A)$ & $(\forall x A \to \exists y B) \to \exists y(\forall x A \to B)$

$\qquad\qquad\qquad\qquad\qquad\qquad\qquad\qquad$ (y not free in A)

AC-NS $\forall x \exists X A(x, X) \to \exists Y \forall x A(X, (Y)_x)$.

THEOREM (de Jongh-Smorynski, 1974).
(i) **HAS** + AC-NS + IP_0 + M + UP! + UP^c *is consistent, and in fact conservative over* **HAS**^{c} + AC-NS (**HAS**^{c}: *classical second-order arithmetic*) *w.r.t. arithmetical sentences.* (*Similarly if* AC-NS *is dropped on both sides.*)
(ii) **HAS**^{1} + AC-NS + AC_{01} + UP! + UP^c_1 + IP_0 + M *is consistent.*

THEOREM (van Dalen, 1974). **HAS**^{1} + KS + CONT_0! + CONT^c_0 + UP + AC_{01} + UP! + UP^c_1 *is consistent.*

THEOREM (Troelstra, 1973A).
(i) **HAS** + CT_0 + IP_0 + M + UP + AC-NS *is consistent*
$\quad$ **HAS** + IP + CT_0 + AC-NS + UP *is consistent.*
(ii) **HAS** + CONT_1 + IP_0 + M + UP_1 + AC-NS *is consistent*
$\quad$ **HAS** + CONT_1 + IP + AC-NS + UP_1 *is consistent.*

(Here too we can in fact obtain appropriate conservative extension results.)

In the models constructed by de Jongh-Smorynski, and van Dalen, UP_1 (with parameters) is false.

5.6. As to open problems connected with UP and UP_1, we have first of all

(i) *Problem.* Explore mathematical consequences of UP, UP_1 (not already obtainable by continuity postulates). We mention in this connection that as a result of UP, the sets of natural numbers do not permit a non-trivial apartness relation. (We call an apartness relation $\#$ non-trivial iff $\exists x \exists y (x \# y)$.) To see this, let $X \# Y$ for some X, Y; by the properties of an apartness relation $\forall Z (X \# Z \vee Y \# Z)$ or equivalently $\forall Z \exists x ((X \# Z \ \& \ x = 0) \vee (Y \# Z \ \& \ x \neq 0))$, hence by UP $\forall Z (X \# Z) \vee \forall Z (Y \# Z)$; each disjunct is falsifiable, taking X, Y for Z respectively. From the viewpoint of metamathematics there is a problem of obvious interest.

(ii) *Problem.* Suppose we introduce 'uniform' species quantification satisfying UP next to ordinary species quantification. Can we define an elimination mapping and prove elimination theorems similar to the choice-quantifier elimination in the preceding example?

5.7. *Generalizing* UP

It is natural to propose for suitable families $\mathcal{F}_A$

$$\{X : A(X)\}$$

of sets of natural numbers a relativized version of UP

UP_A $\qquad \forall X [A(X) \to \exists x B(X, x)] \to \exists x \forall X [A(X) \to B(X, x)].$

UP_A reduces to UP for $A(X) \equiv (0 = 0)$. It is quite obvious that A has to fulfill certain requirements: UP_A is obviously false for the family of decidable sets, i.e. for $A(X) \equiv \forall x (Xx \vee \neg Xx)$. On the other hand, UP_A is trivially implied by UP for all families $\mathcal{F}_A$ of the form

$$A(X) \equiv \exists Y \forall x [Xx \leftrightarrow Yx \wedge Cx]$$

(i.e. $\mathcal{F}_A$ is the power set of $\{x : Cx\}$) since $\forall X [A(X) \to \exists y B(X, y)]$ is equivalent to

$$\forall Y \exists y B(Y \cap \lambda x. Cx, y);$$

with UP

$$\exists y \forall Y B(Y \cap \lambda x . Cx, y)$$

hence $\exists y \forall X[A(X) \to B(X, y)]$ (B in $L(\mathbf{HAS})$). In Troelstra (1973A), § 4 it is shown that $\mathbf{HAS} + \mathrm{UP}_A$ is consistent if $A(X) \equiv \neg C(X)$ for suitable C.

Let us consider the property '$\mathscr{F}_A$ is amalgamated', expressed by

$$\mathrm{Amal}\,(A) \equiv \forall X \in \mathscr{F}_A \forall Y \in \mathscr{F}_A \neg \forall Z \in \mathscr{F}_A (X \neq Z \vee Y \neq Z).$$

Parallel to the remark contained in 5.6 (i), it is easy to see that

$$\mathrm{UP}_A \to \mathrm{Amal}\,(A).$$

Conversely, it seems plausible to propose the general schema

$\mathrm{UP}_G \qquad \mathrm{Amal}\,(A) \to \mathrm{UP}_A.$

This schema cannot properly be termed a generalization of UP, since for the power set of N we cannot prove that it is amalgamated, so we cannot obtain UP as a special case of UP_G.

Note that if $\mathscr{F}_A$ is such that in each $X, Y \in \mathscr{F}_A$ the whole collection of Z_α defined by

$$Z_\alpha = \{x : (Xx \wedge \exists y (\alpha y = 0)) \vee (Yx \wedge \forall y (\alpha y \neq 0))\}$$

belongs to $\mathscr{F}_A$, and CONT_0 holds, then $\mathrm{Amal}\,(A)$ readily follows (compare the argument in 5.3, and 5.6 (i)). On the other hand, $\mathrm{Amal}\,(A)$ expresses a much stronger property than the assertion that $\mathscr{F}_A$ does not admit an apartness relation. In fact, if we take the $\mathscr{F}_A$ given by

$$A(X) \equiv (X = \{0\}) \vee (X0 \wedge X1)$$

then for $B(X, y) \equiv (y = 0 \wedge \neg X1) \vee (y = 1 \wedge X1)$ we have $\forall X \in \mathscr{F}_A \exists y B(X, y)$, but obviously not $\exists y \forall X \in \mathscr{F}_A B(X, y)$, although $\mathscr{F}_A$ does not admit an apartness relation.

6. Example IV: Subcountability of Discrete Sets

6.1. Our next example concerns a schema (indicated by SCDS) which has been investigated hardly at all – its logical relationship to UP and its mathematical consequences are unexplored.

Roughly, the principle of subcountability of Discrete Sets states

"A discrete collection (i.e. equality between elements of the collection is decidable) can be indexed by a subset of the natural numbers".

A stronger formulation requires in addition that the elements of the collection are distinct iff their indices are distinct.

In this generality, the principle is false: if we admit lawless sequences (e.g., Troelstra, 1976, § 2) as legitimate objects, we have $\varepsilon = \eta \vee \varepsilon \neq \eta$ for lawless sequences ε, η, but we cannot find any predicate A containing (hidden or explicitly) only finitely many lawless parameters such that

$$\forall \varepsilon \exists x \forall y A(x, y, \varepsilon y) \wedge \forall xyz[A(x, y, z) \to \exists! u A(x, y, u)].$$

(A countable model of the theory of lawless sequences does exist, but it is not countable in the language of the theory itself, cf. Troelstra, 1970).

For families of sets of natural numbers the principle can be expressed in a weak form

SCDS′ $\quad \forall XY(A(X) \wedge A(Y) \to X = Y \vee X \neq Y)$

$$\to \exists Z \forall X(A(X) \to \exists x(X = (Z)_x)).$$

Here $(Z)_x = \{y: Zj(x, y)\}$. In SCDS′, the family of sets considered is $\{X: A(X)\}$; the premiss assumes this family to be discrete. The conclusion asserts indexing by a subset of natural numbers, but not the stronger assertion that the index is uniquely determined (so as to make the index set itself 'mirror' the discreteness of $\{X: A(X)\}$). This stronger assertion can be expressed as

SCDS $\quad \forall XY(A(X) \wedge A(Y) \to X = Y \vee X \neq Y)$

$$\to \exists Z \forall X(A(X) \to \exists! x(X = (Z)_x)).$$

6.2.   In a sense, SCDS and SCDS′ start where (a generalization of) UP stops. Under suitable syntactical restrictions on A we can show consistency of UP(A), as noted in 5.7.

Suppose now UP(A) holds for a class $\mathscr{C}$ of predicates A such that (modulo provable equivalence)

$$A(X) \in \mathscr{C} \Rightarrow A((X)_y) \in \mathscr{C}$$

$$A(X), B(X) \in \mathscr{C} \Rightarrow A(X) \wedge B(X) \in \mathscr{C}.$$

The class of negated formulae is such a class $\mathscr{C}$; and it is provably consistent (see 5.7) to assume UP(A) for such a class.

Assume for an $A \in \mathscr{C}$ $\forall XY(A(X) \wedge A(Y) \to X = Y \vee X \neq Y)$. This is equivalent   to $\forall X(A((X)_0) \wedge A((X)_1) \to (X)_0 = (X)_1 \vee (X)_0 \neq (X)_1)$. Application of UP$(B)$ for $B(X) \equiv A((X)_0) \wedge A((X)_1)$ yields

$$\forall XY(A(X) \wedge A(Y) \to X = Y)$$
$$\vee \forall XY(A(X) \wedge A(Y) \to X \neq Y).$$

The second case is excluded unless $\mathscr{F}_A$ is empty. The first case expresses that $\mathscr{F}_A$ is a singleton. So either $\mathscr{F}_A$ is a singleton, or $\mathscr{F}_A$ is empty; in both cases SCDS is trivially fulfilled.

6.3.   A not entirely trivial instance of SCDS′ can be obtained as follows. Assume $A(x, X) \equiv \neg B$, and let

$$\forall XY(\exists x A(x, X) \wedge \exists y A(y, Y) \to X = Y \vee X \neq Y).$$
Then
$$\forall xy XY(A(x, X) \wedge A(y, Y) \to X = Y \vee X \neq Y).$$

With the help of UP(B) for $B \equiv A(x, (X)_0) \wedge A(y, (X)_1)$ we obtain

$$\forall xy\{\forall XY(A(x, X) \wedge A(y, Y) \to X = Y)$$
$$\vee \forall XY(A(x, X) \wedge A(y, Y) \to X \neq Y)\}.$$

This shows that for any x $\mathscr{F}_A$ is either empty or a singleton. So we can satisfy SCDS′ taking for Z

$$(Z)_x = \{y: \exists X(A(x, X) \wedge Xy)\}.$$

The argument as it stands does not yield SCDS. Of course this is only an example, it should not be hard to find other cases.

6.4. SCDS *for sets of sequences*

(This section, dealing with SCDS for families of sequences, is a digression which may be skipped by the reader.)

D. van Dalen observed the following

PROPOSITION. *Let* **FIM** *be the system for intuitionistic analysis of Kleene and Vesley, 1965. Let* $B(\delta, n)$ *be defined as*

$$B(\delta, n) \equiv (\delta n = 1 \wedge \forall m < n(\delta m = 0)).$$

Then the following form of SDCS *is consistent relative to* **FIM**:

SDCS* $\forall \alpha \beta [A\alpha \wedge A\beta \rightarrow \alpha = \beta \vee \alpha \neq \beta]$
$$\rightarrow \exists \gamma \forall \alpha [A\alpha \rightarrow \exists! n(B(\gamma, n)$$
$$\wedge \alpha \in n \wedge \forall \alpha' \in n(A\alpha' \rightarrow \alpha = \alpha'))].$$

Here A is almost negative (i.e. does not contain $\vee$*, and* $\exists$ *only in front of prime formulae). In addition,* **EL*** *and* **FIM** *and many systems in between are closed under the rule corresponding to SDCS* (i.e.* $\Rightarrow$ *replacing the main* $\rightarrow$*).*

Proof. **FIM** + GC is consistent (Troelstra, 1973, 3.3.11) where GC is the schema (*A* almost negative)

GC $\forall \alpha [A'\alpha \rightarrow \exists \beta C(\alpha, \beta)] \rightarrow \exists \gamma \forall \alpha [A'\alpha \rightarrow !\gamma \mid \alpha \wedge C(\alpha, \gamma \mid \alpha)].$

Assume $\forall \alpha \beta [A\alpha \wedge A\beta \rightarrow \alpha = \beta \vee \alpha \neq \beta]$ and apply GC to

$$A'\alpha \equiv A(j_1 \alpha) \wedge A(j_2 \alpha)$$
$$C(\alpha, \beta) \equiv (j_1 \alpha = j_2 \alpha \wedge \beta 0 = 0) \vee (j_1 \alpha \neq j_2 \alpha \wedge \beta 0 = 1),$$

then we find a γ' such that

(1) $\forall \alpha \beta [A\alpha \wedge A\beta \rightarrow !\gamma'(\alpha, \beta)$
$$\wedge \{(\gamma'(\alpha, \beta) = 0 \rightarrow \alpha = \beta) \,\&\, (\gamma'(\alpha, \beta) \neq 0 \rightarrow \alpha \neq \beta)\}]$$

(notations as in Troelstra, 1973). Let γ be such that

$$\forall \alpha (\gamma'(\alpha, \alpha) \cong \gamma(\alpha)),$$

then for B as defined above

$$\forall \alpha [A\alpha \to \exists! n (B(\gamma, n) \wedge \alpha \in n \wedge \forall \alpha' \in n(A\alpha' \to \alpha = \alpha'))].$$

To see this, note that if $A\alpha$, then $A\alpha \wedge A\alpha$, hence for the γ' of (1) $\gamma'(\alpha, \alpha) = 0$, so $\gamma(\alpha) = 0$; i.e. $\gamma(\bar{\alpha}x) = 1 \wedge \forall y < x(\gamma(\bar{\alpha}y) = 0)$ for a suitable x etc. etc. ■

To establish the *rule* corresponding to SCDS*, one should use closure of **FIM*** under the rule GCR corresponding to GC (cf. Troelstra, 1973, 3.7.9); in other respects the argument is similar.

Parallel to van Dalen's observation, we can establish the following more general, but weaker result:

PROPOSITION. *For any $A\alpha$, there is a predicate $C(\gamma, \alpha, m)$, such that the following schema*

SCDS'* $\forall \alpha \beta [A\alpha \wedge A\beta \to \alpha = \beta \vee \alpha \neq \beta]$

$$\to \exists \gamma \forall \alpha [A\alpha \to \exists m (C(\gamma, \alpha, m) \wedge \forall \alpha'(C(\gamma, \alpha', m) \to \alpha = \alpha'))]$$

is consistent relative to **FIM**.

Proof. We argue in **FIM** + GC, and assume results and notations in Troelstra (1973, § 3.3). Assume $\forall \alpha \beta [A\alpha \wedge A\beta \to \alpha = \beta \vee \alpha \neq \beta]$. In **FIM** + GC

$$\exists \delta (\delta \mathbf{r}^1 (A\alpha \wedge A\beta)) \leftrightarrow A\alpha \wedge A\beta,$$

where $\delta \mathbf{r}^1 (A\alpha \wedge A\beta)$ is almost negative. Now apply GC to $\forall \alpha \beta \, \delta [\delta \mathbf{r}^1 (A\alpha \wedge A\beta) \to \alpha = \beta \vee \alpha \neq \beta]$, then we find a γ such that

$$(1) \qquad \forall \alpha \beta \, \delta [\delta \mathbf{r}^1 (A\alpha \wedge A\beta) \to !\gamma(\delta, \alpha, \beta)$$

$$\wedge (\gamma(\delta, \alpha, \beta) = 0 \to \alpha = \beta) \wedge (\gamma(\delta, \alpha, \beta) = 1 \to \alpha \neq \beta)].$$

Let ν_3 be a coding of N^3 onto N, with inverses j_1^3, j_2^3, j_3^3; and let k_i^3, $i = 1, 2, 3$ be such that

$$k_i^3 0 = 0, \qquad k_i^3(n * \hat{x}) = k_i^3 n * \langle j_i^3 x \rangle, \qquad i = 1, 2, 3.$$

We put

$$C(\gamma, \alpha, m) \equiv \exists \delta(\nu_3(\delta, \alpha, \alpha) \in m \wedge \delta \mathbf{r}^1(A\alpha \wedge A\alpha) \wedge \gamma m = 1).$$

Assume $A\alpha$; then for some δ, $\delta \mathbf{r}^1(A\alpha \wedge A\alpha)$, and thus $\gamma(\delta, \alpha, \alpha) = 0$, i.e. for a suitable m $C(\gamma, \alpha, m)$. If also $C(\gamma, \alpha', m)$ it follows that for some δ' $\delta' \mathbf{r}^1(A\alpha' \wedge A\alpha')$, $\nu_3(\delta', \alpha', \alpha') \in m$, $\gamma m = 1$; and thus δ, $\delta' \in k_1^3 m$. For $\delta'' = j(j_1\delta, j_2\delta')$ it follows that $\delta'' \in k_1^3 m$, $\delta'' \mathbf{r}^1(A\alpha \wedge A\alpha')$; also $\alpha, \alpha' \in k_2^3 m = k_3^3 m$. Hence $\nu_3(\delta'', \alpha, \alpha') \in m$, and so $\gamma(\delta'', \alpha, \alpha') = 0$, i.e. $\alpha = \alpha'$ (since $\delta' \mathbf{r}^1(A\alpha' \wedge A\alpha')$ implies $A\alpha'$). ∎

6.5. It is easy to see that SCDS' also refutes certain consequences of classical logic: take for $A(X) \equiv \forall YZ(Y = Z \vee Y \neq Z)$. Then obviously $A(X) \& A(Y) \rightarrow X = Y \vee X \neq Y$. The conclusion of SCDS' is obviously false if we assume $A(X)$ for some X, because then $\forall YZ(Y = Z \vee Y \neq Z)$; hence $\forall XA(X)$, but then for some Z $\forall X\exists x(X = (Z)_x)$ which is easily refuted by a diagonal argument.

6.6. As noted by Friedman (1975) SSDC is a special case of the assertion: for every metric space there is a subcountable set of points dense in the space. This is seen by regarding $\{X: A(X)\}$ with $A(X) \& A(Y) \rightarrow X = Y \vee X \neq Y$ as a space with a discrete metric $(\rho(X, Y) = 1$ if $X \neq Y$, 0 otherwise); it follows that $\{X: A(X)\}$ must be subcountable.

6.7. As to the intuitive plausibility arguments for SCDS (and similarly for the assertion on metric spaces[5]) we can hardly say more than that all examples of discrete collections we know are in fact subcountable (if we bypass lawless sequences etc.).

7. Example v: generalizations of SCDS

7.1. The idea behind UP was: a mapping of something very diffuse (sets) into something very discrete (the natural numbers) must be constant; and for UP$_1$: a mapping of something very diffuse into something separable is constant. For SCDS': a discrete subfamily of something very diffuse is subcountable. The analogy might now suggest:

SCDS$_1'$ $\forall XY[A(X) \wedge A(Y) \rightarrow \exists \alpha \{(\forall x(\alpha x = 0) \rightarrow X = Y)$

$\wedge (\neg \forall x(\alpha x = 0) \rightarrow X \neq Y)\}] \rightarrow \exists Z \forall X[A(X) \rightarrow \exists \alpha (X = (Z)_\alpha)]$

For the special case where $\{X : A(X)\}$ can be provided with a metric the premiss is true; the hypothesis that every metric space has a subcountable set of points dense in the space then justifies the conclusion. (Let $\{X_i : i \in I\}$, $I \subseteq N$, be dense in $\{X : A(X)\}$, and let ρ be a metric on $\{X : A(X)\}$; then $X(\alpha, \beta) \equiv \exists X(\lim_{n \to \infty} \rho(X_{\alpha n}, X) = 0 \wedge A(X) \wedge X_\beta \wedge$

$\forall n(\alpha n \in I))$ satisfies the conclusion.) However, the premise of SCDS$_1'$ might hold also in cases where $\{X : A(X)\}$ cannot be provided with a metric, at least this possibility is not excluded.

7.2. Note that reinforcing $\exists \alpha$ to $\exists ! \alpha$ does not give us an acceptable strengthening of SCDS$_1'$ to SCDS, since e.g. $\mathbb{R}$, a connected space, can be represented as $\{X : A(X)\}$; and $\exists ! \alpha$ would imply a mapping into (totally disconnected) Baire space, in strong contrast with continuity theorems. The following weakened variant seems to be more plausible:

SCDS$_1$ $\forall XY[A(X) \wedge A(Y) \rightarrow$

$\exists \alpha \{(\forall x(\alpha x = 0) \rightarrow X = Y) \wedge (\neg \forall x(\alpha x = 0) \rightarrow X \neq Y)\}]$
$\rightarrow \exists \gamma Z \forall X[A(X) \rightarrow$
$\exists \alpha \{(Z)_\alpha = X \wedge \forall \alpha \beta ((Z)_\alpha = (Z)_\beta \leftrightarrow \forall x(\gamma(\bar{\alpha} x, \bar{\beta} x) = 0))\}].$

8. RULES CORRESPONDING TO THE AXIOMS

8.1. The various axioms mentioned of varying degrees of plausibility, suggest that we should look, for familiar systems, for closure under the corresponding rule (where the main implication is replaced by metamathematical $\Rightarrow$: when the premise is derivable, then so is the conclusion). In the case of the first three examples, this means closure under Church's rule, Continuity rule, and the Uniformity rule. Closure under these rules has indeed been established for many systems; for some examples see Troelstra (1973, 1973A) and de Jongh-Smorynski (1974).

As to SCDS, for the usual systems (e.g. **HAS**) we cannot possibly expect to establish the corresponding *rule:*

SCDS-Rule

$$\vdash\forall XY(A(X)\,\&\,A(Y)\rightarrow X=Y\vee X\neq Y)$$
$$\Rightarrow\vdash\exists Z\forall X(A(X)\rightarrow\exists!\,x((Z)_x=X)).$$

For if we take

$$A(X)\equiv\forall YZ(Y=Z\vee Y\neq Z)$$

then the premise is certainly provable in **HAS**, but the conclusion can only be true if $\neg\exists XA(X)$, i.e. we should be able to show $\neg\forall XY(X=Y\vee X\neq Y)$ which is impossible since $\textbf{HAS}\subseteq\textbf{HAS}^c$ (i.e. **HAS** is consistent with classical logic).

8.2. Only by reinforcing **HAS** we can expect the rule to hold; in our example, **HAS** needs to be reinforced by $\neg\forall XY(X=Y\vee X\neq Y)$. A similar remark applies, mutatis mutandis, to rules corresponding to: every metric space has a subcountable set dense in the space. (Requiring $A(X)$ to be non-empty is not sufficient as a restriction: take e.g. $A(X)\equiv[X=\{2n: n\in N\}\vee(X\subseteq\{2n+1: n\in N\}\ \&\ \forall YZ(Y=Z\vee Y\neq Z))]$.)

9. Completeness of intuitionistic predicate logic

9.1. For intuitionistic propositional logic, and certain fragments of predicate logic, we can establish completeness relative to validity in structures, if we permit the domains and relations of those structures to contain lawless sequences (Troelstra, 1976, § 7). The usefulness of such results is limited, since lawless sequences as such do not play an important role in (intuitionistic) mathematical practice. So we are also interested in completeness results for structures where the parameters range over other classes of objects with properties more relevant to mathematical practice.

9.2.   The assumption of CT_0 yields incompleteness (Kreisel; see van Dalen (1973)). Similarly, the usual continuity axioms (e.g. $CONT_1$), although they can be used, just as CT_0, to refute many well-known theorems from classical predicate logic, are not sufficient to refute even all unprovable propositional schemata. For example, Kleene (1965) noted that in his formalization of intuitionistic analysis

$$IP_v \qquad (\neg A \to B \lor C) \to (\neg A \to B) \lor (\neg A \to C)$$

is special-realizable, hence not provably refutable. This holds even in case we assume $\forall \alpha \neg \neg \exists x \in R(\alpha = \{x\})$ (R set of the Gödel numbers of total recursive functions) and CT_0; we refrain from giving a proof here.

Similarly, $UP + IP_v$ is consistent with **HAS** (Troelstra, 1973A, § 4). Thus we are led to ask for significant (sub-)classes of formulae of intuitionistic predicate logic for which the axioms mentioned in our examples yield completeness; this problem is still wide open.

10.   FINAL REMARKS

10.1.   The examples presented are by no means the only ones which have been mentioned or proposed; for variants of the continuity axioms (local continuity) see e.g. Beeson (B).

10.2.   Our examples are all of the same general type; they consist of an implication where the premise because of the constructive ('strong') interpretation of $\exists$ and $\forall \exists$ permits a conclusion which is stronger than in the classical case.

10.3.   One should not overestimate the impact of the axioms in our examples on the *existing* practice of constructive mathematics. For example, since in practice the theory of metric spaces *is* restricted to separable ones, one should not expect a great impact of the axiom that all complete metric spaces are separable. No doubt a lot of inventiveness and imagination will be needed to find areas of mathematics where these axioms do have an impact (compare this with the fact that in classical mathematics one hardly ever uses strong forms of impredicative comprehension).

Mathematical Institute,
University of Amsterdam

NOTES

[1] These properties are in themselves neither necessary nor sufficient to guarantee the 'constructive' character of a theory; see Troelstra (1973A, 1.8).

[2] We refrained from discussing e.g. lawless sequences, for which an informally rigorous justification of the axioms can be given (Troelstra, 1976, § 2), but concentrated on 'plausible' axiom schemata.

[3] The corresponding uniformity *rule* for intuitionistic predicate logic had already attracted attention before – see e.g. Kreisel (1971, p. 145).

[4] This generalizes a remark by van Dalen (1974A, 1975A).

[5] This possible new axiom seems to have suggested itself to a number of people, and may perhaps be said to belong to 'folklore'. I do not know where it was first formulated explicitly; in any case, the Appendix A of Bishop (1967) is very suggestive in this respect.

BIBLIOGRAPHY

Beeson, M. J.: 1972, *Metamathematics of Constructive Theories of Effective Operations*, Thesis, Stanford University. (See also Beeson, 1975.)

Beeson, M. J.: 1975, 'The Nonderivability in Intuitionistic Formal Systems of Theorems on the Continuity of Effective Operations', *Journ. Symbolic Logic* 40, 321–346.

Beeson, M. J.: A, 'The Unprovability in Constructive formal Systems of the Continuity of Effective Operations on the Reals'. To appear.

Beeson, M. J.: B, 'Principles of Continuous Choice and Continuity of Functions in Formal Systems for Constructive Mathematics'. To appear.

Bishop, E.: 1967, *Foundations of Constructive Analysis*, McGraw Hill, New York.

Čeitin, G. S.: 1959, 'Algorithmic Operators in Constructive Complete Separable Metric Spaces' (Russian), *Doklady Akad. Nauk* 128, 49–52.

Dalen, D. van: 1973, 'Lectures on Intuitionism', in A. R. D. Mathias, H. Rogers (eds.), *Cambridge Summer School in Mathematical Logic*, Springer-Verlag, Berlin, pp. 1–94.

Dalen, D. van: 1974A, 'Choice Sequences in Beth models', *Notes on Logic and Computer Science* 20, Dept. of Mathematics, Rijksuniversiteit, Utrecht (The Netherlands). Revised in 'An Interpretation of Intuitionistic Analysis', to appear in *Annals of Math. Logic*.

Dalen, D. van: 1975, 'Experiments in Lawlessness', *Notes on Logic and Computer Science* 28, Dept. of Mathematics, Rijksuniversiteit, Utrecht (The Netherlands). Revised in 'An Interpretation of Intuitionistic Analysis', to appear in *Annals of Math. Logic*.

Dalen, D. van: 1975A, 'The Use of Kripke's Schema as a Reduction Principle', *Preprint* **11**, Dept. of Mathematics, University of Utrecht.

Dragalin, A. G.: 1973, 'Constructive Mathematics and Models of Intuitionistic Theories', in P. Suppes, L. Henkin, Gr. C. Moisil, A. Joja (eds.), *Logic, Methodology and Philosophy of Science* IV, North-Holland Publ. Co., Amsterdam, 111–128.

Friedman, H.: 1975, '*Set Theoretic Foundations for Constructive Analysis and the Hilbert Program*, Manuscript, Dept. of Mathematics, State University of New York at Buffalo, Buffalo, N.Y.

Jongh, D. H. J. de and Smorynski, C. A.: 1974, 'Kripke Models and the Theory of Species', *Report* 74-03, Dept. of Mathematics, Univ. of Amsterdam. Appeared in: *Annals of Mathematical Logic* **9** (1976), 157–186.

Kleene, S. C.: 1965, 'Logical Calculus and Realizability', *Acta Philosophica Fennica* **18**, 71–80.

Kleene, S. C. and Vesley, R. E.: 1965, *The Foundations of Intuitionistic Mathematics, Especially in Relation to Recursive Functions*, North-Holland Publ. Co., Amsterdam.

Kreisel, G.: 1968, 'Lawless Sequences of Natural Numbers', *Compositio Math.* **20**, 222–248.

Kreisel, G.: 1971, 'A Survey of Proof Theory II', in J.-E. Fenstad (ed.) *Proceedings of the Second Scandinavian Logic Symposium*, North-Holland Publ. Co., Amsterdam, pp. 109–170.

Kreisel, G.: 1972, 'Which Number-Theoretic Problems can be Solved in Recursive Progressions on $\prod_1^1$-Paths Through 0?', *J. Symbolic Logic* **37**, 311–344.

Kreisel, G. and Troelstra, A. S.: 1970, 'Formal Systems for Some Branches of Intuitionistic Analysis', *Annals of Math. Logic* **1**, 229–387.

Moschovakis, J. R.: 1973, 'A Topological Interpretation of Second-order Intuitionistic Arithmetic', *Compositio Math.* **26**, 261–275.

Moschovakis, Y. N.: 1964, 'Recursive Metric Spaces', *Fund. Math.* **55**, 215–238.

Orevkov, V. P.: 1963, 'A Constructive Map of the Square into itself, which Moves every Constructive Point' (Russian), *Doklady Akad. Nauk SSSR* **152**, 55–58; translated in *Soviet Mathematics* **4** (1963), 1253–1256.

Orevkov, V. P.: 1964, 'On Constructive Mappings of a Circle into itself' (Russian), *Trudy Mat. Inst. Steklov* **72**, pp. 437–461; translated in *Translations AMS* **100**, 69–100.

Troelstra, A. S.: 1969, *Principles of Intuitionism*, Springer-Verlag, Berlin.

Troelstra, A. S. (ed.): 1973, *Metamathematical Investigation of Intuitionistic Arithmetic and Analysis*, Springer-Verlag, Berlin.

Troelstra, A. S.: 1973A, 'Notes on Intuitionistic Second-Order Arithmetic', in A. R. D. Mathias, H. Rogers (eds.), *Cambridge Summer School in Mathematical Logic*, Springer-Verlag, Berlin, 171–205.

Troelstra, A. S.: 1976, *Choice Sequences, a Chapter of Intuitionistic Mathematics*, Oxford University Press, Oxford. To appear.

II

FOUNDATIONS OF
MATHEMATICAL THEORIES

JAMES E. BAUMGARTNER[1]

INEFFABILITY PROPERTIES OF CARDINALS II

0. INTRODUCTION

This paper applies the methods of [1] to several classes of cardinals other than ineffables. The central point is the same as [1], namely that many 'large cardinal' properties are better viewed as properties of normal ideals than as properties of cardinals alone, and that in order to understand these properties fully it is necessary to consider the associated normal ideals.

We begin in Section 2 with a treatment of the weakly compact ideal on a weakly compact cardinal. This ideal is easily defined in terms of Π_1^1-indescribability (see [8] and [1]), but not in terms of the usual combinatorial definitions of weak compactness. We take three such common definitions (in terms of partitions, trees, and ultrafilters in fields of sets) and show how to strengthen them to obtain combinatorial properties which easily yield the existence of the weakly compact ideal. The existence of the weakly compact ideal easily implies strong versions of the usual theorems that weakly compact cardinals are 'large' (i.e. Mahlo, hyperMahlo, etc.), so one by-product of this approach is a purely combinatorial proof that weakly compact cardinals are large. Other such proofs are known, and some of them are much shorter (see Kunen [6], for example), but none of them yield the weakly compact ideal.

The weakly compact ideal has also been exploited in [3], where it is shown that the non-stationary ideal on a weakly compact cardinal κ is not κ^+-saturated.

In Section 3 we consider the result of iterating into the transfinite the 'operation' which originally produced ineffable sets from stationary sets. The eventual results about this iteration, however, do not seem as interesting as the method which is developed to treat stages α of the

[1] Preparation of this paper was partially supported by National Science Foundation grant number GP-38026.

Butts and Hintikka (eds.), Logic, Foundations of Mathematics and Computability Theory, 87–106.
Copyright © 1977 by D. Reidel Publishing Company, Dordrecht-Holland. All Rights Reserved.

iteration on a cardinal κ such that $\kappa \leqslant \alpha < \kappa^+$. The method involves 'canonical' sequences of functions ordered by eventual dominance. Several results about these sequences are obtained, including a new characterization of ineffables in L. It seems quite likely that this eventual dominance method may be used to handle other unrelated iterations as well.

Section 4 is devoted to extending the results of [1] to Erdös and Ramsey cardinals. We begin by associating normal ideals with Erdös cardinals and with the α-partition cardinals of Drake [4]. This leads naturally to the definition of new classes of cardinals, the α-*Erdös cardinals*, which generalize the properties of Erdös and α-partition cardinals. Moreover, this definition yields an interesting class of cardinals even when α is a successor ordinal.

Jensen and Kunen [5] showed that ineffable cardinals are Π^1_2-indescribable and subtle, but that the converse does not hold. Perhaps the most striking illustration that properties like ineffability, indescribability, and subtlety are really properties of normal ideals is the result in [1] that a cardinal κ is ineffable iff the ideal of non-subtle sets and the ideal of non-Π^1_2-indescribable sets together generate a non-trivial κ-complete normal ideal, which then coincides with the ideal of non-ineffable sets. This result is extended in Section 4 to Ramsey and ineffably Ramsey cardinals. An analogue of subtle and inaccessible cardinals, the *pre-Ramsey* cardinals, is defined and a version of the theorem above is obtained in which ineffable, subtle, and Π^1_2-indescribable are replaced by Ramsey, pre-Ramsey, and Π^1_1-indescribable, respectively.

1. NOTATION AND TERMINOLOGY

Our set-theoretical usage is standard. If x is a set, then $\mathcal{P}(x)$ is the power set of x and $|x|$ is the cardinality of x. If x and y are sets then ${}^x y$ is the set of all functions from x into y. If x is a set and λ is a cardinal, then $[x]^\lambda = \{y \subseteq x : |y| = \lambda\}$ and $[x]^{<\lambda} = \{y \subseteq x : |y| < \lambda\}$.

We assume the axiom of choice throughout, so cardinals are identified with intial ordinals.

A *tree* is a partially ordered set $(T, \leqslant_T)$ such that for all $t \in T$, $\{s \in T : s \leqslant_T t\}$ is well-ordered by $\leqslant_T$, and such that T has a unique minimal element, the *root* of the tree. A *branch through* a tree $(T, \leqslant_T)$ is a linearly ordered subset; its *length* is its order type. The αth *level* of the tree $(T, \leqslant_T)$ is $\{t \in T : \{s \in T : s \leqslant_T t\}$ has order type $\alpha + 1\}$.

The set of all sets of rank less than α is denoted by V_α. A second order formula φ is Π_n^1 if $\varphi = (Q_1 X_1) \cdots (Q_n X_n)\psi$, where ψ is first-order, each X_i is a second order variable, and Q_i is $\forall$ if i is odd and $\exists$ if i is even. A cardinal κ is Π_n^1-*indescribable* iff for any Π_n^1-sentence φ and relations $R_1, \ldots, R_m$ on V_κ, if $(V_\kappa, \varepsilon, R_1, \ldots, R_m) \vDash \varphi$, then there is $\alpha < \kappa$ such that $(V_\alpha, \varepsilon, R_1 | V_\alpha, \ldots, R_m | V_\alpha) \vDash \varphi$ (here $R_i | V_\alpha$ denotes the restriction of R_i to V_α).

A function f defined on a set of ordinals A is *regressive on* A iff $f(\alpha) < \alpha$ for all $\alpha \in A$ such that $\alpha > 0$. An ideal I on a cardinal κ is κ-*complete* if I is closed under unions of size $< \kappa$; I is *normal* if for any $A \subseteq \kappa$, if $A \notin I$ and f is regressive on A, then there is $\alpha < \kappa$ such that $f^{-1}(\{\alpha\}) \notin I$. Fodor's Theorem asserts that the non-stationary ideal on a regular uncountable cardinal is normal.

If f is a function on $[X]^n$ for some $n < \omega$, then we say a set $Y \subseteq X$ is *homogeneous for* f if f is constant on $[Y]^n$. The notation $\kappa \to (\lambda)_\mu^n$ means that for any $f : [\kappa]^n \to \mu$, there is $X \subseteq \kappa$ such that $|X| = \lambda$ and X is homogeneous for f. The notation $\kappa \to (\alpha)_\lambda^{<\omega}$ means that for any $f : [\kappa]^{<\omega} \to \lambda$ there is $X \subseteq \kappa$ such that X has order type α and for each $n < \omega, X$ is homogeneous for $f | [\kappa]^n$.

If X is a set of ordinals, then a function f defined on $[X]^n$ (or on $[X]^{<\omega}$) is *regressive* if $f(x) < \min (x)$ for all $x \in [X]^n$ (resp.: all $x \in [X]^{<\omega}$). A cardinal κ is *subtle* if for any regressive f on $[\kappa]^2$ and any closed unbounded set $C \subseteq \kappa$, there is $A \subseteq C$ such that $|A| \geqslant 3$ and A is homogeneous for f. The definitions of *almost ineffable* and *ineffable* are obtained from the definition above by replacing '$|A| \geqslant 3$' by '$|A| = \kappa$' and 'A is stationary', respectively. All three notions have reformulations in the following style: e.g. κ is ineffable iff for any $\langle S_\alpha : \alpha > \kappa \rangle$, if $S_\alpha \subseteq \alpha$ for all α then there is $A \subseteq \kappa$ such that A is stationary and for all $\alpha, \beta \in A$, if $\alpha < \beta$ then $S_\alpha = S_\beta \cap \alpha$. In the latter situation we say A is *homogeneous* for $\langle S_\alpha : \alpha < \kappa \rangle$. The fact that both types of formulations exist is used in Section 3, where we state without proof the analogous fact for a generalization of ineffability. The proof is the same (see [5]).

2. THE WEAKLY COMPACT IDEAL

The purpose of this section is to show that every weakly compact cardinal bears a natural normal ideal, the *weakly compact* ideal.

There are a great number of equivalent definitions of weak compactness (see [10]) but the following four are probably the most frequently used:

Assume that κ is strongly inaccessible and uncountable. Then the following are equivalent.

(a) κ is Π_1^1-indescribable.

(b) $\kappa \to (\kappa)_2^2$.

(c) κ has the tree property, i.e. if $(T, \leqslant)$ is a tree of cardinality κ and every element of T has $<\kappa$ immediate successors, then T has a branch of length κ.

(d)   If $S \subseteq \mathscr{P}(\kappa)$ is a κ-complete field of sets and $|S| = \kappa$, then there is a non-principal κ-complete ultrafilter on S.

If κ is weakly compact then κ is not the first strongly inaccessible cardinal but this result is obvious only from definition (a) above. For to say that κ is strongly inaccessible is a Π_1^1 statement about (V_κ, ε); hence there is $\alpha < \kappa$ such that α is inaccessible. Apparently, then, the combinatorial definitions (b), (c), and (d) are not as powerful as (a). We would like to argue that the reason is that there is a normal ideal on κ naturally definable from (a), but not from (b), (c), or (d).

Let us say that $A \subseteq \kappa$ is Π_1^1-*indescribable* iff for any Π_1^1-sentence φ and any relations $R_1, \ldots, R_n$ on V_κ, if φ is true in $(V_\kappa, \varepsilon, R_1, \ldots, R_n)$ then for some $\alpha \in A$, φ is true in $(V_\alpha, \varepsilon, R_1 | V_\alpha, \ldots, R_n | V_\alpha)$. Let $I_1 = \{A \subseteq \kappa: A$ is not Π_1^1-indescribable$\}$. It is easy to see that I_1 is an ideal. Moreover, I_1 contains all nonstationary subsets of κ. For if C is closed and unbounded in κ then $(V_\kappa, \varepsilon, C) \vDash \varphi$, where φ says C is closed and unbounded in κ. If $\alpha < \kappa$ and $(V_\alpha, \varepsilon, C \cap V_\alpha) \vDash \varphi$, then clearly $\alpha \in C$. Hence $\kappa - C \in I_1$.

In fact, I_1 is normal, as will follow from Theorem 2.7 below. This result is essentially due to Levy [8]. It is also proved in [1].

It turns out, however, that if we replace definitions (b), (c), and (d) with (b′), (c′), and (d′) below, then the ideal is readily obtainable from each of them, and so are the results about the size of κ.

(b') $\kappa \to (\kappa,$ stationary set$)^2$. i.e. if $f: [\kappa]^2 \to 2$, then there is $A \subseteq \kappa$ such that either $|A| = \kappa$ and f is constantly 0 on $[A]^2$, or else A is stationary and f is constantly 1 on $[A]^2$.

(c') Suppose $(T, \leqslant_T)$ is a tree and $T = \kappa$. If $\alpha \leqslant_T \beta$ implies $\alpha \leqslant \beta$ and the set of immediate successors of each $\alpha \in T$ is non-stationary, then T has a branch of length κ.

(d') If S is a κ-complete field of sets, F is a collection of regressive functions on κ, $f^{-1}(\{\alpha\}) \in S$ for each $f \in F$ and $\alpha \in \kappa$, and $|S| = |F| = \kappa$, then there is a non-principal κ-complete ultrafilter U on S such that every $f \in F$ is constant on a set in U.

Note that in each case we have obtained the primed version from the unprimed version by substituting the ideal of non-stationary sets for the ideal of sets of cardinality less than κ at an appropriate point.

Let $P_1 = \{X \subseteq \kappa: X$ is Π_1^1-indescribable$\}$.

Let P_2 be the set of all $X \subseteq \kappa$ such that $X \to (\kappa,$ stationary set$)^2$, i.e. such that in (b') we can always find A so that $A \subseteq X$.

Let P_3 be the set of all $X \subseteq \kappa$ such that every tree $(T, \leqslant_T)$ as in (c') with universe $T = X$ has a branch of length κ.

Let P_4 be the set of all $X \subseteq \kappa$ such that, given F and S as in (d') with $X \in S$, there is U as in (d') with $X \in U$.

THEOREM 2.1.   *Assume κ is strongly inaccessible. Then $P_1 = P_2 = P_3 = P_4$.*

The proof follows from Lemmas 2.2–2.5. Assume κ is strongly inaccessible.

LEMMA 2.2. $P_1 \subseteq P_2$.

Proof. Suppose $X \in P_1$. Let $f: [X]^2 \to 2$. For each $\alpha \in X$ define a sequence $\langle s_\xi^\alpha, \xi < \delta_\alpha \rangle$ inductively as follows. Given s_ξ^α for all $\xi < \eta$, let s_η^α be the least $s \in X$ such that $s_\xi^\alpha < s < \alpha$ and $f(\{s_\xi^\alpha, s\}) = f(\{s, \alpha\}) = 0$ for all $\xi < \eta$, provided such s exists; otherwise let $\delta_\alpha = \eta$. Now let $h(\alpha) = \sup\{s_\xi^\alpha: \xi < \delta_\alpha\}$.

Case 1. $\{\alpha: h(\alpha) < \alpha\}$ is stationary in κ. Then by Fodor's Theorem there is stationary S such that h is constant on S. But now, using the fact that κ is strongly inaccessible, there is stationary $S' \subseteq S$ such that $\langle s_\xi^\alpha: \xi < \delta_\alpha \rangle = \langle s_\xi^\beta: \xi < \delta_\beta \rangle$ for all $\alpha, \beta \in S'$. But then if $\{\alpha, \beta\} \in [S']^2$ it is clear that $f(\{\alpha, \beta\}) = 1$.

Case 2. $\{\alpha: h(\alpha) < \alpha\}$ is non-stationary, hence belongs to I_1. Thus we may as well assume that $h(\alpha) = \alpha$ for all $\alpha \in X$. Let φ be the assertion that there is no cofinal subset A of X such that f is constantly 0 on $[A]^2$. Then φ is a Π_1^1 assertion about $(V_\kappa, \varepsilon, f, X)$ so, since $X \in P_1$, if φ is true in $(V_\kappa, \varepsilon, f, X)$ then there is $\alpha \in X$ such that φ is true in $(V_\alpha, \varepsilon, f \,|\, V_\alpha, X \,|\, V_\alpha)$, and the latter is false since $\{s_\xi^\alpha: \xi < \delta_\alpha\}$ is a counterexample. ∎

LEMMA 2.3. $P_2 \subseteq P_3$.

Proof. Suppose $X \in P_2$. Let $(T, \leqslant_T)$ be a tree such that $X = T$, $\alpha \leqslant_T \beta$ implies $\alpha \leqslant \beta$, and the set of immediate successors of each $\alpha \in T$ is non-stationary. For each $\alpha \in T$, let p_α be the sequence obtained by writing $\{\beta: \beta \leqslant_T \alpha\}$ in $\leqslant_T$-increasing order with the minimal element omitted. If $\alpha, \beta \in T$, $\alpha < \beta$, let $f(\{\alpha, \beta\}) = 0$ if p_β lexicographically (with respect to the usual ordering on ordinals) precedes p_α; $f(\{\alpha, \beta\}) = 1$ otherwise.

Case 1. There is $A \subseteq X$ such that $|A| = \kappa$ and f is 0 on $[A]^2$. If $\alpha, \beta \in A$, $\alpha < \beta$, then p_β lexicographically precedes p_α so there must be some α_0 such that if $\beta > \alpha_0$ then the least element of p_β is always the same, say x_0. Repeating the argument, we can find $\alpha_1 > \alpha_0$ so that the next element of each p_β for $\beta > \alpha_1$ is always the same, say x_1. Proceeding inductively, and using the fact that κ is regular, we can find x_ξ and α_ξ for each $\xi < \kappa$. But then $\{x_\xi: \xi < \kappa\}$ is a branch of length κ.

Case 2. There is stationary $A \subseteq X$ such that f is 1 on $[A]^2$. If $\alpha, \beta \in A$, $\alpha < \beta$, then p_α lexicographically precedes p_β. For each $\alpha \in A$ let $g(\alpha)$ be the least element of p_α. Then g is regressive on A so g is constant on a stationary set. Hence there is $\alpha_0 < \kappa$ so that if $\beta > \alpha_0$ then the least element of p_β is always the same, say x_0. Now proceed inductively as in Case 1. ∎

LEMMA 2.4. $P_3 \subseteq P_4$.

Proof. Let S and F be as in (d'). Let $h: \kappa \to S \cup F$ be one-to-one and onto. For each α, we define a partition D_α of κ inductively as follows: Let $D_0 = \{\kappa\}$. If $h(\alpha) = A \in S$ then $D_{\alpha+1} = \{A \cap Y: Y \in D_\alpha\} \cup \{Y - A: Y \in D_\alpha\}$. If $h(\alpha) = f \in F$ then $D_{\alpha+1} = \{f^{-1}(\{\beta\}) \cap Y: Y \in D_\alpha, \beta \in \kappa\}$. If α is a limit ordinal then D_α is the collection of all sets of the form $\cap \{X_\beta: \beta < \alpha\}$ where $X_\beta \in D_\beta$. Note that if $\alpha < \beta$ then D_β refines D_α.

Now suppose $X \in P_3$. We define a tree $(T, \leqslant_T)$ with $T = X$ as follows. Let T_α denote the αth level of T. We determine T_α by induction on α. Given T_β for all $\beta < \alpha$, let $T_\alpha = \{\xi \in X : (\exists Y \in D_\alpha) \xi$ is the least element of $Y - \bigcup_{\beta < \alpha} T_\beta\}$.

If $\xi \in T_\alpha$, $\eta \in T_\beta$, $\beta < \alpha$, $\xi \in Y \in D_\alpha$, $\eta \in Z \in D_\beta$, put $\eta <_T \xi$ iff $Z \subseteq Y$. It is easy to see that if $h(\alpha) \in S$ then every member of T_α has at most two immediate successors, and if $h(\alpha) = f \in F$ then f is one-to-one and regressive on the immediate successors of each element of T_α. Hence the set of immediate successors of each element of T is non-stationary. Since $X \in P_3$ there is a branch $B \subseteq T$ of length κ. Let $U = \{A \in S : |B - A| < \kappa\}$. It is easy to see that U works. ∎

LEMMA 2.5. $P_4 \subseteq P_1$.

Proof.   Let $X \in P_4$. Let φ be Π_1^1 and assume $(V_\kappa, \varepsilon, R_1, \ldots, R_n) \vDash \varphi$. Let $\mathcal{M} = (V_\kappa, \varepsilon, <_1, R_1, \ldots, R_n)$ where $<_1$ is a well-ordering of V_κ. Let G be the set of all functions from κ into V_κ which are definable from parameters in $\mathcal{M}$, and let F be the set of all regressive functions in G. Let S be the κ-complete field of sets generated by X together with all the subsets of κ which are definable from parameters in $\mathcal{M}$. Let U be a non-principal κ-complete ultrafilter over S such that $X \in U$ and every $f \in F$ is constant on a set in U.

Now form the 'definable ultrapower' of $\mathcal{M}$ modulo U. That is, if f, $g \in G$, let $f \equiv g$ iff $\{\alpha : f(\alpha) = g(\alpha)\} \in U$ and let $[f]$ be the equivalence class of f. Define relations R_i' on $V' = \{[f] : f \in G\}$ by $R_i'([f_1], \ldots, [f_k])$ iff $\{\alpha : \mathcal{M} \vDash R_i(f_1(\alpha), \ldots, f_k(\alpha))\} \in U$. $<_1'$ and ε' may be defined similarly. Let $\mathcal{M}' = (V', \varepsilon' <_1', \ldots, R_n')$. Since $<_1$ is a well-ordering, Skolem functions for $\mathcal{M}$ are definable, and so we have a version of Łos's Theorem:

$$\mathcal{M}' \vDash \psi([f_1], \ldots, [f_k]) \quad \text{iff} \quad \{\alpha : \mathcal{M} \vDash \psi(f_1(\alpha), \ldots, f_k(\alpha))\} \in U,$$

where ψ is any first-order formula appropriate for $\mathcal{M}$. It follows that the embedding $i : \mathcal{M} \to \mathcal{M}'$ given by $i(x) = [f_x]$ is elementary, where f_x is the constant function with value x. Since U is κ-complete, $\mathcal{M}'$ is well-founded, hence is isomorphic to a structure $\mathcal{N} = (N, \varepsilon, <_2, S_1, \ldots, S_n)$ where N is a transitive set. We identify $\mathcal{M}'$ with $\mathcal{N}$, By standard arguments the κ-completeness of U implies that i is the identity on V_κ so $V_\kappa \subseteq N$. Moreover, since each regressive function in G is constant on a set in U, it follows that [id] is identified with κ, where id is the identity function on κ.

If $\varphi = \forall X \psi(X)$, let $\bar{\varphi}(\kappa)$ be $\forall x (x \subseteq V_\kappa \to \bar{\psi}(x))$, where $\bar{\psi}$ is the same as ψ except that all quantifiers are restricted to V_κ. Then $\bar{\varphi}(\kappa)$ is first order, and since $\mathcal{M} \vDash \varphi$ and $V_\kappa \subseteq N$ (so $V_\kappa \in N$), we have $\mathcal{N} \vDash \bar{\varphi}(\kappa)$. Since $\kappa = [\mathrm{id}]$, we have $Y = \{\alpha : \mathcal{M} \vDash \bar{\varphi}(\alpha)\} \in U$ by Łos's Theorem. Now $\mathcal{M} \vDash \bar{\varphi}(\alpha)$ iff $(V_\alpha, \varepsilon, R_1 \mid V_\alpha, \ldots, R_n \mid V_\alpha) \vDash \varphi$, and since $X \in U$ we have $X \cap Y \neq 0$. Thus $X \in P_1$. ∎

COROLLARY 2.6.   *If κ is strongly inaccessible then (b'), (c'), and (d') are equivalent formulations of weak compactness.*

Proof.   To say that κ is weakly compact is the same as saying $\kappa \in P_1$, so we are done by Theorem 2.1. ∎

In view of Theorem 2.1, it seems reasonable to say that weak compactness is really a property of subsets of a cardinal κ. We refer to the set $P = P_i (1 \leq i \leq 4)$ as the set of *weakly compact* subsets of κ. The ideal $I = \{X \subseteq \kappa : X \notin P\}$ is the *weakly compact ideal*, and the filter $\{X \subseteq \kappa : \kappa - X \in I\}$ is the *weakly compact filter*.

THEOREM 2.7.   *The weakly compact ideal is normal.*

Proof.   The proof can be given equally well from any of the definitions (a), (b'), (c'), or (d'). We illustrate with (b'). It will suffice to show that if f is regressive on $X \in P_2$ then for some α, $f^{-1}(\{\alpha\}) \in P_2$. Suppose not. For each α, let $g_\alpha : [f^{-1}(\{\alpha\})]^2 \to 2$ be a counterexample to $f^{-1}(\{\alpha\}) \in P_2$. Define $g : [X]^2 \to 2$ as follows. If $\alpha, \beta \in X$, $\alpha \neq \beta$, let $g(\{\alpha, \beta\}) = 0$ iff $\exists_\gamma f(\alpha) = f(\beta) = \gamma$ and $g_\gamma(\{\alpha, \beta\}) = 0$.

Case 1.   There is $A \subseteq X$ such that $|A| = \kappa$ and g is 0 on $[A]^2$. Then clearly $\exists \gamma g(\alpha) = \gamma$ for all $\alpha \in A$ and g_γ is 0 on $[A]^2$, contradiction.

Case 2.   There is stationary $A \subseteq X$ such that g is 1 on $[A]^2$. Since f is regressive on A, by Fodor's Theorem there is γ and stationary $B \subseteq A$ such that f is constantly γ on B. But then g_γ is 1 on $[B]^2$, contradiction. ∎

It follows from Theorem 2.7 and the proof of Lemma 2.2 that a much stronger version of (b') holds for weakly compact cardinals, namely, $\kappa \to (\kappa, \text{weakly compact set})^2$. The proof is left to the reader. In fact, for this partition property one may even begin with a partition of $[A]^2$, where A is weakly compact, rather than with a partition of $[\kappa]^2$.

The fact that weakly compact cardinals are large is perhaps best illustrated by the following two theorems.

THEOREM 2.8. *If κ is weakly compact, then $\{\alpha < \kappa : \alpha$ is strongly inaccessible$\}$ belongs to the weakly compact filter.*

Proof. The easiest proof uses Π_1^1-indescribability. Since we have asserted that the combinatorial definitions (b'), (c'), and (d') readily yield information about the size of κ, however, we illustrate with (b'). First note that since κ is strongly inaccessible, $\{\alpha < \kappa : \alpha$ is a strong limit cardinal$\}$ is closed and unbounded in κ, hence must belong to the weakly compact filter. It will suffice, therefore, to show that $\{\alpha < \kappa : \alpha$ is regular$\}$ belongs to the weakly compact filter. Suppose not. Since the cofinality function is regressive on $\{\alpha < \kappa : \alpha$ is singular$\}$, we may find λ and $A \in P_2$ such that $cf\alpha = \lambda$ for all $\alpha \in A$.

For $\alpha \in A$, let $\langle s_\xi^\alpha : \xi < \lambda \rangle$ be an increasing sequence cofinal in α. If $\alpha, \beta \in A$, $\alpha < \beta$, let $f(\{\alpha, \beta\}) = 0$ if $\langle s_\xi^\alpha : \xi < \lambda \rangle$ lexicographically precedes $\langle s_\xi^\beta : \xi < \lambda \rangle$, and $f(\{\alpha, \beta\}) = 1$ otherwise. Now arguing as in the proof of Lemma 2.3, we can find a homogeneous set $X \subseteq A$, and s_ξ and α_ξ for each $\xi < \lambda$ such that if $\alpha > \alpha_\xi$ and $\alpha \in X$ then $s_\xi^\alpha = s_\xi$. But this is a contradiction, since then $\langle s_\xi^\alpha : \xi < \lambda \rangle = \langle s_\xi : \xi < \lambda \rangle$ whenever $\alpha \in X$, $\alpha > \sup\{\alpha_\xi : \xi < \lambda\}$. ∎

Similarly, one can prove

THEOREM 2.9. *If κ is weakly compact and $S \subseteq \kappa$ is stationary, then $\{\alpha < \kappa : S \cap \alpha$ is stationary in $\alpha\}$ belongs to the weakly compact filter.*

It follows immediately that $\{\alpha < \kappa : \alpha$ is Mahlo$\}$ belongs to the weakly compact filter, and so forth. Theorem 2.9 has also been utilized in [3] to show that if κ is weakly compact, then the ideal of non-stationary subsets of κ is not κ^+-saturated. In fact, the result yields that if $X \subseteq \kappa$ is weakly compact, then the non-stationary ideal restricted to X is not κ^+-saturated.

It is natural to ask whether a purely combinatorial treatment of the theory of weakly compact cardinals can be given. To be more specific, can one eliminate the role of Π_1^1-indescribability and exhibit a direct combinatorial path from definitions (b), (c), and (d) to (b'), (c'), and (d′)? The answer is affirmative, and one approach is in Theorem 2.10 below. In this connection, it should be remarked that in Theorem 2.1 a direct proof that $P_4 \subseteq P_2$ can be given.

Note that the proof of Theorem 2.10 is really the same as the proof that a measurable cardinal bears a normal ultrafilter.

THEOREM 2.10.   (d) *implies* (d′).

Proof. Let S and F be as in (d′). By enlarging S and F, if necessary, we may assume that

(i) F is closed under composition of functions
(ii)   if $X \in S$ and $f \in F$ then $f^{-1}(X) \in S$
(iii)   if $f, g \in F$ then $\{\alpha : f(\alpha) < g(\alpha)\} \in S$.

By (d), there is a non-principal κ-complete ultrafilter U on S. If every $f \in F$ is constant on a set in U then we are done, so suppose not. For $f, g \in F$, let $f < g$ iff $\{\alpha : f(\alpha) < g(\alpha)\} \in U$. Since U is κ-complete, it follows that $<$ well-orders F (provided we identify functions f and g such that $\{\alpha : f(\alpha) = g(\alpha)\} \in U$. Let f be the $<$-least member of F such that f is not constant on a set in U. For $X \in S$, let $X \in D$ iff $f^{-1}(X) \in U$. It is easy to check that D is a κ-complete non-principal ultrafilter on S. Suppose $g \in F$. Then clearly $gf < f$ so gf is constant on a set in U. Say $\{\alpha : gf(\alpha) = \gamma\} \in U$. But then $\{\alpha : g(\alpha) = \gamma\} \in D$ so we are done.   ∎

If one is only interested in a purely combinatorial proof that weakly compact cardinals are large, then considerably shorter proofs are available. See for example Kunen's proof in [6]. However, the shorter arguments do not seem to yield the weakly compact ideal, and therefore results such as Theorems 2.8 and 2.9 are unobtainable.

3.   EVENTUAL DOMINANCE AND A HIERARCHY OF INEFFABLES

The ineffable sets may be regarded as having been produced from the stationary sets by an operation involving a combinatorial propery. In this section we consider the result of iterating this operation transfinitely. First, however, it is necessary to investigate briefly the notion of eventual dominance of functions.

If κ is a cardinal, D is a filter of subsets of κ, and $f, g \in {}^{\kappa}\kappa$, then we write $f < g$ (mod D) iff $\{\alpha : f(\alpha) < g(\alpha)\} \in D$. We say g *eventually dominates* f *mod D*. $f \leqslant g$ is defined similarly. Note that if D is countably complete, then $<$ (mod D) is well-founded. We will usually be interested in the case when D is the filter generated by the closed unbounded subsets of a regular uncountable cardinal κ. In this case we write simply $f < g$.

Now let κ be regular and uncountable. Let us call a sequence $\langle f_\alpha : \alpha < \kappa^+ \rangle$ of elements of ${}^\kappa\kappa$ a *canonical sequence* iff $\alpha < \beta$ implies $f_\alpha < f_\beta$ and for any other sequence $\langle g_\alpha : \alpha < \kappa^+ \rangle$ well-ordered by $<$, we have $f_\alpha \leqslant g_\alpha$ for all α. Note that if $\langle f_\alpha : \alpha < \kappa^+ \rangle$ and $\langle g_\alpha : \alpha < \kappa^+ \rangle$ are both canonical, then for each $\alpha, \{\xi : f_\alpha(\xi) = g_\alpha(\xi)\}$ contains a closed unbounded set.

The following theorem is part of the folklore.

THEOREM 3.1. *If κ is regular and uncountable, then there is a canonical sequence of elements of ${}^\kappa\kappa$.*

Proof. The f_α can be determined inductively, but it will be convenient to have a direct construction. If $\alpha < \kappa$ let $f_\alpha(\xi) = \alpha$ for all $\xi < \kappa$. Suppose $\kappa \leqslant \alpha < \kappa^+$. Let $h_\alpha : \kappa \to \alpha$ be an isomorphism, and let $<_\alpha$ be the relation on κ induced by the usual ordering on α. For each $\xi < \kappa$, let $f_\alpha(\xi)$ be the order-type of $(\xi, <_\alpha \mid \xi)$.

Now let $\alpha < \beta$. We must show $f_\alpha < f_\beta$. Fix η_0 so that $h_\alpha(\eta_0) \geqslant \alpha$. Let $C_{\alpha\beta} = \{\xi < \kappa : \eta_0 < \xi$ and if $\eta < \xi$ then $h_\beta^{-1} h_\alpha(\eta) < \xi\}$. $C_{\alpha\beta}$ is closed and unbounded, and if $\xi \in C_{\alpha\beta}$ then $(\xi, <_\alpha \mid \xi)$ is carried into a proper initial segment of $(\xi, <_\beta \mid \xi)$ by $h_\beta^{-1} h_\alpha$. Thus $f_\alpha(\xi) < f_\beta(\xi)$ and $f_\alpha < f_\beta$.

To complete the proof, it will suffice to show that if $\alpha < \kappa^+$ and $\{\xi : g(\xi) < f_\alpha(\xi)\}$ is stationary, then there is $\beta < \alpha$ such that $\{\xi : g(\xi) \leqslant f_\beta(\xi)\}$ is stationary. We leave to the reader the case when $cf\alpha < \kappa$.

Suppose $cf\alpha = \kappa$. Let $\langle \alpha_\xi : \xi < \kappa \rangle$ be an increasing sequence with limit α, and let $C_\xi = C_{\alpha_\xi \alpha}$, defined as above. Also, if $\eta < \kappa$, let $j(\eta)$ be the least ξ such that $h_\alpha(\eta) < \alpha_\xi$. Let C be the diagonal intersection of the C_ξ, i.e. $C = \{\xi : \forall \eta < \xi) \xi \in C_\eta\}$. Let $\bar{C} = C \cap \{\xi : \text{if } \eta < \xi \text{ then } j(\eta) < \xi\}$. Then $\bar{C}$ is closed unbounded, and it is easy to check that if $\xi \in \bar{C}$ then $f_\alpha(\xi) = \sup_{\eta < \xi} f_{\alpha_\eta}(\xi)$. By Fodor's Theorem, there is η so that $\{\xi : g(\xi) \leqslant f_{\alpha_\eta}(\xi)\}$ is stationary. ∎

Kunen [7] has shown that if D is a countably complete filter on a cardinal κ, then there is an ordinal $\alpha < (2^\kappa)^+$ such that all subsets of ${}^\kappa\kappa$

well-ordered by $<$ (mod D) have order type less than α. The identity of the least such α is independent of the usual axioms for set theory, as Kunen also shows. Next we observe that two familiar combinatorial 'axioms' settle the question in different ways.

Let κ be a regular uncountable cardinal. Then $\Diamond''_\kappa$ is the following assertion: There is $\langle S_\alpha : \alpha < \kappa \rangle$ such that $S_\alpha \subseteq \mathscr{P}(\alpha)$ for each α, $|S_\alpha| = |\alpha|$, and for any $X \subseteq \kappa$, $\{\alpha : X \cap \alpha \in S_\alpha\}$ contains a closed unbounded set.

Jensen [5] has shown that in L, κ is ineffable iff κ is regular and $\Diamond''_\kappa$ fails.

THEOREM 3.2. $\Diamond''_\kappa$ implies that there is a sequence $\langle f_\alpha : \alpha \leq \kappa^+ \rangle$ well-ordered by $<$, and such that $f_\alpha(\xi) < |\xi|^+$ for all α and all ξ.

Proof. Let $\langle f_\alpha : \alpha < \kappa^+ \rangle$ be canonical. By the proof of Theorem 3.1, we may take $f_\alpha(\xi) < |\xi|^+$ for all $\alpha < \kappa^+$. By $\Diamond''_\kappa$, there is $\langle S_\xi : \xi < \kappa \rangle$ such that $S_\xi \subseteq \mathscr{P}(\xi \times \xi)$, $|S_\xi| = |\xi|$ and for any $X \subseteq \kappa \times \kappa$, $\{\xi : X \cap (\xi \times \xi) \in S_\xi\}$ contains a closed unbounded set. For each ξ, let $f_{\kappa^+}(\xi) = \sup \{\gamma + 1 :$ some $Z \in S_\xi$ is a well-ordering of type $\gamma\}$. Then $f_{\kappa^+}(\xi) < |\xi|^+$. If $\alpha < \kappa^+$ then there is a closed unbounded set C_α such that for all $\xi \in C_\alpha$, $<_\alpha \mid \xi \in S_\alpha$. Hence $C_\alpha \subseteq \{\xi : f_\alpha(\xi) < f_{\kappa^+}(\xi)\}$. ∎

Chang's Conjecture is equivalent to the partition relation

$$\omega_2 \to [\omega_1]^{<\omega}_{\omega_1, \omega},$$

which is translated as follows: for any $f : [\omega_2]^{<\omega} \to \omega_1$, there is $A \in [\omega_2]^{\omega_1}$ such that the range of f on $[A]^{<\omega}$ is countable. In unpublished work, Silver has shown Chang's Conjecture to be consistent relative to the existence of certain large cardinals.

THEOREM 3.3. Chang's Conjecture implies that there is no sequence $\langle f_\alpha : \alpha \leq \omega_2 \rangle$ of elements of $^{\omega_1}\omega_1$ well-ordered by $<$.

Proof. Suppose not. We will construct $f : [\omega_2]^{<\omega} \to \omega_1$ inductively. At the nth stage we determine $f \mid [\omega_2]^n$. Let $f \mid [\omega_2]^n$ be arbitrary for $n = 0, 1$. For $\alpha < \beta < \omega_2$, let $C_{\alpha\beta}$ be a closed unbounded set such that for all $\xi \in C_{\alpha\beta}$, $f_\alpha(\xi) < f_\beta(\xi) < f_{\omega_2}(\xi)$. Now if $n \geq 2$ and $x \in [\omega_2]^n$, let $f(x)$ be the least element of $\cap \{C_{\alpha\beta} : \alpha, \beta \in x, \alpha < \beta\}$ which is greater than $f(y)$ for all $y \in [x]^{n-1}$.

By Chang's Conjecture, there is $A \in [\omega_2]^{\omega_1}$ such that f has countable range on $[A]^{<\omega}$. Let γ be the least element of ω_1 not in the range of f. Suppose $\delta < \gamma$, $\delta \in$ range f. Say $\delta = f(x)$. Let $\alpha, \beta \in A$, $\alpha < \beta$. Choose $y \supseteq x$, $y \neq x$ such that $\alpha, \beta \in \gamma$. Then $f(y) \in C_{\alpha\beta}$, $f(y) > \delta$. Hence $C_{\alpha\beta}$ is cofinal in γ, so $\gamma \in C_{\alpha\beta}$. But then $f_\alpha(\gamma) < f_\beta(\gamma) < f_{\omega_2}(\gamma)$ whenever $\alpha, \beta \in A$ and $\alpha < \beta$, contradicting the fact that A is uncountable. ∎

The proof of Theorem 3.3 can easily be generalized to other cardinals.

THEOREM 3.4. *If κ is ineffable, then there is no sequence $\langle f_\alpha : \alpha \leq \kappa^+ \rangle$ of elements of $^\kappa\kappa$ such that $\alpha < \beta$ implies $f_\alpha < f_\beta$, and $f_\alpha(\xi) < |\xi|^+$ for all ξ.*

Proof. Suppose not. We may assume that $\langle f_\alpha : \alpha < \kappa^+ \rangle$ is as constructed in the proof of Theorem 3.1. For each ξ, let $S_\xi \subseteq \xi \times \xi$ be a well-ordering of ξ in order type $f_{\kappa^+}(\xi)$. Since κ is ineffable, there is $S \subseteq \kappa \times \kappa$ such that $\{\xi : S \cap (\xi \times \xi) = S_\xi\}$ is stationary. But if S has order type α, then $\{\xi : f_\alpha(\xi) = f_{\kappa^+}(\xi)\}$ is stationary, contradiction. ∎

Combining Theorems 3.2 and 3.4 with the result of Jensen mentioned earlier, we obtain:

COROLLARY 3.5. *Assume $V = L$ and κ is regular. Then κ is ineffable iff there is no sequence $\langle f_\alpha : \alpha \leq \kappa^+ \rangle$ such that $\alpha < \beta$ implies $f_\alpha < f_\beta$, and $f_\alpha(\xi) < |\xi|^+$ for all $\xi < \kappa$.*

Now we turn to the process of iterating the operation that produced ineffables. We define In_α^κ by induction on κ. Let $\text{In}_0^\kappa = \{X \subseteq \kappa : X$ is stationary in $\kappa\}$. If $\text{In}_\alpha^\kappa \neq 0$, let $\text{In}_{\alpha+1}^\kappa$ be the set of all $X \subseteq \kappa$ such that if $S_\xi \subseteq \xi$ for all $\xi \in X$, then there exists $S \subseteq \kappa$ such that $\{\xi \in X : S \cap \xi = S_\xi\} \in \text{In}_\alpha^\kappa$. If $\text{In}_\alpha^\kappa = 0$, let $\text{In}_{\alpha+1}^\kappa = 0$. If α is a limit ordinal, let $\text{In}_\alpha^\kappa = \cap \{\text{In}_\beta^\kappa : \beta < \alpha\}$. Note that In_1^κ is the set of all ineffable subsets of κ. Also, $\alpha < \beta$ implies $\text{In}_\beta^\kappa \subseteq \text{In}_\alpha^\kappa$.

It is not difficult to see that if α is a successor ordinal, or if $\text{cf}\,\alpha > \kappa$, then $\mathscr{P}(\kappa) - \text{In}_\alpha^\kappa$ is a κ-complete normal ideal. Also, as in the case of ineffables, there are other characterizations of $\text{In}_{\alpha+1}^\kappa$ in terms of In_α^κ. For example, $X \in \text{In}_{\alpha+1}^\kappa$ iff for any regressive $f : [X]^2 \to \kappa$ there is $Y \in \text{In}_\alpha^\kappa$ which is homogeneous for f.

The In_α^κ for α finite were studied in [1]. Here we are interested in α infinite, particularly when $\kappa \leq \alpha < \kappa^+$. Our goal is to show (Theorem

3.7 below) that if $\beta < \kappa^+$ and $\mathrm{In}_\beta^\kappa \neq 0$ then for all $\alpha < \beta$, $\mathrm{In}_{\alpha+1}^\kappa - \mathrm{In}_\alpha^\kappa \neq 0$. Our basic tool is the notion of a canonical sequence.

For each α, let $D_\alpha^\kappa = \{X \subseteq \kappa : \kappa - X \notin \mathrm{In}_\alpha^\kappa\}$.

THEOREM 3.6.   *Let $\langle f_\alpha : \alpha < \nu \rangle$ be a sequence of members of $^\kappa\kappa$ such that if $\alpha < \beta$ then $f_\alpha < f_\beta$ (mod D_β^κ). Then for all $\alpha < \nu$ and all $X \subseteq \kappa$, if $\{\xi : X \cap \xi \in \mathrm{In}_{f_\alpha(\xi)}^\xi\} \in \mathrm{In}_\alpha^\kappa$ then $X \in \mathrm{In}_\alpha^\kappa$.* ∎

Proof.   By induction on α. If $\alpha = 0$ this follows from the well-known fact that if $\{\xi : X \cap \xi \in \mathrm{In}_0^\xi\} \in \mathrm{In}_0^\kappa$, then $X \in \mathrm{In}_0^\kappa$.

Suppose $\alpha = \beta + 1$. Then $Y = \{\xi : X \cap \xi \in \mathrm{In}_{f_\alpha(\xi)}^\xi\} \cap \{\xi : f_\beta(\xi) < f_\alpha(\xi)\} \in \mathrm{In}_\alpha^\kappa$. Now let $\langle S_\eta : \eta \in X \rangle$ be such that $S_\eta \subseteq \eta$. If $\xi \in Y$, then there exists T_ξ such that $\{\eta < \xi : T_\xi \cap \eta = S_\eta\} \in \mathrm{In}_{f_\beta(\xi)}^\xi$. Also, since $Y \in \mathrm{In}_\alpha^\kappa$, there is T such that $\{\xi \in Y : T \cap \xi = T_\xi\} \in \mathrm{In}_\beta^\kappa$. But now if $Z = \{\eta : T \cap \eta = S_\eta\}$, we have $\{\xi : Z \cap \xi \in \mathrm{In}_{f_\beta(\xi)}^\xi\} \in \mathrm{In}_\beta^\kappa$, so by inductive hypothesis $Z \in \mathrm{In}_\beta^\kappa$. Hence $X \in \mathrm{In}_\alpha^\kappa$.

Suppose α is limit and $\beta < \alpha$. Then since $\{\xi : f_\beta(\xi) < f_\alpha(\xi)\} \in D_\alpha^\kappa$, we have $\{\xi : X \cap \xi \in \mathrm{In}_{f_\beta(\xi)}^\xi\} \in \mathrm{In}_\alpha^\kappa \subseteq \mathrm{In}_\beta^\kappa$, so by inductive hypothesis $X \in \mathrm{In}_\beta^\kappa$. Hence $X \in \cap \{\mathrm{In}_\beta^\kappa : \beta < \alpha\} = \mathrm{In}_\alpha^\kappa$. ∎

THEOREM 3.7.   *If $\beta < \kappa^+$ and $\mathrm{In}_\beta^\kappa \neq 0$, then for all $\alpha < \beta$, $\mathrm{In}_{\alpha+1}^\kappa - \mathrm{In}_\alpha^\kappa \neq 0$.*

Proof.   Let $\langle f_\alpha : \alpha < \kappa^+ \rangle$ be a canonical sequence as constructed in the proof of Theorem 3.1. Let $X_\alpha = \{\xi : \xi \notin \mathrm{In}_{f_\alpha(\xi)}^\xi\}$. The theorem follows from Lemmas 3.8 and 3.9. ∎

LEMMA 3.8.   *For all $\alpha \leq \beta$, $X_\alpha \in \mathrm{In}_\alpha^\kappa$.*

Proof.   The proof is by induction on κ. If $\{\xi : \xi \in \mathrm{In}_{f_\alpha(\xi)}^\xi\} \notin \mathrm{In}_\alpha^\kappa$ we are done, so assume otherwise. But now if $\xi \in \mathrm{In}_{f_\alpha(\xi)}^\xi$ then by inductive hypothesis $X_\alpha \cap \xi \in \mathrm{In}_{f_\alpha(\xi)}^\xi$. This uses the particular definition of $f_\alpha(\xi)$ as the order-type of $<_\alpha \mid \xi$ (the case $\alpha < \kappa$ is left to the reader), since then it is clear that there is a canonical sequence $\langle g_\eta : \eta < \xi^+ \rangle$ of elements of $^\xi\xi$ such that $g_{f_\alpha(\xi)} = f_\alpha \mid \xi$. Hence $\{\xi : X_\alpha \cap \xi \in \mathrm{In}_{f_\alpha(\xi)}^\xi\} \in \mathrm{In}_\alpha^\kappa$, so by Theorem 3.6, $X_\alpha \in \mathrm{In}_\alpha^\kappa$. ∎

LEMMA 3.9.   *For all α, $X_\alpha \notin \mathrm{In}_{\alpha+1}^\kappa$.*

Proof.   Suppose $X_\alpha \in \mathrm{In}_{\alpha+1}^\kappa$. For each $\xi \in X_\alpha$ there is an ordinal $\zeta < f_\alpha(\xi)$ and a counterexample $\langle S_\eta^\xi : \eta < \xi \rangle$ to $\xi \in \mathrm{In}_{\xi+1}^\xi$. Let R_ξ be a well-ordering of ξ of order-type ζ (we leave the argument to the reader in case $\alpha < \kappa$ or $\zeta < \xi$). Since $X_\alpha \in \text{In}_{\alpha+1}^\kappa$, there are $\langle S_\eta : \eta < \kappa \rangle$ and R such that $E = \{\xi : \langle S_\eta : \eta < \xi \rangle = \langle S_\eta^\xi : \eta < \xi \rangle$ and $R \cap (\xi \times \xi) = R_\xi\} \in \text{In}_\alpha^\kappa$. Then R is a well-ordering of κ in type β, say, where $\beta < \alpha$.

Thus it will suffice to show by induction on α that there are no E, $\langle S_\eta : \eta \in E \rangle$, and $\beta < \alpha$ such that $E \in \text{In}_\alpha^\kappa$ and

(1) for all $\xi \in E$, $\langle S_\eta : \eta \in E \cap \xi \rangle$ is a counterexample to $E \cap \xi \in \text{In}_{f_\alpha(\xi)}^\xi$, and

(2) for all $\xi \in E$ no subset of $E \cap \beta$ homogeneous for $\langle S_\eta : \eta \in E \cap \beta \rangle$ lies in $\text{In}_{f_\beta(\xi)}^\xi$.

First suppose $\alpha = 1$. For $\xi \in E$, let $C_\xi \subseteq \xi$ be closed unbounded such that $C_\xi \cap \{\eta : S_\xi \cap \eta = S_\eta\} = 0$. Since $E \in \text{In}_1^\kappa$, there are C and S such that $E' = \{\xi \in E : C \cap \xi = C_\xi, S \cap \xi = S_\xi\}$ is stationary. Choose $\eta, \xi \in E' \cap C$, $\eta < \xi$. Then $\eta \in C_\xi$ since $C_\eta \subseteq C_\xi$ and C_ξ is closed. But also $S_\eta = S_\xi \cap \eta$, contradiction.

Now suppose $\alpha > 1$. For $\xi \in E$, let $T_\xi = \{\eta : S_\xi \cap \eta = S_\eta\}$, and let $\langle T_\eta^\xi : \eta \in T_\xi \rangle$ be a counterexample to $T_\xi \in \text{In}_{f_\beta(\xi)}^\xi$. Choose $\zeta < f_\beta(\xi)$ such that no homogeneous set for $\langle T_\eta^\xi : \eta \in T_\xi \rangle$ lies in In_ζ^ξ, and let R_ξ be a well-ordering of ξ in order type ζ (again we leave to the reader the argument when $\zeta < \xi$). Since $E \in \text{In}_\alpha^\kappa$, there are S, T, $\langle T_\eta' : \eta \in T \rangle$, and R such that if $E' = \{\xi \in E : S \cap \xi = S_\xi, T \cap \xi = T_\xi, \langle T_\eta' : \eta \in T_\xi \rangle = \langle T_\eta^\xi : \eta \in T_\xi \rangle$, and $R \cap (\xi \times \xi) = R_\xi\}$, then $E' \in \text{In}_\beta^\kappa$. Let γ be the order type of R. Then $\gamma < \beta$. If we can show that $E' \subseteq T$ then (1) and (2) will be satisfied by E', $\langle T_\xi' : \xi \in E' \rangle$, and $\gamma < \beta$, so we can apply the inductive hypothesis. Let $\eta \in E'$. If $\xi \in E'$ and $\eta < \xi$, then $S_\xi \cap \eta = S_\eta$. Hence $\eta \in T_\xi \subseteq T$. ∎

It is clear that for some α, $\text{In}_\alpha^\kappa = \text{In}_{\alpha+1}^\kappa$. If when this happens $\text{In}_\alpha^\kappa \neq 0$, we call κ *completely ineffable*. If κ is measurable and U is a normal measure ultrafilter on κ, then it is easy to see by induction that $U \subseteq \text{In}_\alpha^\kappa$ for all α. Hence κ is completely ineffable. The least completely ineffable, however, is much smaller than the least measurable.

One may also ask which is the least α such that $\text{In}_\alpha^\kappa = \text{In}_{\alpha+1}^\kappa$ when κ is completely ineffable. By Theorem 3.7, $\alpha \geq \kappa^+$. In recent unpublished work, Kleinberg has determined the least such α.

Finally, note that if κ is completely ineffable, $\text{In}_\alpha^\kappa = \text{In}_{\alpha+1}^\kappa$, $X \in \text{In}_\alpha^\kappa$, $n < \omega$, and $f : [X]^n \to 2$, then there is $Y \in \text{In}_\alpha^\kappa$ which is homogeneous for f. The proof is easy, and is left to the reader.

4. Erdös and Ramsey cardinals

At the end of [1] it is asserted that the methods of that paper can be applied to Ramsey cardinals as well as to ineffable cardinals. In particular, there is an inaccessibility-like property which, when combined appropriately with Π_1^1-indescribability, yields the Ramsey property. This section is devoted to a brief treatment of Erdös and Ramsey cardinals using the techniques of normal ideals and regressive partition functions.

Let us denote by $\kappa(\alpha)$ the least cardinal κ such that $\kappa \to (\alpha)_2^{<\omega}$, where α is an infinite ordinal. The cardinals $\kappa(\alpha)$ are usually called *Erdös cardinals*, particularly when α is a limit ordinal. Drake [4] has extended this notion to what he calls α-*partition cardinals*, where $\alpha \geqslant \omega$ and α is a limit ordinal. κ is an α-partition cardinal iff for some $\lambda < \kappa$, κ is the least cardinal such that $\kappa \to (\alpha)_\lambda^{<\omega}$. It is shown in [4] that this notion successfully generalizes the notion of an Erdös cardinal.

We propose a further generalization. If $\alpha \geqslant \omega$, let us call κ an α-*Erdös cardinal* iff for every regressive function f on $[\kappa]^{<\omega}$ and every closed unbounded subset C of κ, there is $A \subseteq C$ such that A has order type α and for all $n < \omega$, A is homogeneous for $f \mid [\kappa]^n$. The latter condition is abbreviated by saying that A is homogeneous for f.

Some of the results below on α-Erdös cardinals were obtained independently by J. Henle.

In [9] Schmerl shows that if α is a limit ordinal then $\kappa(\alpha)$ is an α-Erdös cardinal. Virtually the same proof extends to show that every α-partition cardinal is an α-Erdös cardinal.

The α-Erdös cardinals form a natural class for several reasons. It is clear from the characterization of subtle cardinals via regressive functions that every α-Erdös cardinal is subtle, and hence is strongly inaccessible, Mahlo, hyperMahlo, etc. Also, the α-Erdös cardinals are closed under 'stationary limits':

THEOREM 4.1. *If κ is a cardinal and $\{\xi < \kappa : \xi$ is an α-Erdös cardinal$\}$ is stationary in κ, then κ is an α-Erdös cardinal.*

Proof. Let $f : [\kappa]^{<\omega} \to \kappa$ be regressive and let $C \subseteq \kappa$ be closed unbounded. Choose $\xi < \kappa$ so that ξ is α-Erdös and ξ is a limit point of C. Now apply the fact that ξ is α-Erdös to $f \mid [\xi]^{<\omega}$ and $C \cap \xi$.

This notion naturally leads to normal ideals. Let us say that $X \subseteq \kappa$ is α-Erdös iff for any regressive f on $[X]^{<\omega}$ and any closed unbounded $C \subseteq \kappa$ there is $A \subseteq X \cap C$ such that A has type α and is homogeneous for f. The α-Erdös ideal on κ is $\{X \subseteq \kappa : X$ is not α-Erdös $\}$. The α-Erdös filter on κ is $\{X \subseteq \kappa : \kappa - X$ is not α-Erdös$\}$. ■

THEOREM 4.2. *The α-Erdös ideal on κ is a normal κ-complete ideal.*

Proof. We check only normality. Suppose $X \subseteq \kappa$, $h : X \to \kappa$ is regressive, and for all α, $h^{-1}(\{\alpha\})$ is not α-Erdös. Let f_α and C_α witness that $h^{-1}(\{\alpha\})$ is not α-Erdös. Let $p : \kappa \times \kappa \to \kappa$ be one-one and onto. Let $C = \{\beta : (\forall \alpha < \beta) \beta \in C_\alpha\} \cap \{\beta : \gamma, \delta < \beta$ implies $p(\gamma, \delta) < \beta\}$. Define regressive f on $[X \cap C]^{<\omega}$ by $f(\{x\}) = p(h(x), f_\alpha(\{x\}))$, where $\alpha = h(x)$, and $f(\{x_1, \ldots, x_n\}) = f_\alpha(\{x_1, \ldots, x_n\})$, where $n > 1$ and $\alpha = h(x_1)$. Clearly there is no homogeneous $A \subseteq X \cap C$ of type α. Hence X is not α-Erdös. ■

THEOREM 4.3. *Suppose $\omega \leq \alpha < \beta$ and κ is β-Erdös. Then for any κ, $\{\xi < \kappa : \xi$ is α-Erdös$\}$ lies in the β-Erdös filter on κ.*

Proof. Let $X = \{\xi < \kappa : \xi$ is not α-Erdös$\}$ and suppose X is β-Erdös. For each $\xi \in X$, let f_ξ and C_ξ witness that ξ is not α-Erdös. Define f on $[X]^{<\omega}$ by $f(\{x_1, \ldots, x_n\}) = f_{x_n}(\{x_1, \ldots, x_{n-1}\})$, where $x_1 < \cdots < x_n$. Let A be a subset of X of type β which is homogeneous for f. Moreover, by using a pairing function p as in the proof of Theorem 4.2, we may assume that A is also homogeneous for a function g, where g is constructed so that A is homogeneous for g iff A is homogeneous for $\langle C_\xi : \xi \in X \rangle$. It is easy to see that if x is the αth member of A, then the first α members of A are homogeneous for f_x. But since A is also homogeneous for $\langle C_\xi : \xi \in X \rangle$, we have $C_\xi \cap \eta = C_\eta$ whenever $\eta, \xi \in A$ and $\eta < \xi$, and therefore $\eta \in C_\xi$. Hence the first α members of A lie in C_x, and this contradicts the choice of f_x and C_x. ■

It is worth emphasizing that Theorem 4.3 holds for successor ordinals as well as limit ordinals. The usual definition of Erdös cardinals does not yield essentially new cardinals for successor ordinals for, as Galvin [2] has shown, if α is limit and $1 \leq n < \omega$ then $\kappa(\alpha + n) = (2^\kappa_{n-1}(\alpha))^+$, where $2^\kappa_0 = \kappa$ and $2^\kappa_{m+1} = 2^\lambda$, where $\lambda = 2^\kappa_m$. Thus $\kappa(\alpha + 1)$ is never an $(\alpha + 1)$-Erdös cardinal.

It is clear that if κ is α-Erdös then $\kappa \to (\alpha)_\lambda^{<\omega}$ for all $\lambda < \kappa$. In general there are α-Erdös cardinals which are not α-partition cardinals. If α is limit and κ is $(\alpha + 1)$-Erdös, for example, then by Theorem 4.3 κ is a limit of α-Erdös cardinals, and no limit of α-Erdös cardinals can be an α-partition cardinal.

Now we consider Ramsey cardinals. A cardinal κ is *Ramsey* iff $\kappa \to (\kappa)_2^{<\omega}$, i.e. iff κ is κ-Erdös. If κ is Ramsey then we replace the term 'κ-Erdös' by 'Ramsey', and speak of the Ramsey ideal, filter, etc.

In [1] it was shown that a cardinal κ is ineffable iff the subtle ideal and the Π_2^1 ideal (defined using Π_2^1-indescribability in the same way the weakly compact ideal is defined using Π_1^1-indescribability) together generate a non-trivial ideal, which then coincides with the ineffable ideal. Moreover, one must speak of ideals; the least cardinal which is simultaneously subtle and Π_2^1-indescribable is not ineffable.

Similar results hold for Ramsey cardinals. The analogue of subtlety is as follows. A set $X \subseteq \kappa$ is *pre-Ramsey* iff for any regressive $f: [X]^{<\omega} \to \kappa$ and any closed unbounded $C \subseteq \kappa$, there is $\alpha \in X \cap C$ and $A \subseteq X \cap C \cap \alpha$ such that A is cofinal in α, A is homogeneous for f, and f is constant on $[A \cup \{\alpha\}]^1$. The *pre-Ramsey ideal* on κ is $\{X \subseteq \kappa: X$ is not pre-Ramsey$\}$. Of course κ is a *pre-Ramsey cardinal* iff κ is pre-Ramsey as a subset of itself.

As in Theorems 4.1 and 4.2, we see that the pre-Ramsey ideal is normal, and that the pre-Ramsey cardinals are closed under stationary limits. Moreover, pre-Ramsey-ness and α-Erdös-ness, like inaccessibility and subtlety, can be described by Π_1^1-formulas. Since Ramsey cardinals are weakly compact and hence Π_1^1-indescribable, the least pre-Ramsey cardinal is smaller than the least Ramsey cardinal.

Note that if κ is pre-Ramsey then κ is α-Erdös for all $\alpha < \kappa$. In fact, the α-Erdös filter is a subset of the pre-Ramsey filter for each $\alpha < \kappa$, so by normality of the pre-Ramsey ideal $\{\xi < \kappa: \xi$ is α-Erdös for all $\alpha < \xi\}$ belongs to the pre-Ramsey filter. Thus the pre-Ramsey cardinals do not coincide with the cardinals κ which are α-Erdös for all $\alpha < \kappa$.

THEOREM 4.4. *A cardinal κ is Ramsey iff the pre-Ramsey ideal and the weakly compact ideal generate a non-trivial ideal. In that case, the ideal generated is the Ramsey ideal.*

Proof. Let I be the ideal on κ generated by the pre-Ramsey ideal and the weakly compact ideal. It will suffice to show $X \notin I$ iff X is Ramsey.

First suppose $X \notin I$. Let $f : [X]^{<\omega} \to \kappa$ be regressive, and let C be closed unbounded. Since I is normal, $C \cap X \notin I$. Suppose there is no cofinal $A \subseteq C \cap X$ which is homogeneous for f. Then this fact can be expressed by a Π^1_1-sentence φ over $(V_\kappa, \varepsilon, C, X, f)$ so $Y = \{\alpha < \kappa : (V_\alpha, \varepsilon, C \cap \alpha, X \cap \alpha, f \,|\, [\alpha]^{<\omega}) \vDash \varphi\}$ belongs to the weakly compact filter on κ. Since $X \cap C \notin I$, $X \cap C$ is not the union of a set in the pre-Ramsey ideal and a set in the weakly compact ideal. Hence $Y \cap X \cap C$ is pre-Ramsey. But if $\alpha \in Y \cap X \cap C$ and $B \subseteq Y \cap X \cap C \cap \alpha$ is homogeneous for f and cofinal in α, this contradicts $\alpha \in Y$. Hence X is Ramsey.

For the other direction, it will suffice to show that if X is Ramsey then X is weakly compact and X is pre-Ramsey. If X is Ramsey then X is almost ineffable, so (see [1, Section 7]) X is weakly compact. For the other part, suppose $f : [X]^{<\omega} \to \kappa$ is regressive and C is closed unbounded. Since the Ramsey ideal is normal, there is $Y \subseteq X$ such that f is constant on $[Y]^1$ and $Y \cap C$ is Ramsey, so there is $Z \subseteq Y \cap C$ such that $|Z| = \kappa$ and Z is homogeneous for f. Since $Y \cap C$ is stationary, there is $\alpha \in Y \cap C$ which is a limit point of Z. But then $Z \cap \alpha$ and α satisfy the definition of pre-Ramsey. ∎

As in the case of ineffable cardinals, we have:

THEOREM 4.5. *If κ is Ramsey, then $\{\alpha < \kappa : \alpha$ is pre-Ramsey and weakly compact$\}$ belongs to the Ramsey filter. Hence the least cardinal which is pre-Ramsey and weakly compact is not Ramsey.*

Proof. It will suffice to show that $\{\alpha < \kappa : \alpha$ is weakly compact$\}$ and $\{\alpha < \kappa : \alpha$ is pre-Ramsey$\}$ belong to the Ramsey filter. The first set belongs to the subtle filter on κ (see [1]) and the subtle filter is a subset of the Ramsey filter. The second set belongs to the weakly compact filter since the pre-Ramsey property can be expressed with a Π^1_1-formula, and by Theorem 4.4 the weakly compact filter is a subset of the Ramsey filter. ∎

One may ask whether the pre-Ramsey ideal must be considered in Theorem 4.4. Would the α-Erdös ideal for some $\alpha < \kappa$ do as well? The answer is negative. In fact, if κ is the least $(\alpha + 1)$-Erdös cardinal then for some $\xi < \kappa$ the α-Erdös ideal on ξ and the weakly compact ideal generate a non-trivial ideal. The proof is left to the reader. It would be interesting to have a combinatorial, ideal-free characterization of such cardinals ξ, but we do not know of one.

Finally we mention briefly the notion of *ineffably Ramsey*, which is defined like Ramsey except that the homogeneous set is required to be stationary.

Let us say that κ is *pre-ineffably Ramsey* iff for any regressive $f: [\kappa]^{<\omega} \to \kappa$ and any closed unbounded $C \subseteq \kappa$, there is $\alpha \in C$ and $A \subseteq C \cap \alpha$ such that A is homogeneous for f, f is constant on $[A \cup \{\alpha\}]^1$, and A is stationary in α. We leave to the reader the theory of the ineffably Ramsey and pre-ineffably Ramsey ideals, etc.

It is not difficult to show that if κ is pre-ineffably Ramsey, then $\{\alpha < \kappa: \alpha$ is Ramsey$\}$ belongs to the pre-ineffably Ramsey filter. Also, κ is ineffably Ramsey iff the pre-ineffably Ramsey and Π_2^1 ideals generate a non-trivial ideal, which is then the ineffably Ramsey ideal.

Dartmouth College

BIBLIOGRAPHY

[1] J. Baumgartner, 'Ineffability Properties of Cardinals I', in *Colloq. Math. Soc. János Bolyai* **10**, 'Infinite ,and Finite Sets', Keszthely, Hungary, 1973, pp. 109–130.
[2] J. Baumgartner and F. Galvin, to appear.
[3] J. Baumgartner, A. Taylor, and S. Wagon, 'On Splitting Stationary Subsets of Large Cardinals', *J. Symbolic Logic*, to appear.
[4] F. R. Drake, 'A Fine Hierarchy of Partition Cardinals', *Fund. Math.* **81**, (1974), 271–277.
[5] R. Jensen and K. Kunen, 'Some Combinatorial Properties of L and V', mimeographed.
[6] K. Kunen, 'Combinatorics', in J. Barwise (ed.), *A Handbook of Mathematical Logic*, to appear.
[7] K. Kunen, 'Inaccessibility Properties of Cardinals', Ph.D. dissertation, Stanford University, 1968.
[8] A. Levy, 'The Sizes of Indescribable Cardinals', in *Axiomatic Set Theory*, Proc. of Symposia in Pure Math., Vol XIII, Part 1, Amer. Math. Soc., pp. 205–218.
[9] J. Schmerl, 'On κ-like Structures which embed Stationary and Closed Unbounded Subsets', *Annals Math. Logic*, to appear.
[10] J. Silver, 'Some Applications of Model Theory in Set Theory', *Annals Math. Logic* **3**, (1971), 45–110.

W. A. J. LUXEMBURG*

NON-STANDARD ANALYSIS

1. As early as 1934 it was pointed out by Thoralf Skolem (see [17]) that there exist proper extensions of the natural number system which have, in some sense, 'the same properties' as the natural numbers. The title of Skolem's paper indicates that the purpose of it was to show that no axiomatic system specified in a formal language, in Skolem's case the lower predicate calculus, can characterize the natural numbers categorically. At that time, however, Skolem did not concern himself with the properties of the structures whose existence he had established. In due course these structures became known as non-standard models of arithmetic. For nearly thirty years since the appearance of Skolem's paper non-standard models were not used or considered in any sense by the working mathematician. Robinson's fundamental paper, which appeared in 1961 under the title 'Non-standard Analysis', (see [11]) changed this situation dramatically. In this paper Abraham Robinson was the first to point out that this highly abstract part of model theory could be applied fruitfully to a theory so far removed from it as the infinitesimal calculus. As a result Robinson obtained a firm foundation for the non-archimedian approach to the calculus based on a number system containing infinitely small and infinitely large numbers, in a manner almost identical to that suggested by Leibniz some three centuries ago, and which predominated the calculus until the middle of the nineteenth century when it was rejected as unsound and replaced by the ε, δ-method of Weierstrass.

Once Abraham Robinson had shown by the example of the calculus in which sense non-standard models can be used in mathematics other applications to other fields were soon found by Robinson and his followers (see [4], [6], [8], [13]). So it turned out that the new non-standard methods not only could be made to work for arithmetic and the calculus but for many other branches of mathematics as well. Its range of applications today seems almost unlimited and includes such areas as mathematical economics (see [3]).

Butts and Hintikka (eds.), *Logic, Foundations of Mathematics and Computability Theory*, 107–119.
Copyright © 1977 by D. Reidel Publishing Company, Dordrecht-Holland. All Rights Reserved.

The purpose of the present exposition is to show, by means of a number of examples, how Abraham Robinson used non-standard models and made them into such effective tools in mathematics.

2. In order to make the paper self-contained we shall begin with a brief description of the concept of a non-standard model and the concept of an enlargement. For details we will have to refer the reader to the pertinent literature on this subject (see [9], [10], [13], [18]).

Let M be a mathematical super-structure based on an infinite set X whose elements will be called the individuals of the structure. In symbols,

$$M = \bigcup_{n=0}^{\infty} M_n \quad \text{where} \quad M_0 = X \quad \text{and,}$$

$$M_{n+1} = p\left(\bigcup_{k=0}^{n} M_k\right) \quad (n = 1, 2, \ldots),$$

where $p(\cdot)$ denotes the power set operation. In the examples below X may be the field of real and complex numbers or more generally a topological space.

The elements of M will be called the *entities* of the structure, and so the individuals of M are also entities of M. Clearly $M_n \in M_{n+1}$ for all $n = 0, 1, 2, \ldots$ and also $M_n \subset M_{n+1}$ for $n = 1, 2, \ldots$. The elements of the difference set $M_n \setminus M_{n-1}$ are called entities of type n $(n = 0, 1, 2, \ldots)$ of M. Observe also that if $x, y \in M_n$ $(n \geq 0)$, then the ordered pair $(x, y) = \{\{x\}, \{x, y\}\} \in M_{n+2}$. Consequently, if x and y are entities of M, then the ordered pair (x, y) is an entity of M and similarly for k-tuples of entities of M. Hence, M contains as entities all k-ary relations between entities of M.

Let K be the set of all statements which hold in M. We shall assume that these statements are expressed in a formal language L. The language L is supposed to include the logical connectives, variables, the quantifier symbols ($\forall$) 'for all' and ($\exists$) 'there exists', and sufficiently many extralogical constants to denote the entities of M of all finite types. The reader should take note that the main reason for introducing the super-structure M is that this allows us to extend quantification over sets of individuals to quantification over sets of entities of any type of M.

Assume now that $*M$ is a super-structure which is a proper extension of M, that is, $*M$ is a mathematical super-structure based on a set of individuals $*X$ which contains X properly and such that to every entity a of M there corresponds a unique entity $*a$ of $*M$ of the same type and which is denoted by the same symbol of L. The entities of $*M$ which are elements of $*M_n$ ($n = 0, 1, 2, ...$) will be called the *internal* entities of $*M$. All the other elements of $*M$ will be referred to as the *external* elements of $*M$.

On this basis any statement W of K can be reinterpreted in $*M$, where we assign the usual meaning to the connectives and quantification over sets of individuals, but where quantification is involved over sets of entities of higher types of $*M$ we shall only allow such quantification over internal entities of $*M$. Under these hypotheses we shall refer to $*M$ as a non-standard model of M whenever all statements of K under this new interpretation also hold in $*M$, the so-called *transfer principle*. For instance, if $p(X)$ denotes the entity of M of all subsets of X, then the internal entity $*p(*X)$ of $*M$ of the internal subsets of $*X$ has the same properties in $*M$ as $p(X)$ in M as far as they can be expressed by sentences of L belonging to K. As we will see later this fact can be used to show that not every set of entities of $*X$ is internal.

The existence of non-standard models for a given structure is an immediate consequence of the so-called compactness principle of model theory (see [13]). They can also be constructed in the form of an ultrapower (see [8]).

From a proof-theoretic point of view Robinson's method of examining the properties of M in $*M$ via the transfer principle is not based on an axiomatic theory of the structures $*M$. In fact, the structures $*M$ are not in any sense uniquely determined by the given structure M. Rather it consists of a set of instructions determined by the particular structure $*M$ which are superimposed on M and which themselves cannot be categorically axiomatized. The effectiveness of Robinson's method depends on the use of these instructions in combination with internal notions, to which the transfer principle applies, and the external notions. It is obvious that we should get nothing new for M if we considered only internal notions of *M. After all, they are by defini-tion, the notions transferred from M and they have the same proper-ties in *M as in M. For this reason we also have to introduce external notions relative to M and notions derived from them. This will enlarge our vocabulary and enhance our deductive processes. As we shall see below in the calculus the external notions are precisely the notions of infinitely small and infinitely large. In another direction the use of non-standard models leads to a greater syntactical simplicity in that it allows us to eliminate certain quantifiers by replacing them by certain appropriate external predicates.

As we have indicated earlier we shall illustrate this new kind of reasoning by a few examples which have played a fundamental role in the development of the non-standard methods.

3. For our first example we have selected a beautiful result of A. Robinson (see [12]) which is now known as Robinson's lemma. The proof of this lemma will clearly show the features of the new method discussed in such general terms in the previous section.

Since the lemma deals with the properties of a certain family of internal sequences of real numbers we shall begin with a brief intro-duction of a non-standard model of the reals.

By R we shall denote as usual the field of real numbers. The super-structure determined by R will be denoted by $M(R)$; and the set of all statements which hold in $M(R)$ and which are formulated in a formal language L in the sense of Section 2 will be denoted by $K = K(R)$. Let us now consider a non-standard model *M(*R) of $M(R)$. Since R is a totally ordered field it follows from the transfer principle that the new set of individuals *R under the interpretation of the algebraic and order relation of R in *M, is also a totally ordered field. More generally, the *-mapping injects M into *M thereby extending all the entities of $M(R)$ to entities of *M(*R) with the same properties as long as they are listed by K. In particular, the set of natural numbers $N = \{1, 2, ...\}$ of R extends to a set *N of individuals of *R which has the same properties as N formulated by sentences of K. For example, K contains the sentence $(\forall x)(\forall y)(0 < x < y \in R \Rightarrow (\exists n)$ $(n \in N$ and $nx > y))$ expressing the archimedian property of R. From the transfer principle it follows that *R has the Archimedian property in the following sense. For every pair of positive numbers $x, y \in {}^*R$ there exists a natural number $n \in {}^*N$ such that $nx > y$. But ${}^*R$ being a proper extension of R is not Archimedian if we mean by the Archimedian property the metamathematical postulate that for any pair of real numbers $0 < x < y$ there exists a natural number n in the metalanguage such that $x + x + \cdots + x$ (n-times) $> y$.

It is customary to call the elements of ${}^*R$ *real numbers* and the elements of ${}^*R$ which are the images under the *-mapping of the real numbers of R *the standard real numbers* of ${}^*R$. Also any entity of ${}^*M({}^*R)$ of the form ${}^*a$, where $a \in M(R)$ will be called a *standard entity* of ${}^*M$. For the sake of convenience we shall not use the star-notation ${}^*a$ for the standard real numbers of R. However, for the internal entities of ${}^*M({}^*R)$ of type $n \geqslant 1$ which are standard we shall use the star-notation ${}^*A$.

A real number $a \in {}^*R$ is called *finite* whenever $|a| < r$ for some positive standard real number r. A real number which is not finite will be called *infinite*. A real number $h \in {}^*R$ will be called *infinitely small* whenever $|h| < \varepsilon$ for all positive standard real numbers ε. The set of finite real numbers of ${}^*R$ is a ring denoted by R_0 and the set of infinitesimals denoted by R_1, is a maximal ideal in R_0. It can be shown that the quotient using R_0/R_1 is isomorphic to R. The algebraic and order-homomorphism of R_0 onto R with kernel R_1 is denoted by 'st' and assigns to every finite real number $a \in {}^*R$ a unique standard number $st(a)$ which is infinitely close to a. In general, we shall say that a_1 is infinitely close to a_2, and we write $a_1 \simeq a_2$, whenever $a_1 - a_2 \in R_1$. Thus, in particular, $a \simeq st(a)$ for all finite reals of ${}^*R$. For any standard real number $r \in R$ the set of all real numbers which are infinitely close to r will be called the *monad* of r and will be denoted by $\mu(r)$. For $r = 0$, the monad $\mu(0)$ is just the set of all infinitesimals of ${}^*R$. Within this framework Robinson and his followers (see [5], [7], [13], [18]) developed the calculus in a manner as was proposed by Leibniz about three centuries ago.

It is now easy to see that the notions of the set of infinitesimals of ${}^*R$ and the notions of the set of all finite real numbers are typical examples of external notions. Let us examine this for a moment for the notion of the set of infinitesimals. If the set $\mu(0)$ is an internal entity of ${}^*M({}^*R)$, then, since $\mu(0)$ is not empty and bounded in ${}^*R$, it follows from the transfer principle that $\mu(0)$ must have a least upper bound, say a. From $0 \in \mu(0)$ it follows that $a \geq 0$. Furthermore, $a \notin \mu(0)$, because in that case, $a/2$ would also be an upper bound of $\mu(0)$, contradicting the definition of a. If $a \in \mu(0)$, then $2a \in \mu(0)$ again leads to a contradiction, showing that $\mu(0)$ is an external element of $*M(*R)$. Similarly one can show that the set of standard reals is external. Another interesting example is given by the set $*N - N$ of infinitely large natural numbers. Indeed, if $*N - N$ is internal, then, by the transfer principle, $*N - N$ must have a first element ω_0 contradicting that fact that the natural number $\omega_0 - 1$ is also infinitely large. Consequently, the set of standard natural numbers is also external.

Let us now denote by S the set of all real sequences, that is, S is the set of all mappings $\{a_n\}$, $n \in N$ of N into R. It is obvious that $S \in M(R)$. As we pass from M to $*M$ the entity S of M extends to the entity $*S$ of $*M$ of all internal real sequences, that is, of internal mappings of $*N$ into $*R$. In particular, a standard real sequence $\{a_n\}$, $n \in N$, extends, in passing from M to $*M$ to a real sequence defined over $*N$. In the standard language we express the fact $\lim_{n \to \infty} a_n = a$ by the sentence $(\exists a)(\forall \varepsilon > 0)(\exists n)(\forall m)(m \geq n \Rightarrow |a_m - a| < \varepsilon)$ while it was shown by Robinson that the non-standard expression is simply: "For all infinitely large values of the index n, a_n is infinitely close to a" replacing the quantifiers by the new external predicates and thereby achieving a greater syntactical simplicity referred to earlier.

Robinson's lemma deals with internal sequences which have an external property namely that its values are infinitely small for each standard values of its index. The conclusion is then that this property extends beyond the finite range of the set of indices. In particular showing that a sequence $\{a_n\}$ $(n \in *N)$ satisfying $a_n \cong 0$ for all standard n and $a_n = 1$ for all infinite n is not internal. The precise formulation of the lemma and its proof is as follows.

ROBINSON'S LEMMA. *If an internal sequence $\{a_n\}$, $n \in *N$, has the property that $a_n \simeq 0$ for all finite n, then there exists an infinitely large natural number ω such that $a_n \simeq 0$ for all $n \leq \omega$.*

Proof. The sequence $\{n |a_n|\}$, $n \in *N$, is also an internal sequence with the property that for finite n its value is infinitesimal. From the transfer principle it follows that the set of indices $A = \{n : n \in *N$ and

$k |a_k| \leq 1$ for all $k \leq n\}$ is internal. From the hypothesis it follows that $N \subset A$. Since A is internal and N is not it follows that $N \neq A$, and so, A must contain an infinitely large natural number, say ω_0. Then for $k \leq \omega_0$ and k infinitely large $k |a_k| \leq 1$ implies $|a_k| \leq 1/k \simeq 0$, that is, $a_k \simeq 0$; and the proof is finished. ∎

From the proof of this lemma, which appears frequently in applications, we see that the success of the new methods is based on the complimentary use of the transfer principle and the external notions.

4. In our next example we shall turn our attention to the theory of non-standard topology as developed by A. Robinson and the present author (see [8], [13]). For this purpose we need to adopt a slightly more general point of view.

Let $M = M(X)$ be a structure generated in the sense of Section 2 by an infinite set X of individuals. A binary relation $\Phi(x, y)$ which is an entity of M will be called *concurrent* whenever it has the following property. For any finite set $\{a_1, ..., a_n\}$ of entities in the domain of the first argument of Φ there exists an entity b such that $\Phi(a_k, b)$ holds in M for all $k = 1, 2, ..., n$. We shall call a non standard model $*M(*X)$ of $M(X)$ an *enlargement* whenever $*M$ has the following property. For any concurrent binary relation $\Phi \in M$ there exists an internal entity $b \in *M$ such that $*\Phi(*a, b)$ holds in $*M$, where now $*a$ ranges over *all* the standard entities which make up the first domain of Φ. The entity b is usually referred to as a *bound* of Φ.

An example of a concurrent binary relation is furnished by the relation Φ whose domain of the first argument is the set of all finite subsets of X and the domain of the second argument of Φ or the range of Φ is the set X and $\Phi(y, x)$ holds if $x \notin y$. In terms of an enlargement of $M(X)$ this means that there is an individual which is not standard. We can also turn this around whereby the domain of Φ is the set X and the range of Φ is the set of all finite subsets of X and define $\Phi(x, y)$ to mean $x \in y$. In passing from $M(X)$ to $*M(*X)$, Φ extends to $*\Phi$ and it follows immediately from the transfer principle that the domain of $*\Phi$ is the set $*X$ of individuals. To characterize the range of $*\Phi$ we shall introduce the following bit of additional terminology. An internal set A of entities of $*M$ will be called *-*finite* whenever there exists a natural number $n \in *N$ such that A can be brought into one-to-one correspondence by an internal mapping with the set $\{1, 2, ..., n\}$.

With this definition it follows that the range of $^*\Phi$ is the internal entity of all *-finite subsets of $^*X$. A bound B of Φ is now a *-finite subset of $^*X$ which contains all the standard individuals or, in symbols, $X \subset B$.

Another more concrete example of a concurrent binary relation is furnished by the order relation $<$ between real numbers. Then any positive infinite real number $b \in {}^*R$ is a bound.

The notion of an enlargement is due to Abraham Robinson and their existence was shown by him to follow from the compactness theorem of model theory (see [13]). The existence of enlargements in the form of ultrapowers was shown independently by several mathematicians including the author (see [8]).

The present author has shown that within the framework of an enlargement the theory of filters can be fully developed (see [8]). Since we will need some of the concepts of this theory we shall recall briefly some of the pertinent definitions and results of the so-called theory of monads. For the sake of simplicity we shall restrict the discussion entirely to filters over a given set X of individuals.

For a filter $\mathcal{F}$ of subsets of X we define its monad $\mu(\mathcal{F})$ as the intersection of the extensions $^*A$ of all the elements A of $\mathcal{F}$. In symbols, $\mu(\mathcal{F}) = \cap(^*A : A \in \mathcal{F})$. In an enlargement $^*M(^*X)$ of $M(X)$ it can be easily shown that the monad of a filter is not empty. To this end, we introduce the binary relation Φ whose domain and range are $\mathcal{F}$ and $\Phi(x, y)$ holds whenever $x \supset y$. The fact that Φ is concurrent follows from the basic property of a filter that if $x_1, \ldots, x_n \in \mathcal{F}$, then $x_1 \cap \cdots \cap x_n \in \mathcal{F}$. Consequently, in an enlargement the bounds of Φ are entities of $^*\mathcal{F}$ contained in the monad of $\mathcal{F}$, and so, in an enlargement the monads are non-empty. In [8] examples are given to show that there also exist non-standard models in which certain filters may have empty monads.

The monads derive their importance from the fact that they reflect faithfully all the properties of filters. For instance, monads determine filters uniquely, in the sense, that two filters are equal if and only if they have the same monad. Furthermore, the structure of the set of filters ordered by inclusion can be completely analyzed in terms of the Boolean operation properties of the set of their monads. The guiding principle here is that a filter $\mathcal{F}_1$ is finer than a filter $\mathcal{F}_2$ if and only if $\mu(\mathcal{F}_1) \subset \mu(\mathcal{F}_2)$. Consequently, the intersection of any family of monads is either empty or a monad, namely the monad of the filter generated by the given family. The monad of an ultrafilter is the smallest filter monad, more precisely, they are the atoms of the set of monads ordered by inclusion. Another important aspect of the notion of a monad is that it is an external notion. We have already come across this fact when we showed that the set of infinitesimals is external. Indeed, the set of infinitesimals is the filter monad of the neighborhood filter of the zero element of R. In general, it was first shown by the author (see [8], Theorem 2.2.6) that a monad of a filter is internal if and only if the filter is a principal filter, and in that case the monad is the extension of a standard set. For further information about the theory of monads we have to refer the reader to [8] and [10].

We shall now turn to a discussion of some aspects of the non-standard versions of topological concepts. Let (X, τ) be a topological space whose topological structure is determined by the set τ of subsets of X declared open, and let $M(X, \tau)$ be the superstructure generated by the set X. Then τ is an entity of M of type 2. We shall now assume that $^*M(^*X, ^*\tau)$ is an enlargement of M in the sense defined above. The entities contained in the standard entity $^*\tau$ of $^*M$ will be called the internal open point sets of $^*X$. The extensions $^*O$ of the standard open point sets of X will be called the standard open point sets of $^*X$. For a non-standard version of the theory of the topological space X we first need to reformulate the most important standard notions in terms of the notions of $^*M$. To this end, we introduce for each point $x \in X$ the monad $\mu_\tau(x)$ of the τ-neighborhood filter of the point x. The family of monads $\mu_\tau(x)$, $x \in X$ is the most fundamental external notion in non-standard topology. We shall illustrate this by quoting the following two theorems.

A topological space (X, τ) is a Hausdorff space if and only if for each pair x, y of distinct points of X their monads are disjoint.

A point $a \in X^*$ will be called nearstandard whenever there exists a standard point $x \in X$ such that $A \in \mu_\tau(x)$. In terms of the external notion of 'nearstandardness' Robinson was able to give the following non-standard characterization of compactness.

A topological space (X, τ) is compact if and only if every point of $^*X$ is nearstandard.

The proof of this important result rests entirely on the theory of monads. Assume first that X is compact and that $a \in {}^*X$. We observe that the filter $\mathscr{F}_a = \{A : A \subset X \text{ and } a \in {}^*A\}$ is an ultrafilter of subsets of X, and so, by the compactness of X converges to a point $x \in X$. Hence, $\mathscr{F}_a$ is finer than the neighborhood filter of x. Consequently, $a \in \mu(\mathscr{F}_a) \subset \mu_\tau(x)$ shows that a is nearstandard. If we assume that every point of ${}^*X$ is nearstandard, then, by considering the monad $\mu(\mathscr{U})$ of an ultrafilter $\mathscr{U}$, we conclude from the fact that $\mu(\mathscr{U})$ is non-empty that there is a point $a \in {}^*X$ such that $a \in \mu(\mathscr{U})$. Since a is nearstandard it follows that $a \in \mu_\tau(x)$ for some $x \in X$. Then $a \in \mu_\tau(x) \cap \mu(\mathscr{U})$ and $\mu(\mathscr{U})$ being an ultrafilter monad implies that $\mu(\mathscr{U}) \subset \mu_\tau(x)$. Hence, $\mathscr{U}$ is convergent to x; and the proof is complete.

Robinson's non-standard version of compactness has been fruitfully applied in many areas of topology and analysis. To be applicable in a special case, however, one needs first to characterize the nearstandard points of the special topological space under consideration. For instance, the nearstandard points of the n-dimensional Euclidean space $R^n (n = 1, 2, ...)$ are those points $(a_1, ..., a_n) \in {}^*R^n$ whose coordinates are finite real numbers. A more sophisticated example is furnished by the space $C[0, 1]$ of the real continuous functions on the unit interval and with the topology of uniform convergence. In an enlargement of $C[0, 1]$ the nearstandard points of ${}^*C[0, 1]$ can be characterized as those internal continuous ${}^*R$-valued functions f defined on the unit interval ${}^*[0, 1]$ of ${}^*R$ such that $f(x)$ is finite for all $x \in {}^*[0, 1]$ and $f(x_1) \simeq f(x_2)$ for all $x_1, x_2 \in {}^*[0, 1]$ satisfying $x_1 \simeq x_2$. From this result the famous characterization of Ascoli of the compact subsets of $C[0, 1]$ follows now readily (see [8], Theorem 3.8.2).

Continuous mappings between topological spaces map nearstandard points into nearstandard points. This result embodies the non-standard version of the well-known fact that the continuous image of a compact space is compact.

In the theory of normed linear spaces the compact linear transformations can be characterized non-standardly as those linear transformations which map points with finite norm into nearstandard points. This characterization played a fundamental role in the proof of the celebrated result of Robinson and Bernstein that every polynomial compact linear transformation of a Banach space into itself has a non-trivial invariant subspace. (See [1], [2]). This theorem settled a question raised by Halmos and has led to a renewed interest in the general invariant subspace problem.

5. An example of a slightly different nature which illustrates the interplay between internal and external notions can be found in field theory. Suppose that W is a sentence which is formulated in the lower predicate calculus in terms of equality, addition and multiplication holds for all fields of characteristic zero, then it is also true for fields of characteristic $p \geqslant p_0$, where p_0 depends on W. A quick proof of this result of Tarski's (see [14]) using internal and external notions runs as follows. Assume that the statement is false. Then the sentence: For every natural number n there exists a prime number $p > n$ such that W is false in a field of characteristic p holds for the natural numbers N. Hence, by the transfer principle, this statement holds for the extension $*N$ in a non-standard model of set theory. Choosing now $n \in *N$ to be infinitely large we obtain that there is a field F of infinite characteristic $p > n$ such that W does not hold in F. But when we look at the field F from an external point of view it must be a field of characteristic zero since it is not of any finite characteristic. This completes the proof of the theorem.

6. There seems little doubt that we may expect further progress with the use of non-standard methods by the introduction of additional external notions. A glimpse of what the future may hold may perhaps be gotten from studying a recent non-standard version of the famous Siegel–Mahler theorem obtained by Abraham Robinson, just before his tragic death, in collaboration with P. Roquette (see [16]). The theorem referred to above in its original form due to Siegel can be formulated as follows.

Any plane algebraic curve Γ defined by an absolute irreducible algebraic equation $f(x, y) = 0$ whose coefficients are contained in an algebraic number field K of finite degree and whose genus $g > 0$ has only finitely many points (x, y) whose coordinates are algebraic integers in K.

It was shown earlier by Robinson (see [15]) that if a point $(\alpha, \beta) \in *K \times *K$ is a non-standard point of $*\Gamma$, then $K(\alpha, \beta)$ is an algebraic function field over K which can be K-isomorphically injected into $*K$ such that $K \subset K(\alpha, \beta) \subset *K$. Furthermore, the injection of $K(\alpha, \beta)$ into

*K is an external injection. The importance of the existence of this external injection consists in the fact that it shows that the elements of the function field $K(\alpha, \beta)$ can be represented in an external sense by numbers in *K. Returning now to the theorem of Siegel we observe that a proof by contradiction would necessitate the assumption that there are infinitely many points on Γ whose coordinates are algebraic integers in K. This assumption implies that, by passing from K to *K, that there exists a non-standard point (α, β) in *K on *Γ. This brings us exactly at the situation described above and examined earlier by Robinson. Now in this case it can be shown that the genus of the algebraic function field $K(\alpha, \beta)$ is equal to the genus of Γ. This in turn leads to the following non-standard version of Siegel's Theorem.

*Let F be an algebraic function field of one variable over K, and assume that F is K-isomorphic (externally) to a subfield of *K in such a manner that $K \subset F \subset$ *K, whereby F again denotes its external copy in *K. If F has genus $g > 0$, then every non-constant function $x \in F$ admits at least one non-standard prime divisor of *K in its denominator.*

The proof of this result can be found in [16]. It is a combination of standard and non-standard methods and an excellent illustration of the nature of the non-standard methods we have tried to explain in this paper.

California Institute of Technology, Pasadena

NOTE

* Work on this paper was supported in part by NSF Grant MPS 74-17845.

BIBLIOGRAPHY

[1] Bernstein, Allen R. and Robinson, Abraham: 1966, 'Solution of an Invariant Subspace Problem of K. T. Smith and P. R. Halmos', *Pacific J. Math.* **16**, 421–431.
[2] Bernstein, Allen R.: 1967, 'Invariant Subspaces of Polynomially Compact Operators on a Banach Space', *Pacific J. Math.* **21**, 445–464.
[3] Brown, Donald J. and Robinson, Abraham: 1972, 'A Limit Theorem on the Cores of Large Standard Exchange Economics', *Proc. Nat. Acad. Sci. U.S.A.* **69**, 1258–1260.

[4] Henson, C. Ward: 1974, 'The Isomorphism Property in Nonstandard Analysis and its Use in the Theory of Banach Spaces', *J. Symbolic Logic* **39**, 717–731.

[5] Keisler, H. Jerome: 1975, *Elementary Calculus, An Approach Using Infinitesimals*, Prindle, Weber and Schmidt.

[6] Loeb, Peter A.: 1971, 'A Nonstandard Representation of Measurable Spaces and L_∞', *Bull. Am. Math. Soc.* **77**, 540–544.

[7] Luxemburg, W. A. J.: 1962, *Nonstandard Analysis, Lectures on A. Robinson's Theory of Infinitesimals and Infinitely Large Numbers*, Calif. Inst. of Technology, Pasadena 1962 and revised edition, 1964.

[8] Luxemburg, W. A. J. (ed.): 1969, 'A General Theory of Monads', in *Applications of Model Theory to Algebra, Analysis and Probability* (Proc. Symposium on Nonstandard Analysis, Calif. Inst. of Techn., 1967), Holt-Rinehart and Winston, New York, pp. 123–137.

[9] Luxemburg, W. A. J.: 1973, 'What is Nonstandard Analysis?', *Am. Math. Monthly* **80**, 38–67.

[10] Machover, Moshe and Hirschfeld, Jòram: 1969, *Lectures on Nonstandard Analysis* (*Lecture Notes in Mathematics* **94**), Springer-Verlag, Berlin.

[11] Robinson, Abraham: 1961, 'Non-Standard Analysis', *Kon. Nederl. Akad. Wetensch. Amsterdam Proc.* **A64** (= *Indag. Math.* **23**), 432–440.

[12] Robinson, Abraham: 1964, 'On Generalized Limits and Linear Functionals', *Pacific J. Math.* **14**, 269–283.

[13] Robinson, Abraham: 1966, *Non-standard Analysis, Studies in Logic*, North-Holland Publ. Co. Amsterdam, second edition, 1975.

[14] Robinson, Abraham: 1967, 'Non-standard Arithmetic', *Bull. Am. Math. Soc.* **73**, 818–843.

[15] Robinson, Abraham: 1973, 'Nonstandard Points on Algebraic Curves', *J. Number Theory* **5**, 301–327.

[16] Robinson, Abraham and Roquette, Peter J.: 1975, 'On the Finiteness Theorem of Siegel and Mahler Concerning Diophantine Equations', *J. Number Theory* **7**, 121–176.

[17] Skolem, Thoralf: 1934, Über die Nicht-charakterisierbarkeit der zahlenreihe Mittels endlich oder abzählbar unendlich vieler Aussagen mit ausschliesslich Zahlenvariablen, *Fund. Math.* **23**, 150–161.

[18] Stroyan, K. D. and Luxemburg, W. A. J.: 1967, *Introduction to the Theory of Infinitesimals*, Academic Press, New York.

YU. V. MATIJASEVIČ

SOME PURELY MATHEMATICAL RESULTS INSPIRED BY MATHEMATICAL LOGIC

It is well-known that mathematical logic has greatly contributed to mathematics proper. As examples, one can mention famous results connected with the continuum hypothesis and numerous results about algorithmical unsolvability of many important decision problems in mathematics. However, these results are not purely mathematical since their very formulations involve some logical notions such as the notion of axiomatic theory or that of algorithm.

In my present talk I would like to consider some questions of methodological interest connected with purely mathematical theorems whose original proofs had been obtained with the aid of mathematical logic. The talk is by no means a survey of all such theorems, on the contrary, I will restrict myself to theorems which are in the scope of my personal interest. Namely, we shall deal with some recent investigations connected with Hilbert's tenth problem. For all technical and historical details the reader is referred to my joint paper with Davis *et al.* (1976) and the exhaustive bibliography given there.

A *Diophantine* equation is an equation of the form

$$(1) \qquad P(x_1, ..., x_\mu, a_1, ..., a_\nu) = 0$$

where P is a polynomial with integer coefficients; *parameters* $a_1, ..., a_\nu$ and *unknowns* $x_1, ..., x_\mu$ will always be restricted to natural numbers. Sets of the form

$$(2) \qquad \{\langle a_1, ..., a_\nu \rangle \mid \exists x_1 \cdots x_\mu [P(x_1, ..., x_\mu, a_1, ..., a_\nu) = 0]\}$$

are also called *Diophantine*.

The negative solution of Hilbert's tenth problem was obtained on the base of the following main result:

$$(3) \qquad \textit{The class of all Diophantine sets coincides with the class of}$$
$$\textit{all recursively enumerable sets.}$$

Butts and Hintikka (eds.), Logic, Foundations of Mathematics and Computability Theory, 121–127.

This result provides a bridge between the number-theoretical notion of Diophantine equation and the logical notion of recursive enumeration and thus gives a way for applying methods of mathematical logic in the theory of numbers.

We start from the following example. Let us consider, for simplicity, one parameter equation

(4) $P(x_1, ..., x_\mu, a) = 0.$

The set

(5) $\{a \mid \exists x_1 \cdots x_\mu [P(x_1, ..., x_\mu, a) = 0]\}$

is recursively enumerable and hence is on the list of all recursively enumerable sets

(6) $\mathfrak{M}_0, \mathfrak{M}_1, \ldots.$

The set of couples

(7) $\{\langle a, k \rangle \mid a \in \mathfrak{M}_k\}$

is also recursively enumerable and hence it is Diophantine, that is, for some polynomial U we have

(8) $a \in \mathfrak{M}_k \Leftrightarrow \exists z_1 \cdots z_\lambda [U(z_1, ..., z_\lambda, a, k) = 0].$

Now let κ be the Gödel number of (5) in the enumeration (6). Finally we have: the original Equation (4) is solvable if and only if the equation

(9) $U(z_1, ..., z_\lambda, a, \kappa) = 0$

is solvable.

What has been gained in this transition from Diophantine equations to enumerable sets and back to Diophantine equations? It is a reduction in the number of unknowns. In (4) μ may be arbitrarily large, but

λ does not depend on P at all. Thus we get the following number-theoretical theorem.

(10)   *There is an absolute constant λ such that for every one parameter Diophantine equation there is another one parameter Diophantine equation in λ unknowns which has solutions for exactly the same values of the parameter.*

It is interesting to trace the history of this theorem. It was in the early 50's that Martin Davis posed (3) as a conjecture. Some mathematicians did not believe in the possibility of a universal bound for the number of unknowns in Diophantine equations and considered this unbelievable consequence as informal evidence against Davis' conjecture. Nevertheless in 1970 the conjecture was shown to be true and we got a number-theoretical theorem by means of mathematical logic.

At first, the logical character of the proof seemed to be unavoidable. The boundary to the number of unknowns arises just when we pass from individual sets (6) to the universal set (7) and no number-theoretical counterpart was seen to this logical step.

Today the situation is different. A large amount of work has been done by Julia Robinson and me to get a small value of λ. Our starting point in 1970 was $\lambda = 35$. Four years ago I reported to the IV International Congress of Logic, Methodology and Philosophy of Science that λ can be as small as 14. Since then our joint paper appeared with $\lambda = 13$ (see Matijasevič and Robinson, 1975). At present we can get λ even as small as 9 but it is not the exact number of unknowns which is of interest for us now. The cutting down of the number of unknowns was obtained as a result of a long series of partial improvements and tracing now these improvements back I can say that all of them had the same nature: logical ideas and methods were replaced by number-theoretical ones. Our joint paper is published in a number-theoretical journal and we say in the introduction that our proof is an arithmetical one. Formally it is so, but the reader familiar with investigations on Hilbert's tenth problem will find influence of the earlier logical ideas. However, the new proof for $\lambda = 9$ is practically free from the initial logical ideas and really is a number-theoretical proof.

Thus today we can say that the theorem (10) is an example of a purely mathematical theorem, having a purely mathematical proof, but which was originally unbelievable to mathematicians and was *inspired* by mathematical logic.

Let us consider some other purely mathematical results obtained as consequences of the Diophantine. nature of recursively enumerable sets.

A trivial example concerns Fermat's last theorem which states that equation

$$(11) \qquad (x+1)^{w+3} + (y+1)^{w+3} = z^{w+3}$$

has no solution in non-negative integers w, x, y, z. Again the set of quadruples $\langle w, x, y, z \rangle$ satisfying (11) is recursively enumerable and hence Fermat's theorem is equivalent to the unsolvability of some Diophantine equation with additional variables

$$(12) \qquad F(v_1, ..., v_\mu, w, x, y, z) = 0.$$

Thus, we have removed the unknowns from the exponent for here F is an ordinary polynomial.

More interesting results can be obtained for another famous problem, namely, for Goldbach's conjecture. It states that every even integer greater than 2 can be represented as the sum of two prime numbers:

$$(13) \qquad \forall a\{a > 1 \Rightarrow \exists p_1 p_2 [2a = p_1 + p_2 \ \& \ p_1, p_2 \ are \ primes]\}.$$

Once again we notice that the set of natural numbers satisfying the condition in the braces in (13) is recursively enumerable and thus it can be represented in the form

$$(14) \qquad \{a \mid \exists x_1 \cdots x_\mu [G(x_1, ..., x_\mu) = a]\}$$

where G is a particular polynomial (a representation of the type (2) in case $\nu = 1$ can be easily transformed into a representation of the same set of the type (14)). In terms of this polynomial Goldbach's conjecture can be reformulated as follows:

(15)   *Every natural number is a value of polynomial G.*

The new formulation has the advantage of not containing any mention of primes at all.

Goldbach's conjecture can be reformulated in one more way connected with Diophantine equations. Let us consider this possibility at first in a general setting.

Suppose we are interested in a problem of the form

(16) $\forall a \mathcal{A}(a)$

where $\mathcal{A}$ is a decidable predicate. Then both $\mathcal{A}$ and its negation are recursively enumerable and hence they have Diophantine representations:

(17) $\mathcal{A}(a) \Leftrightarrow \exists t_1 \cdots t_\mu [P(t_1, ..., t_\mu) = a],$

(18) $]\mathcal{A}(a) \Leftrightarrow \exists s_1 \cdots s_\mu [Q(s_1, ..., s_\mu) = a].$

Finally, we have:

(19)

$$\forall a \mathcal{A}(a) \Leftrightarrow \forall a \exists t_1 \cdots t_\mu [P(t_1, ..., t_\mu) = a]$$
$$\Leftrightarrow]\exists q s_1 \cdots s_\mu [Q(s_1, ..., s_\mu) = q].$$

In other words, every problem of the form (16) can be reformulated in two ways: first, as an assertion that a particular polynomial represents every natural number or, second, as an assertion that a particular Diophantine equation has no solution.

Coming back to Goldbach's conjecture we see that it is of the form (16). Hence the conjecture is essentially a conjecture about the unsolvability of a particular Diophantine equation.

Many problems, not necessarily in number theory, are of the form (16). For example, the famous four color conjecture states that every planar map can be properly colored in four colors. One can take for $\mathcal{A}$ the property "every planar map with exactly a countries has a coloring".

In some cases it is not so evident that a problem is of the form (16). Let us consider in this connection another famous problem, namely, the outstanding Riemann hypothesis about the ζ-function. The function is defined by

$$(20) \qquad \zeta(z) = \sum_{n=1}^{\infty} n^{-z} \, .$$

The series convergers for all z with $\mathbf{Re}\,(z) > 1$ and can be extended to an analytic function defined on the whole complex plane with the exception of $z = 1$. One among many equivalent formulations of Riemann's conjecture states that ζ has no zeros inside the strip $0.5 < \mathbf{Re}\,(z) < 1$.

We can cover the strip by an infinite set of rectangles lying with their boundaries inside the strip. Then, roughly speaking, one can take for $\mathcal{A}$ the property "ζ has no zero inside the ath rectangle". This property is decidable because ζ is an analytic function and the number of zeros inside the rectangle with the boundary R is given by

$$(21) \qquad \frac{1}{2\pi i} \int_R \frac{\zeta'(z)}{\zeta(z)} \, dz.$$

The value of (21) is always an integer and to decide whether it is equal to zero or not we need only to calculate it with accuracy, say, 0.1, which can be done by numerical methods. (In fact, the situation is more involved: *a priori* ζ may have a zero inside the strip lying on the boundary of a chosen rectangle and in such a case the formula (21) cannot be applied. The difficulty can be overcome by a more sophisticated method taking advantage of the fact that a non-zero analytic function has at most finite number of zeros inside a bounded region.)

Thus we see that one formulation of Riemann's hypothesis has the same universal form: a particular Diophantine equation has no solution. Evidently, the proof of this purely number-theoretical result is by no means traditional, on the contrary, it takes advantage of such general logical ideas as the general notion of computability and of arithmetization of computations.

We have considered a number of examples of famous problems which can be effectively reduced by quite general methods to particular problems on Diophantine equations. However we know that there is no general method for solving these Diophantine problems. So the question arises whether there is anything behind these reductions to Diophantine equations or it is merely a kind of mathematical curiosity? For example, can we learn anything about the ζ-function by looking at the corresponding Diophantine equation described above? I daresay "Not". We can hardly expect to find out something new about specific problems by applying to them universal methods of reduction. But still it seems to me that another approach may prove to be fruitful. Namely, now that we know that a certain problem can be reduced to Diophantine equation by a general universal procedure, we can look for another reduction based on specific properties of the problem under consideration. When such a specific reduction is obtained, we may hope to have further progress in the problem itself.

Looking from this point of view at the famous problems considered above we can mention that reduction of Fermat's last theorem to a Diophantine equation can be easily obtained in a purely mathematical way. The same is true for replacing $p_1 + p_2$ by a polynomial in Goldbach's conjecture. As for reductions of the four color problem and Riemann's hypothesis, some more sophisticated methods are described in Davis *et al.* (1976) which take advantage of some specific properties of the problems. However they cannot be considered as purely mathematical ones.

Academy of Sciences of the USSR.

BIBLIOGRAPHY

Davis, M., Matijasevič, Yu., and Robinson, J.: 1976, 'Hilbert's Tenth Problem. Diophantine Equations: Positive Aspects of a Negative Solution', *Proceedings of Symposia in Pure Mathematics*, vol. 28. American Mathematical Society.
Matijasevič, Yu. V. and Robinson, J.: 1975, 'Reduction of an Arbitrary Diophantine Equation to One in 13 Unknowns', *Acta Arithmetica* **27**, 521–553.

L. W. SZCZERBA

INTERPRETABILITY OF ELEMENTARY THEORIES

The word 'theory' is often understood as dependent on the language used to formalise it (cf. e.g. Henkin *et al.*, 1971, p. 44). As a consequence, the theory of groups formalised without a neutral element symbol is a proper subtheory of that formalised in a language with a symbol for this element. We may observe an even more striking situation in geometry: the systems presented in Pieri (1908) and Tarski (1959) using different primitive notions are completely different theories, though they are intuitively closely related. The series of papers (e.g. Beth and Tarski, 1956; Scott, 1956; Tarski, 1956; R. Robinson, 1959; Royden, 1959) has been devoted to the study of possible systems of primitive notions of Euclidean geometry. Each of these systems may be used to express the same facts in a completely different manner. The difference may even be in the elements of the universes: Tarski (1959) uses points only, Hilbert (1930) – points, lines, planes and angles, Schwabhäuser and Szczerba (1975) – lines only, and Tarski (1929) – open discs or balls. Each of these formalisations may be included in any other by means of proper definitions, the procedure used commonly by most mathematicians. In fact if a lecturer speaking about some mathematical topic needs a new definable notion, no matter whether relation or object, he usually just introduces it, not bothering to change the language; and from that moment he works with a different theory.

To conform with this practice we should understand theories as closed not only under rules of inference, but under rules of definition too. As Tarski has pointed out to me, if we would introduce rules of definition by analogy with rules of inference, then the theory would depend on the order of use of the rules of definition. He felt this property unpleasant: the present notion of theory does not depend on the order of use of rules of inference. The aim of this paper is therefore a little less ambitious. An equivalence of elementary theories

Butts and Hintikka (eds.), *Logic, Foundations of Mathematics and Computability Theory*, 129–145.
Copyright © 1977 *by D. Reidel Publishing Company, Dordrecht-Holland. All Rights Reserved.*

will be introduced in such a way that all geometries mentioned above will be equivalent. In fact this notion has been used (implicitly) in several papers to prove the representation theorem (cf. e.g. Tarski, 1959; Hartshorne, 1967).

A theory T will be said to be interpretable in T' if it is obtainable by means of some definitions from T'. The problem is which schemas of definition should be admitted. Tarski (Tarski *et al.*, 1953) admits only defining relations by means of first order formulas. Corresponding equivalence, called mutual interpretability (Szmielew and Tarski, 1952), bilateral interpretability (Montague, 1957), synonymity (Bouvère, 1972), does not cover the examples of geometries: it requires the universe to remain unchanged. For this reason such equivalence cannot be used to obtain the metamathematical properties of elementary geometry from the analogous properties of the arithmetic of real numbers, as is done by Tarski (1951). It seems that defining relations by means of first order formulas is sufficient as a schema of introducing new relations. It is necessary to add some schemas for the introduction of new objects. In proofs of representation theorems for some geometries, authors frequently use sequences of elements as new elements, members of a new universe (e.g. points may be pairs of real numbers as in the case of the Cartesian plane), universes might be restricted to definable subsets, and moreover new elements might be equivalence classes with respect to some definable equivalence relation. (The projective point is an equivalence class of triples of real numbers (with triple $(0, 0, 0)$ excluded) with respect to the relation of proportionality.) It is obvious that all these definitional schemas are elementary, but it is only my belief that there is no definitional schema essentially different to that mentioned above which we would be willing to call elementary. In particular, definitions with parameters might be elementary in some inessential extensions, but the passage to such extensions are, at least for me, non-elementary.

The equivalence relation corresponding to the interpretability relation will be called, after Szmielew and Tarski (1952), mutual interpretability. It is obvious that if two theories are mutually interpretable then they are isomorphic in the sense of Hanf (1972) and definitionally equivalent in the sense of Schreiber (1974). The question whether there are elementary theories which are equivalent in the sense of Hanf or Schreiber, but not mutually interpretable, remains open.

It would be very difficult to give proper credit to everybody who influenced this paper by remarks, comments or suggestions, but at least I should mention, in addition to those quoted above, A. Mostowski, H. Herre, and K. Prażmowski. Their contributions go far beyond those explicitly credited to them in the text.

The results of this paper are given without proofs. Only some hints are given, where possible. All the expressions written in logical symbols are treated, not as expressions of a formalised theory, but as their metamathematical descriptions. Bold type will be used consistently to represent variables, predicates and other symbols of formalised languages. The corresponding light face type will denote corresponding elements of mathematical structures, relations between them and so on. Let φ be a formula with at most $\mathbf{v}_0, \mathbf{v}_1, ..., \mathbf{v}_n$ occurring as free variables. The result of the simultaneous substitution of the variables $\mathbf{v}_{i_0}, \mathbf{v}_{i_1}, ..., \mathbf{v}_{i_n}$ for the variables $\mathbf{v}_0, \mathbf{v}_1, ..., \mathbf{v}_n$ will be denoted by $\varphi(\mathbf{v}_{i_0}, \mathbf{v}_{i_1}, ..., \mathbf{v}_{i_n})$.

The phrase 'φ is satisfied in the structure $\mathfrak{A}$ by the sequence $(a_0, a_1, ..., a_n)$' is assumed to be understood. It will be denoted by $\mathfrak{A} \vDash \varphi(a_0, a_1, ..., a_n)$. Such metamathematical notions as model completeness, categoricity in power, decidability, recursive function etc. are also assumed to be known.

All the theories discussed below are understood as formalised in the first order predicate calculus. It is assumed that the languages of these theories employ only one sort of variable and a finite number of non-logical constants: moreover all of them denote relations, i.e. there are no individual constants or operation symbols. As has been shown (cf., e.g. A. Robinson, 1963, p. 226 and ff.), this assumption is inessential – for it is always possible to eliminate the function symbols from the language. Because of this assumption it is possible to understand a signature of a given language as a finite sequence of natural numbers. If n is ith term of the signature then the ith relation symbol is n-ary.

The set of all formulas built up in the usual way from symbols of the signature σ will be denoted by L_σ. By L_σ^n will be denoted the set of those formulas of L_σ in which at most the variables $\mathbf{v}_0, \mathbf{v}_1, ..., \mathbf{v}_{n-1}$ occur free. Thus L_σ^0 is the set of all sentences in L_σ. The class of corresponding structures (class of structures of the signature σ) will be denoted by $\mathscr{St}_\sigma$. The class of all models of a given set of sentences X

will be denoted by $\mathcal{M}od$ X and the set of all sentences from L_σ true in the class $M \subseteq \mathcal{S}t_\sigma$ (the theory of $\mathcal{M}$) will be denoted by Th $\mathcal{M}$.

Let $R \subseteq A^n$ and $A' \subseteq A$. The relation

$$R \upharpoonright A' = R \cap (A')^n$$
$$= \{(a_0, ..., a_{n-1}) : (a_0, ..., a_{n-1}) \in R \wedge a_0 \in A' \wedge \cdots$$
$$\wedge\, a_{n-1} \in A'\}$$

is called the *restriction* of R to the set A'. Let now $\approx$ be any equivalence relation on A. The equivalence class of an element a is denoted by $[a]/_\approx$. By the *quotient relation* of R we mean the relation $R/_\approx$ where

$$R/_\approx([a_0]/_\approx, ..., [a_{n-1}]/_\approx) \leftrightarrow \exists a_0', ..., \exists a_{n-1}'(a_0'$$
$$\approx a_0 \wedge \cdots \wedge a_{n-1}' \approx a_{n-1} \wedge R(a_0', ..., a_{n-1}')).$$

Let $\mathfrak{A} = (A, R)$ be a structure of signature σ, i.e. $\mathfrak{A} \in \mathcal{S}t_\sigma$, and let $A' \subseteq A$. The structure $\mathfrak{A} \upharpoonright A' = (A', (R_0 \upharpoonright A', ..., R_m \upharpoonright A'))$, where m is the length of σ, is called the *restriction* of $\mathfrak{A}$ to A' and the structure $\mathfrak{A}/_\approx = (A/_\approx, (R_0/_\approx, ..., R_m/_\approx))$ – the *quotient structure* of $\mathfrak{A}$ by an equivalence relation $\approx$.

Let in addition $\mathfrak{B} = (B, P)$ be any structure of the signature τ. Suppose that $A \cap B = \varnothing$. Let $\mathfrak{A} \oplus' \mathfrak{B} = (A \cup B, (A, B)^\cap R^\cap P)$, where $(A, B)^\cap R^\cap P$ denotes the concatenation of the sequences (A, B), R and P. To be able to perform this operation for structures with non-disjoint universes, it is necessary to pass to some isomorphic structures which already have disjoint universa. Let then $\mathfrak{A}^{(i)}$ be an isomorphic image of $\mathfrak{A}$ with respect to the function mapping of an element $a \in A$ onto a pair (a, i). Since the universa of $\mathfrak{A}^{(0)}$ and $\mathfrak{B}^{(1)}$ are disjoint, we may put $\mathfrak{A} \oplus \mathfrak{B} = \mathfrak{A}^{(0)} \oplus' \mathfrak{B}^{(1)}$.

Let now $\sigma = (s_i)_{i<p}$ and $\tau = (t_i)_{i<r}$ be signatures. A sequence c will be called a *code* from σ to τ ($c \in \mathbf{Cod}_\tau^\sigma$) if it is of the form

(1) $\qquad c = (n, \varphi, \psi, \xi_0, ..., \xi_{r-1}),$

where n is a natural number different from zero (and called the *exponent* of the code c), $\varphi \in L_\sigma^n$, $\psi \in L_\sigma^{2n}$, $\xi_i \in L_\sigma^{nt}$. Let $\mathfrak{A} \in \mathscr{St}_\delta$ and let a be the universe of $\mathfrak{A}$. Suppose that

$$A' = \{a: a \in A^n \wedge \mathfrak{A} \vDash \varphi a\},$$

i.e. A' is the set defined in A^n by the formula φ. Similarly let $\approx$ be a binary relation on A^n defined by ψ, and R_i and t_i-ary relation on A^n, defined by the formula ξ_i. We put

$$\Gamma c\mathfrak{A} = (A^n, (R_0, ..., R_{r-1})) \upharpoonright A'/_\approx$$

provided $\approx$ is an equivalence relation. Otherwise we consider $\Gamma c\mathfrak{A}$ undefined. Thus Γc is a partial function from $\mathscr{St}_\sigma$ into $\mathscr{St}_r$. Any one of these functions is called an *interpretation*.

Interpretation maps structures onto structures. The corresponding functions mapping formulas onto formulas are called translations. Let c be a code (1). The *translation* Fc is defined by induction as follows:

$$Fc(v_i = v_j) = \varphi(v_{ni}, ..., v_{ni+n-1}, v_{nj}, ..., v_{nj+n-1})$$
$$Fc(R_i(v_{j_0}, ..., v_{j_{t_i-1}})) = (\exists v_{nq} \cdots \exists v_{n(q+t_i-1)+n-1}$$
$$Fc(v_{j_0} = v_q) \wedge \cdots \wedge Fc(v_{j_{t_i-1}} = v_{q+t_i-1}) \wedge \xi_i(v_{nq}, ..., v_{n(q+t_i-1)+n-1}))$$

where $q = \max\{j_0, ..., j_{t_i-1}\} + 1$

$$Fc(\neg\zeta) = (\neg Fc\zeta)$$
$$Fc(\eta \vee \zeta) = (Fc\eta \vee Fc\zeta)$$
$$Fc(\eta \wedge \zeta) = (Fc\eta \wedge Fc\zeta)$$
$$Fc(\eta \rightarrow \zeta) = (Fc\eta \rightarrow Fc\zeta)$$
$$Fc(\eta \leftrightarrow \zeta) = (Fc\eta \leftrightarrow Fc\zeta)$$
$$Fc(\exists v_i\zeta) = (\exists v_{ni} \cdots \exists v_{ni+n-1}(\varphi(v_{ni}, ..., v_{ni+n-1}) \wedge Fc))$$
$$Fc(\mathcal{M}v_i\zeta) = (\forall v_{ni} \cdots \forall v_{ni+n-1}(\varphi(v_{ni}, ..., v_{ni+n-1}) \rightarrow Fc\zeta)).$$

It is easy to prove

LEMMA 1. Translation lemma. *For any formula* $\zeta \in L_\sigma^m$, *any* c$\in$ Cod$_r^\sigma$, *any* $\mathfrak{A} = (A, R) \in \text{Dom}\Gamma c$ *and any* $a_0, ..., a_{m-1} \in A^n$, *where n is an exponent of* c, *we have*

(i) $Fc\zeta \in L_\tau^{mn}$

(ii) $\Gamma c\mathfrak{A} \vDash \zeta([a_0]/_\approx, \dots, [a_{m-1}]/_\approx) \leftrightarrow \mathfrak{A} \vDash Fc\zeta(a_0 \cap \cdots \cap a_{m-1})$ *where* $\approx$ *is the equivalence relation of* c.

As a simple consequence of translation of the lemma we get

COROLLARY 2.   *If* $\mathfrak{A} \in Dom\Gamma c$ *then* $Th\overleftarrow{\Gamma c}\mathfrak{A} = \overleftarrow{Fc}Th\mathfrak{A}$.
(where $\overleftarrow{Fc}Th\mathscr{A}$ is a coimage of $Th\mathscr{A}$ with respect to Fc) and therefore

COROLLARY 3.   *If* $\mathscr{A} \subseteq Dom\Gamma c$ *then* $Th\overrightarrow{\Gamma c}\mathscr{A} = \overleftarrow{Fc}Th\mathscr{A}$.
(where $\overrightarrow{\Gamma c}\mathscr{A}$ is an image of $\mathscr{A}$ with respect to Γc).

It is possible to prove by the explicit construction of a code

LEMMA 4.   Composition lemma.   *There is a recursive partial operation* $\circ$ *such that for any* $c \in Cod_\tau^\sigma$ *and* $c' \in Cod_\kappa^\tau$, *if* $\mathfrak{A} \in Dom\Gamma c$ *and* $\Gamma c\mathfrak{A} \in Dom\Gamma c'$ *then* $\Gamma c'\Gamma c\mathfrak{A} \cong \Gamma(c \circ c')\mathfrak{A}$.

(The symbol $\cong$ stands for the relation of being isomorphic.) From the composition lemma it follows easily that composition of interpretations is an interpretation.

Similarly we may prove

LEMMA 5.   *There is a recursive partial operation* $\oplus$ *such that for any* $c \in Cod_\tau^\sigma$ *and* $c' \in Cod_\kappa^\sigma$ *if* $\mathfrak{A} \in Dom\Gamma c \cap Dom\Gamma c'$ *then* $\Gamma c\mathfrak{A} \oplus \Gamma c'\mathfrak{A} \cong \Gamma(c \oplus c')\mathfrak{A}$.

Let T and T' are two theories with finite signatures σ and τ respectively. We put

$$T \leqslant T' \leftrightarrow \exists c \quad c \in Cod_\tau^\sigma \wedge \mathcal{M}od\, T' \subseteq Dom\Gamma c \wedge T$$
$$= Th\overrightarrow{\Gamma c} \,\mathcal{M}od\, T'.$$

If $T \leqslant T'$ we say that T is *interpretable* in T'.

From Corollary 3 follows that the notion of interpretability may be characterized in purely syntactical terms:

THEOREM 6.

$$T \leqslant T' \leftrightarrow \exists c \quad c \in Cod_\tau^\sigma \wedge T' \subseteq Codom\, Fc \wedge T = \bar{F}cT'.$$

From the composition lemma it follows that interpretability is transitive. Since it is obviously reflexive, we conclude that the relation of interpretability is a quasi-order relation and therefore the notion of *mutual interpretability:*

$$T \equiv T' \leftrightarrow T \leqslant T' \wedge T' \leqslant T,$$

is an equivalence relation.

Two mutually interpretable theories concern the same problems. But they express these problems in different languages. Therefore there is a suggestion that we should use the word 'theory' to denote an equivalence class with respect to mutual interpretability. The elements of such classes, called now 'theories', should be referred to as 'presentations of a theory'. It is very tempting to adopt this terminology since it mirrors what is usually meant. It seems however, that this terminology has little chance of being accepted: the present meaning of 'theory' is too strongly bred in the bone of people working in the foundations of mathematics. Thus it seems proper to keep, with some regret, the present meaning of 'theory' and to use '*domain*' to denote equivalence classes. We can still say that a theory T represents the domain [T]. We say that a domain [T] is interpretable in a domain [T'], or in symbols [T] $\leqslant$ [T'], is and only if T is interpretable in T'.

The notion of interpretability is often used to infer some metamathematical properties of some theories from analogous properties of some other theory. For example, Tarski (1951) used it to infer the completeness and decidability of elementary Euclidean geometry from the completeness and decidability of elementary arithmetic of real-closed fields. It is possible because of the following

THEOREM 7. Hereditarity theorem. *Let* T $\leqslant$ T' *then*
   (i)   *if* T' *is complete then* T *is complete*
   (ii) *if* T' *is decidable then* T *is decidable*
   (iii) *if* T' *is* $\aleph_0$-*categorical then* T *is* $\aleph_0$-*categorical*
   (iv) *if* T' *is stable then* T *is stable.*

In fact the first two parts of this theorem may be strengthened if we use degrees of incompleteness and undecidability. By *incompleteness degree* of a theory T, in symbols CdT, we mean the number (finite or infinite) of its complete and consistent extensions. Thus $CdT \leqslant CdT'$ means that there are not less complete and consistent extensions of T' than of T. Similarly $DdT \leqslant DdT'$ (the *undecidability degree* of T is less or equal to that of T') means that T is decidable relatively to T' (cf. Post, 1944).

THEOREM 8. *Let* $T \leqslant T'$ *then*
  (i) $CdT \leqslant CdT'$
  (ii) $DdT \leqslant DdT'$.

Parts (i) and (ii) of Theorems 7 and 8 follow from the translation lemma. Part (iii) of Theorem 7 is proved in Prazmowski and Szczerba (1976). Finally part (iv) of Theorem 7 may be proved by the theorem of Shelah 1971 (cf. Chang and Keisler, 1973, p. 424 Theorem 7.1.33).

The categoricity in any infinite power other than $\aleph_0$ is not hereditary with respect to interpretability. It is hereditary with respect to a stronger relation defined as follows:

$$T \leqslant_2 T' \leftrightarrow \exists c \quad c \in Cod_T^\sigma \wedge \mathcal{M}od\,T' \subseteq Dom\vec{\Gamma}c \wedge \mathcal{M}od\,T$$
$$= I\vec{\Gamma}c\mathcal{M}od\,T',$$

here $I\vec{\Gamma}c\mathcal{M}od\,T'$ denotes the closure of $\vec{\Gamma}c\mathcal{M}od\,T'$ under isomorphisms. For the proofs see Prażmowski and Szczerba (1976).

Another example of non-hereditary metamathematical property is axiomatisability. Let T be the theory of rings and T' the theory of all algebras $\mathfrak{F} \oplus \mathfrak{N}$ where $\mathfrak{F}$ is any ring and $\mathfrak{N}$ is the structure of natural numbers. The theory T, being an axiomatic theory, is finitely axiomatisable, but T' is not, nevertheless $T \leqslant T'$.

THEOREM 9. *Let* $T \equiv T'$ *then*
  (i)   *if* T *is axiomatisable then* T' *is axiomatisable*
  (ii) *if* T *is finitely axiomatisable then* T' *is finitely axiomatisable.*

We may say therefore that axiomatisability, finite axiomatisability as well as completeness, decidability, $\aleph_0$-categoricity, incompleteness degree and undecidability degree are properties not only of theories but in fact of equivalence classes of theories i.e. of domains. On the other hand $\aleph_\alpha$-categoricity (with $\alpha > 0$) and model-completeness are properties only of theories themselves. For an example of $\aleph_1$-categorical theory mutually interpretable with a theory which is not $\aleph_1$-categorical, see Prazmowski and Szczerba 1976. P. Tuschik has constructed the following example of two mutually interpretable theories, one of them is model-complete and the other is not. Let T be a binary relation on natural numbers holding for pairs $(3n, 3n+1)$, $(3n+1, 3n+2)$ for any n and for pairs $(3n, 3n+2)$ for even n, and for no other pairs. It is easy to prove that $\mathrm{Th}(\omega, T)$ is model-complete. Consider now the structure with universe being a sum of ω and $\omega \times \omega$, and ternary relation T' holding between a, b, c if and only if a and b are natural numbers and c is the pair (a, b). The theory $\mathrm{Th}(\omega \cup \omega \times \omega, T')$ is not model-complete, nevertheless

$$\mathrm{Th}(\omega, T) \equiv \mathrm{Th}(\omega \cup \omega \times \omega, T').$$

In both cases mutual interpretability may be replaced by an equivalence which seems to be much stronger

$$\mathsf{T} \equiv {}'\mathsf{T}' \leftrightarrow \exists cc'\{c \in \mathrm{Cod}_\tau^\sigma \wedge c' \in \mathrm{Cod}_\sigma^\tau \wedge \mathcal{M}\!od\,\mathsf{T}$$
$$\subseteq \mathrm{Dom}\,\Gamma c \wedge \mathcal{M}\!od\,\mathsf{T}' \subseteq \mathrm{Dom}\,\Gamma c'$$
$$\wedge [\forall \mathfrak{A} \vDash \mathsf{T}\ \ \Gamma c' \Gamma c \mathfrak{A} \cong \mathfrak{A}] \wedge [\forall \mathfrak{A} \vDash \mathsf{T}'\ \ \Gamma c \Gamma c' \mathfrak{A} \cong \mathfrak{A}]\}.$$

It is easy to see that this stronger equivalence relation implies mutual interpretability. I do not know any example of two not mutually interpretable theories T and T' such that $\mathsf{T} \equiv \mathsf{T}'$.

Now we shall turn to some examples of interpretability of elementary theories. Let $\mathfrak{N}$ be the algebra of natural numbers, $\mathfrak{Z}$ – the ring of integers and $\mathfrak{Q}$ – the field of rational numbers. From the well known constructions of integers in natural numbers, and rational numbers in integers, it follows that $\mathrm{Th}\mathfrak{Q} \leqslant \mathrm{Th}\mathfrak{Z} \leqslant \mathrm{Th}\mathfrak{N}$. Since from the construction of natural numbers in the field of the rationals (cf. J. Robinson, 1949) it follows that $\mathrm{Th}\mathfrak{N} \leqslant \mathrm{Th}\mathfrak{Q}$, therefore the arithmetics of natural and rational numbers, as well as the arithmetic of integers are mutually interpretable:

$$\mathrm{Th}\mathfrak{Q} \equiv \mathrm{Th}\mathfrak{Z} \equiv \mathrm{Th}\mathfrak{N}.$$

Similarly, from the introduction of coordinates in a real Euclidean plane, Weierstraß coordinates in projective plane and finally of Beltrami coordinates in Bolyai-Lobatshevski plane, it follows that Euclidean plane geometry E, projective plane geometry P, and Bolyai-Lobatshevski plane geometry BL are all interpretable in the arithmetic of real numbers. On the other hand in each plane the field of real numbers has been constructed (cf. Hilbert, 1930, for the case of Euclidean and Bolyai-Lobatshevski geometries and Hartshorne, 1967, for the case of projective plane geometry). The quoted definitions employ parameters, but it is possible to remove them. When we start with any model of E, P or BL, the corresponding construction gives us some real-closed field. Thus

$$E \equiv P \equiv BL \equiv Th\Re,$$

where $\Re$ is the field of real numbers.

It seems therefore natural to say that 'arithmetic' (no matter whether of natural or rational numbers) is the domain $[Th\Re]$ and that 'geometry' denotes the domain $[E]$. At the first moment it may seem strange to think that the arithmetic of real numbers is a presentation of geometry, but after all the real numbers has been invented as a tool to measure segments.

As K. Hauschild and I. Korec have informed me, they had proved that the elementary arithmetic of real numbers is interpretable in elementary arithmetic of natural numbers. The proof proceeds roughly as follows: It is possible to express finite sequence of natural numbers $(n_0, ..., n_k)$ as a single natural number $2^{n_0} \cdot 3^{n_1} \cdot 5^{n_2} \cdots p_k^{n_k}$, where p_k stands for the kth prime number. Similarly we may code by means of natural numbers all finite sequences of integers, or what amounts to the same, all polynomials with integer coefficients. Thus we may construct the field of real algebraic numbers. Since this field is elementarily equivalent to the field of real numbers (cf. Tarski, 1951) we get

$$Th\Re < Th\mathfrak{N}.$$

The symbol $<$ stands for 'interpretable, but not mutually interpretable': $T < T' \leftrightarrow T \leq T' \wedge \neg T \equiv T'$. To prove that $\neg Th\Re \equiv Th\mathfrak{N}$, or more exactly that $\rightarrow$Th$\mathfrak{N} \leqslant$Th$\mathfrak{R}$ it is sufficient to observe that Th$\mathfrak{R}$ is decidable (cf. Tarski 1951) but Th$\mathfrak{N}$ is not (cf. Gödel, 1931 or Tarski *et al.*, 1953) and use hereditarity Theorem 7(ii).

It is obvious that the elementary arithmetic of complex numbers is interpretable in the elementary arithmetic of real numbers. H. Herre has observed that the arithmetic of complex numbers, being χ_1-categorical, is ω-stable and therefore stable, while the arithmetic of real numbers is not (cf. Chang and Keisler, 1973, Proposition 1.4.10, Lemma 7.1.4 and 7.1.33). Thus by Theorem 7(iv) the arithmetic of real numbers is not interpretable in the elementary arithmetic of complex numbers.

The results of the above considerations may be summarized as follows:

$$\text{Th}\mathfrak{N} \equiv \text{Th}\mathfrak{Z} \equiv \text{Th}\mathfrak{Q} > \text{Th}\mathfrak{R} \equiv E \equiv P \equiv BL > \text{Th}\mathfrak{C}.$$

Any domain may be represented by different theories, and in different languages. It is interesting how much complexity is needed in a particular domain. It turns out that not too much:

THEOREM 10.   *If* T *is any theory of finite signature then there is a theory* T' *of the signature* (2) (*i.e. with just one binary relation symbol*) *such that*

$$T \equiv T'.$$

This theorem justifies the frequent procedure of proving meta-mathematical theorems just in case of one binary relation. It turns out that in case of properties of domains nothing more is to be proved. Moreover, since the mutual interpretability in the Theorem 10 may be replaced by mutual interpretability in the sense of $\leqslant_2$, the same applies to categoricity in any power.

The Theorem 10 may be formulated conveniently by means of the notion of universal theory. We say that a theory T of signature σ is *universal* if for any theory T' there is an extension of the theory T, of the same signature σ, mutually interpretable with T'. The notion of universality used here is different (and in fact stronger) to that used by Hauschild and Rautenberg (1971). The proof of the Theorem 10 does

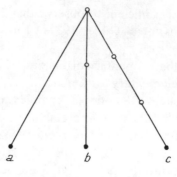

Fig. 1

not differ essentially from that used by Hauschild and Rautenberg. It will be described here in the case of ternary relation only: For any triple (a, b, c) in the relation we define new elements represented in Figure 1 as circlets and binary relation between them represented by lines. It turns out that both theories are mutually interpretable even in the sense of $\leqslant_2$. This proof may be easily modified to prove

THEOREM 11. *The following theories are universal:*
  (i)   *the logic theory of one binary relation*
 (ii)   *the theory of partial order*
(iii)   *the theory of lattices*
 (iv)   *the theory of two unary functions*
  (v)   *the theory of one unary function and one equivalence relation.*

As it follows from Behman (1922) any theory with unary relations symbols only is decidable and therefore it cannot be universal. Similarly from Ehrenfeuht (1959) follows that the theory of one unary function is decidable and therefore, by Theorem 7(ii) it cannot be universal too.

Now we shall turn to the study of properties of interpretability understood as order relations of domains. The bottom part of this ordered set is known (see Figure 2). The smallest one is denoted by D_1. It consists of all theories of structures with one element sets as universes. Since these structures, as well as their theories, are trivial, we call this domain trivial. The remaining domains of the Figure 2 are

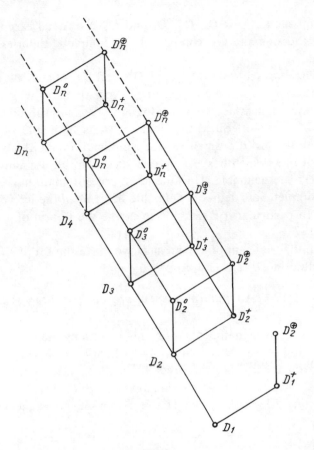

Fig. 2

denoted as follows: all of them are equivalence classes of theories of equality in some family of sets. In case of D_n it is the theory of equality in a one element set or a two element set of... or an n element set. In the case of D_n^+ it is the theory of equality in a two element set or a three element set or... or an $(n+1)$ element set. In the case of D_n^0 it is the theory of equality in a one element set or a two element set or... or an $(n-1)$ element set or infinite set, and finally in the case of $D_n^\oplus$ it is the theory of equality in a two element set or a three element set or... or an n element set or infinite set. There are no other connections than those which follow from the arrangement shown on Figure 2.

None of the domains D_n, D_n^+, D_n^0 and $D_n^\oplus$ are trivial except D_1. For example the domain D_1^+ consists of all complete theories of finite structures.

It is easy to see that D_n, D_n^+, D_n^0 and $D_n^\oplus$ have all the same incompleteness degree, namely n. In all other domains which have the same incompleteness degree n the domain $D_n^\oplus$ is interpretable, but the domain D_{n+1} is not. Since the domain D_1 is definable in terms of interpretability as the smallest one (minimal domain), D_1^+ as the successor of D_1 which has exactly one successor, all domains D_n, D_n^+, D_n^0 and $D_n^\oplus$ are definable. Thus the notion of a domain having a given finite incompleteness degree is definable in terms of the interpretability relation. In particular it is possible to define the notion of a complete domain.

By an initial segment of a domain D we mean the set of all domains interpretable in D. It is easy to prove

THEOREM 12.   *An initial segment of any theory is at most countable.*

In fact, for any signature there is only countably many codes in this signature, and thus only countably many theories may be interpretable from a given theory. On the other hand a co-initial segment of any domain D, i.e. the set of all domains in which the domain D is interpretable, is uncountable. This is a simple consequence of the following

THEOREM 13.   *For any countable sequence of theories* (T_i) *there is a theory* T *such that* T_i *is interpretable in* T *for any i.*

We shall construct T in the case where all theories T_i are complete and of signature (2). By the Theorem 10 the second assumption is inessential. Let $T_i = \text{Th}\mathfrak{A}_i$ and $\mathfrak{A}_i = (\omega, A_i)$ where ω is the set of natural numbers. We define T to be a ternary relation on natural numbers, holding for a triple (i, j, k) if and only if the pair (j, k) is in the relation $\mathfrak{A}_i$. For T we may take $\text{Th}(\omega, +, \cdot, T)$.

Since any undecidability degree is an undecidability degree of some elementary theory (cf. Feferman, 1957), we may take several results from the theory of undecidability degrees and reproof them for the theory of interpretability by means of the Theorem 8(ii). In particular we may prove

THEOREM 14.   *There is a continuum of mutually incomparable domains.*

In fact, there is a continuum of mutually incomparable undecidability degrees (cf. Yates, 1970, Theorem 1.7). Therefore there is a continuum of theories with mutually incomparable degrees. By Theorem 8(ii) all domains of these theories are mutually incomparable. From Theorems 12, 13 and 14 we get

COROLLARY 15.   *There is no maximal domain.*

THEOREM 16.   *The structure of domains with finite incompleteness degree ordered by the interpretability relation is upper semi-lattice.*

Let T and T' be two complete theories. Thus there are structures $\mathfrak{A}$ and $\mathfrak{A}'$ such that $T = Th\mathfrak{A}$ and $T' = Th\mathfrak{A}'$. The elementary theory of $\mathfrak{A} \oplus \mathfrak{A}'$ is a lowest upper bound of T and T'. To extend this proof for arbitrary theories with finite incompleteness degree it is enough to observe that any such theory may be defined as an elementary theory of finite collection of structures, each of which may be singled out among others by means of a single elementary sentence.

University of Warsaw

BIBLIOGRAPHY

Beth, E. W. and Tarski, A.: 1956, 'Equilaterality as the Only Primitive Notion of Euclidean Geometry', *Ind. Math.* **18**, 462–467.
Bouvère, K. de: 1963, 'Synonymous Theories', in J. W. Addison, L. Henkin and A. Tarski (eds.), *The Theory of Models*, North-Holland, Amsterdam, pp. 402–406.
Chang, C. C. and Keisler, H. J.: 1973, *Model Theory*, North-Holland, Amsterdam.
Erenfeuht, A.: 1959, 'Decidability of the Theory of One-Function', *Notices Amer. Math. Soc.* **6**, 268.
Feferman, S.: 1957, 'Degrees of Unsolvability Associated with Classes of Formalised Theories', *J. Symbolic Logic* **22**, 161–175.
Gödel, K.: 1931, 'Uber formal unenscheidbare Sätze der Principia Mathematica und verwandter Systeme I', *Monatshefte Math. u. Phys.* **4**, 173–198.

Hanf, W.: 1972, 'Model-Theoretic Methods in the Study of Elementary Logic', in J. W. Addison, L. Henkin, and A. Tarski (eds.), *The Theory of Models*, North-Holland, Amsterdam, pp. 132–145.

Hartshorne, R.: 1967, *Foundations of Projective Geometry*, New York.

Hauschild, K. and Rautenberg, W.: 1971, 'Interpretierbarkeit in Gruppentheorie', *Algebra Universalis* **1**, 136–151.

Hauschild, K. and Rautenberg, W.: 1971, 'Interpretierbarkeit und Entscheidbarkeit in der Gruppentheorie I', *Z. Math. Logik Grundlagen Math.* **17**, 47–55.

Henkin, L., Monk, J. D., and Tarski, A.: 1971, *Cylindric Algebras*, North-Holland, Amsterdam and London.

Hilbert, D.: 1930, *Grundlagen der Geometrie*, B. G. Teubner, Leipzig and Berlin.

Montague, R. M.: 1957, *Contributions to the Axiomatic Foundations of Set Theory*, Doctoral dissertation, University of California, Berkeley.

Pieri, M.: 1908, *La Geometrie Elementare sulle nozioni di 'punto' e 'sfera'*.

Post, E. L.: 1944, 'Recursively Enumerable Sets of Positive Integers and Their Decision Problems', *Bull. Amer. Math. Soc.* **50**, 284–316.

Prażmowski, K. and Szczerba, L. W.: 1976, 'Interpretability and Categoricity', *Bull. Acad. Polon. Sci. ser. sci. math. astronom. phys.* **24**, 309–312.

Robinson, A.: 1963, *Introduction to Model Theory and to the Metamathematic of Algebra*, North-Holland, Amsterdam.

Robinson, J.: 1949, 'Decidability and Decision Problems in Arithmetic', *J. Symbolic Logic* **14**, 98–114.

Robinson, R.: 1959, 'Binary Relations as Primitive Notions in Elementary Geometry' in: L. Henkin, P. Suppes, and A. Tarski (eds.), *The Axiomatic Method with Special Reference to Geometry and Physics*, North-Holland, Amsterdam, pp. 68–85.

Royden, H. L.: 1959, 'Remarks on Primitive Notions for Elementary Euclidean and Non-Euclidean Plane Geometry', in L. Henkin, P. Suppes, and A. Tarski (eds.), *The Axiomatic Method with Special Reference to Geometry and Physics*, North-Holland, Amsterdam, pp. 86–96.

Scott, D.: 1956, 'A Symmetric Primitive Notion for Euclidean Geometry', *Ind. Math.* **18**, 457–461.

Schreiber, P.: 1974, 'Definitorische Erweiterungen formalisierter Geometrien' in *Grundlagen der Geometrie und algebraische Methoden* (Int. Kolloq. Pädag. Hohsch. 'Karl Liebknecht', Potsdam, 1973) Potsdamer Forschungen, Reihe B, Heft 3. Pädag. Hohsch. 'Karl Liebknecht', Potsdam.

Schwabhäuser, W. and Szczerba, L. W.: 1975, 'Relations on Lines as Primitive Notions for Euclidean Geometry', *Fund. Math.* **82**, 347–355.

Shelah, S.: 1971, 'Stability and the F.P.C.; Model-Theoretic Properties of Formulas in First Order Theories', *Ann. Math. Logic* **3**, 271–362.

Szmielew, W. and Tarski, A.: 1952, 'Mutual Interpretability of Some Essentially Undecidable Theories', in *Proceedings of the International Congress of Mathematicians, Cambridge Mass. 1950*, American Math. Soc. Providence, p. 734.

Tarski, A.: 1929, 'Les fondements de la geometrie des corps', in *Ksiega Pamiatkowa Pierwszego Polskiego Zjazdu Matematyków*, Kraków, pp. 29–33.

Tarski, A.: 1951, *A Decision Method for Elementary Algebra and Geometry*, 2nd ed. revised, Berkeley and Los Angeles.

Tarski, A.: 1956, 'A General Theorem Concerning Primitive Notions of Euclidean Geometry', *Ind. Math.* **18**, 468–474.

Tarski, A.: 1959, 'What is Elementary Geometry?' in L. Henkin, P. Suppes, and A. Tarski (eds.), *The Axiomatic Method With Special Reference to Geometry and Physics*, North-Holland, Amsterdam.

Tarski, A., Mostowski, A., and Robinson, R. M.: 1953, *Undecidable Theories*, North-Holland, Amsterdam.

Yates, C. E. M.: 1970, 'Initial Segments of Degrees of Unsolvability, Part I: A Survey', in Y. Bar-Hillel (ed.), *Mathematical Logic and Foundations of Set Theory*, North-Holland, Amsterdam.

III

CATEGORY THEORY

CATEGORICAL FOUNDATIONS AND FOUNDATIONS OF CATEGORY THEORY*

ABSTRACT. This paper is divided into two parts. Part I deals briefly with the thesis that category theory (or something like it) should provide the proper foundations of mathematics, in preference to current foundational schemes. An opposite view is argued here on the grounds that the notions of operation and collection are prior to all structural notions. However, no position is taken as to whether such are to be conceived extensionally or intensionally.

Part II describes work by the author on a new non-extensional type-free theory $\hat{T}$ of operations and classifications.[1] Its interest in the present connection is that much of 'naive' or 'unrestricted' category theory can be given a direct account within $\hat{T}$. This work illustrates requirements and possibilities for a foundation of unrestricted category theory.

I

The reader need have no more than a general idea of the nature of category theory to appreciate most of the issues discussed below. MacLane [15] gives a succinct account which is particularly related to these questions. Two views are intermixed in [15] as to current set-theoretical foundations, namely that (i) they are *inappropriate* for mathematics as practised, and (ii) they are *inadequate* for the full needs of category theory. The latter is taken up in Part II below. The view (i) evidently derives from the increasingly dominant conception (by mathematicians) of mathematics as the study of abstract structures. This view has been favored particularly by workers in category theory because of its successes in organizing substantial portions of algebra, topology, and analysis. It is perhaps best expressed by Lawvere [13]: "In the mathematical development of recent decades one sees clearly the rise of the conviction that the relevant properties of mathematical objects are those which can be stated in terms of their abstract structure rather than in terms of the elements which the objects were thought to be made of. The question thus naturally arises whether one can give a foundation for mathematics which expresses wholeheartedly this conviction concerning what mathematics is about and in particular

Butts and Hintikka (eds.), Logic, Foundations of Mathematics and Computability Theory, 149–169.
Copyright © 1977 by D. Reidel Publishing Company, Dordrecht-Holland. All Rights Reserved.

in which classes and membership in classes do not play any role."
Further: "A foundation of the sort we have in mind would seemingly
be much more natural and readily usable than the classical one ..."
Lawvere went on in [13] to formulate a (first-order) theory whose
objects are conceived to be arbitrary categories and functors between
them.[2] Each object in Lawvere's theory is thus part of a highly
structured situation which must be prescribed axiomatically by the
elementary theory of categories.

There are several objections to such a view and program,[3] among
which are arguments for what is achieved in set-theoretical foundations
that is not achieved in other schemes (present or projected). I wish to
stress here instead a very simple objection to it which is otherwise
neutral on the question of 'proper foundations' for mathematics: the
argument given is itself not novel.[4]

The point is simply that *when explaining* the general notion of
structure and of particular kinds of structures such as groups, rings,
categories, etc. we implicitly *presume as understood* the ideas of
operation and *collection;* e.g. we say that a group consists of a collection of objects together with a binary operation satisfying such and
such conditions. Next, when explaining the notion of *homomorphism*
for groups or *functor* for categories, etc., we must again understand the
concept of operation. Then to follow category theory beyond the basic
definitions, we must deal with questions of *completeness*, which are
formulated in terms of collections of morphisms. Further to verify
completeness in concrete categories, we must be able to form the
operation of *Cartesian product* over collections of its structures. Thus
at each step we must make use of the unstructured notions of operation and collection to explain the structural notions to be studied. The
logical and *psychological priority* if not primacy of the notions of
operation and collection is thus evident.

It follows that a theory whose objects are supposed to be highly
structured and which does not explicitly reveal assumptions about
operations and collections cannot claim to constitute a foundation for
mathematics, simply because those assumptions are unexamined. It is
evidently begging the question to treat collections (and operations
between them) as a category which is supposed to be one of the objects
of the universe of the theory to be formulated.

The foundations of mathematics must still be pursued in a direct examination of the notions of operation and collection. There are at present only two (more or less) coherent and comprehensive approaches to these, based respectively on the Platonist and the constructivist viewpoints. Only the first of these has been fully elaborated, taking as basis the conception of sets in the cumulative hierarchy. It is distinctive of this approach that it is *extensional*, i.e., collections are considered independent of any means of definition. Further, operations are identified with their graphs.

On the other hand, it is distinctive of the constructive point of view that the basic notions are conceived to be *intensional*, i.e. operations are supposed to be *given by rules* and collections are supposed to be *given by defining properties*.[5] Theories of such are still undergoing development and have not yet settled down to an agreed core comparable to the set theories of Zermelo or Zermelo-Fraenkel. Nevertheless, a number of common features appear to be emerging, for example, in the systems proposed by Scott [20], Martin-Löf [16], and myself [5].

Since neither the realist nor constructivist point of view encompasses the other, there cannot be any present claim to a *universal foundation* for mathematics, unless one takes the line of rejecting all that lies outside the favored scheme. Indeed, *multiple foundations* in this sense may be necessary, in analogy to the use of both wave and particle conceptions in physics. Moreover, it is conceivable that still other kinds of theories of operations and collections will be developed as a result of further experience and reflection. I believe that none of these considerations affects the counter-thesis of this part, namely that foundations for structural mathematics are to be sought in theories of operations and collections (if they are to be sought at all).

In correspondence concerning the preceding Professor MacLane has raised several criticisms which I shall try to summarize and respond to here. Though I had tried to keep the argument as simple as possible, it now seems to me that elaboration on these points will help clarify certain issues; I am thus indebted to Professor MacLane for his timely rejoinder.

First, he says that the program of categorical foundations has made considerable progress (since the papers [13], [15], and [14]) via work on elementary theories of topoi by Lawvere, Tierney, Mitchell, Cole,

Osius, and others. He believes that this makes the discussion above out of date and beside the point.

Second, MacLane thinks that questions of psychological priority are 'exceedingly fuzzy' and subjective. Further, mathematicians are well known to have very different intuitions, and these may be strongly affected by training.

It is necessary to indicate the nature of the work on topoi before going into these points. A good introductory survey is to be found in MacLane [21]. Elementary topoi are special kinds of categories $\mathbb{C}$ which have features strongly suggested by the category of sets. (At the same time there are various other examples of mathematical interest such as in categories of sheaves.) In place of membership, one deals with morphisms $f: 1 \to X$ where 1 is a terminal object of $\mathbb{C}$. The requirements to be a topos include closure under such constructions on objects X, Y of $\mathbb{C}$ as product $X \times Y$ and exponentiation X^Y. The principal new feature is that $\mathbb{C}$ is required to have a 'subobject classifier' Ω whereby every subobject $S \to X$ of an object X corresponds uniquely to a 'member' of Ω^X. Thus Ω generalizes the role of the set of truth-values $\{0, 1\}$ and Ω^X generalizes the role of the power set of X for sets X. By means of additional axioms on topoi one may reflect more and more of the particularities of the category of sets. Indeed, (following Mitchell and Cole) Osius [22] gives two extensions ETS(Z) and ETS(ZF) of the elementary theory of topoi which are equivalent (by translation), respectively, to the theories of sets Z and ZF. (Incidentally, this work clarifies the relationship of Lawvere's theory in [13] to ZF.)

According to MacLane, the technical development just described shows that set-theoretical foundations and categorical foundations are entirely equivalent and hence that one cannot assign any logical priority to the former.

In response:

(i) My use of 'logical priority' refers not to relative strength of formal theories but to order of definition of concepts, in the cases where certain of these *must* be defined before others. For example, the concept of vector space is logically prior to that of linear transformation; closer to home, the (or rather some) notions of set and function are logically prior to the concept of cardinal equivalence. (By contrast, there are cases where there is no priority, e.g., as between Boolean algebras and Boolean rings.)

(ii)   On the other hand, 'psychological priority' has to do with natural order of understanding. This is admittedly 'fuzzy' but not always 'exceedingly' so. Thus one cannot understand abstract mathematics unless one has understood the use of the logical particles 'and', 'implies', 'for all', etc. and understood the conception of the positive integers. Moreover, in these cases formal systems do not serve to explain what is not already understood since these concepts are implicitly involved in understanding the workings of the systems themselves. This is not to deny that formal systems as well as informal discussions can serve to clarify meanings when there is ambiguity, e.g. as between classical and constructive usage.

(iii)   My claim above is that the general concepts of operation and collection have logical priority with respect to structural notions (such as 'group', 'category', etc.) because the latter are defined in terms of the former but not conversely. At the same time, I believe our experience demonstrates their psychological priority. I realize that workers in category theory are so at home in their subject that they find it more natural to think in categorical rather than set-theoretical terms, but I would liken this to not needing to hear, once one has learned to compose music.

(iv)   The preceding has to do with an order between concepts, according to which some of them appear to be more basic than others. There is in consequence an order between theories of these concepts. Namely, we choose certain systems *first* because they reflect our understanding (as well as one can formulate it) of basic conceptions; other systems may be chosen *later* because they are useful and reducible to the former. For example, axioms about real numbers reflect some sort of basic understanding, while axioms for non-standard real numbers are only justified by a relative consistency proof. This is in opposition to the (formalist) idea that equivalence of formal systems makes one just as good a foundation for a certain part of mathematics as another; it is contrary to our actual experience in the development and choice of such systems.

(v)   My neglect of the work on topoi was not due to ignorance but rather based on the conclusion that it was irrelevant to my argument. Indeed, since topoi are just special kinds of categories the objections here to a program for the categorical foundations of mathematics apply all the more to foundations via theories of topoi. It should be added that the axioms of ETS(Z) and ETS(ZF) were clearly obtained by tracing out just what was needed to secure the translations of Z and ZF in the language of topoi. This applies particularly to the replacement scheme and bears out my contention of the priority of set-theoretical concepts.

(vi) To avoid misunderstanding, let me repeat that I am *not* arguing for accepting current set-theoretical foundations of mathematics. Rather, it is that on the platonist view of mathematics something like present systems of set theory must be prior to any categorical foundations. More generally, on any view of abstract mathematics priority must lie with notions of operation and collection.

II

1. *Defects of Present Foundations for Category Theory*

Current set-theoretical foundations do not permit us to meet the following two requirements:

(R1) *form the category of all structures of a given kind, e.g. the category* $\mathbb{G}$ *of all groups, the category* $\mathbb{T}$ *of all topological spaces, the category* $\mathbb{C}$ *of all categories, etc.;*

(R2) *form the category* $\mathbb{B}^{\mathbb{A}}$ *of all functors from* $\mathbb{A}$ *to* $\mathbb{B}$ *when* $\mathbb{A}$, $\mathbb{B}$ *are any given categories.*

This is the main reason that MacLane [15] argues that set-theoretical foundations are inadequate. As described in [15] there are two means at present to reformulate (R1), (R2), and other constructions of category theory in set-theoretically acceptable terms. Briefly, these are:

(i) The Grothendieck method of 'universes'. A *universe U* is a set of sets satisfying strong closure conditions, including closure under exponentiation and under Cartesian product more generally. (The sets of rank $<\alpha$ form a universe when α is inaccessible.) It is assumed that for every universe U there is another U' which contains U as member. With each universe and notion of a particular kind of structure is associated the category of all such structures in U; this is a member of U' when $U \in U'$. For example, we can speak of the category $\mathbb{G}_U$ of all groups in U; $\mathbb{G}_U$ is then an element of U' when $U \in U'$. Thus (R1) is satisfied only in a relative way. For any U and categories $\mathbb{A}, \mathbb{B} \in U$ the category $\mathbb{B}^\mathbb{A}$ also belongs to U. The method of universes provides a reduction of category theory to ZF + 'there exist infinitely many inaccessibles'.[6]

(ii) The Eilenberg-MacLane reduction to the BG theory of sets and classes, via the distinctions between 'small' and 'large'. A category is said to be *small* if it is a set, otherwise *large*. For example, the category $\mathbb{G}$ of all sets which are groups is large. The functor category $\mathbb{B}^\mathbb{A}$ exists only under the hypothesis that $\mathbb{A}$ is small, so (R2) is not satisfied. Also (R1) may be said to be satisfied only partially since, e.g. there is no category of all large groups.[7]

In addition to the inadequacies just explained there is dissatisfaction with the two schemes in that the restrictions employed seem mathematically unnatural and irrelevant. Though bordering on the territory of the paradoxes, it is felt that the notions and constructions involved in (R1), (R2) have evolved naturally from ordinary mathematics and do not have the contrived look of the paradoxes. Thus it may be hoped to find a way which gives them a more direct account.

It must be said that there is no urgent or compelling reason to pursue foundations of unrestricted category theory, since the schemes (i), (ii) (or their variants and refinements) serve to secure all practical purposes. Speaking in logical terms, one can be sure that any statement of set theory proved using general category theory has a set-theoretical proof (to be more precise, a proof in ZF if we follow scheme (ii), since BG is conservative over ZF). The aim in seeking new foundations is mainly as a problem of logical interest motivated largely by aesthetic considerations (or rather by the inaesthetic character of the present solutions).

2. The Search for Foundations of Unrestricted Category Theory

The situation here is analogous to several classical problems of foundations in mathematics, when one employed objects conceived beyond ordinary experience, such as infinitesimals, imaginary numbers, and points at infinity. The restrictions by set-theoretical distinctions of size in category theory are analogous to the employment of the "ε, δ" language for the foundations of the calculus. On the other hand, the consistency proofs for the complex number system and for projective geometry justified a direct treatment of the latter ideas.[8] We have in these cases *extensions* of familiar systems, namely of the real number system and Euclidean geometry, resp. But it is to be noted that in each case a price was paid: certain properties of the familiar objects were sacrificed in the process. When passing from the reals to the complex numbers one must give up the ordering properties; when passing from Euclidean to the projective space, metric properties must be given up.

In the case of category theory the idea would be to formulate a system in which one could obtain familiar collections and carry out familiar operations, such as the following:

(R3)   *form the set N of natural numbers, form ordered pairs* (a, b), *form* $A \cup B$, $A \cap B$, $A - B$, $A \times B$, B^A, $\mathscr{P}(A)$, $\bigcup_{x \in A} B_x$, $\bigcap_{x \in A} B_x$, $\prod_{x \in A} B_x$, *etc.*

Since new 'unlimited' collections would have to be objects of the theory and we would have *self-application* by (R1) we could not expect to carry over all familiar laws. The problem then would be to select an appropriate *part of* (R3) and such familar laws as *extensionality*, to see which should be extended. Here there is no clear criterion for selection, but one wants to preserve ordinary mathematics as much as possible.

In the past few years I have experimented with several formal systems to achieve this purpose. One system based on an extension of Quine's stratification was reported in [4] and a write-up of the work was informally distributed. However, the paper has not been sent for publication for two reasons: (i) the system introduced turned out to be very similar to one that had been discovered independently by Oberschelp [19], and (ii) it was not completely successful for the intended purposes. In particular, though one could carry out (R1), (R2), and (R3) to a certain extent (e.g. to form $A \cup B$, $A \cap B$, $A - B$, $A \times B$, B^A) one could not carry out $\prod_{x \in A} B_x$. However, Cartesian product is a very important operation used to build structures, to verify completeness of concrete categories, etc., and so ought not to be given up.

Last year I found another theory which does somewhat better and also has a more intrinsic plausibility. This was suggested by my earlier work [5] on constructive theories of operations and classifications (collections). The point in the preceding work had been to deal with those notions conceived of as given *intensionally* by presentations of rules respectively defining properties, i.e. in both cases as certain kinds of syntactic expressions. In a universe which contains objects which 'code' syntactic expressions, we can have self-application. Of course, this may be accidental, depending on the coding. The problem is to arrange instead mathematically significant instances of self-application. Where [5] had used *partial operations* and *total classifications*, the new theory [6] achieved its aims by also using *partial classifications*. Intuitively, suppose c is a classification, i.e. an object given by a property φ_c. In testing whether or not $\varphi_c(x)$ holds we are in some cases led into a circle. Write $(x\eta c)$ if it can be *verified* (in some sense) that φ_c holds of x and $(x\bar{\eta}c)$ if it can be *verified* that φ_c does *not* hold of x. Here we conceive of verification as a possibly transfinite process, e.g. $\forall y \varphi(y)$ is verified if for each b, $\varphi(b)$ is verified. Not every attempt to verify leads to a conclusion, e.g. $\varphi_c(c)$ cannot be verified if $\varphi_c(x)$ is $x\eta x$ or $(x\bar{\eta}x)$. Write $\Box\varphi$ if φ is verified. Thus $x\bar{\eta}c \leftrightarrow \Box\neg(x\eta c)$. Pursuing these ideas leads one to a comprehension scheme of the following form:

(*) $\exists c \, \forall x [x\eta c \leftrightarrow \Box\varphi(x)] \wedge [x\bar{\eta}c \leftrightarrow \Box\neg\varphi(x)],$

where no restrictions are placed on φ in (*). Such a scheme was first shown to be consistent by Fitch [7, 8] and a closely related scheme in ordinary predicate calculus was shown consistent by Gilmore [9]. However, to arrange (R1), (R2) and as much as possible of (R3), it appears that somewhat more must be built in. Namely, there must be a theory of operations in terms of which one defines such classifications as b^a; further, these operations should be extensive enough to take both operations and classifications as arguments or values. In the next section I shall describe a theory resulting from this combination of ideas.

3. A Non-extensional Theory of Partial Operations and Classifications

3.1. Let T be any theory in any logic including the classical first order predicate calculus, and which contains a symbol 0 and operation symbols ′ and (,) and which proves:

A1. $x' \neq 0$.

A2. $x' = y' \rightarrow x = y$.

A3. $(x_1, y_1) = (x_2, y_2) \rightarrow x_1 = x_2 \wedge y_1 = y_2$.

We write $\mathscr{L}$ for the language of T.

Our first step is to extend this to a theory of partial operations. We adjoin a three-place predicate symbol App (x, y, z) which is also written $xy \simeq z$; App is then denoted $\simeq$. The intuitive interpretation is that the operation x is defined at y with value z. We write $\mathscr{L}(\simeq)$ for the language extended by the symbol $\simeq$. Note that expressions compounded by xy are not terms of $\mathscr{L}(\simeq)$. Extend the language by a binary operation symbol for application; by a *pseudo-term* t we mean a term of the extended language. The meaning of $t \simeq z$ for any pseudo-term t is defined inductively as a formula of $\mathscr{L}(\simeq)$. Then $(t\!\downarrow)$ is written for $\exists z(t \simeq z)$ and $t_1 \simeq t_2$ for $\forall z[t_1 \simeq z \leftrightarrow t_2 \simeq z]$. Finally, $xy_1 \cdots y_n$ is written for $(\ldots(xy_1)\ldots)y_n$.

The theory $T_{\simeq}$ contains the following axioms:

A4.   (*Unicity*). $xy \simeq z_1 \wedge xy \simeq z_2 \rightarrow z_1 = z_2$.

A5.   (*Explicit definition*). For each pseudo-term t with variables $x, y_1, ..., y_n$

$$\exists f \forall y_1, ..., y_n, x[fy_1 \cdots y_n\!\downarrow \wedge fy_1 \cdots y_n x \simeq t].$$

In addition, $T_{\simeq}$ contains axioms guaranteeing the existence of p_1, p_2, d, and e satisfying:

A6.   (*Projections*) $p_1(x, y) \simeq x \wedge p_2(x, y) \simeq y$.

A7.   (*Definition by cases*)

$$(x = y \rightarrow dxyab \simeq a) \wedge (x \neq y \rightarrow dxyab \simeq b).$$

A8.   (*Quantification*)

$$[ex \simeq 0 \leftrightarrow \exists y(xy \simeq o)] \wedge [ex \simeq 1 \leftrightarrow \forall y \exists z(xy \simeq z \wedge z \neq 0)].$$

If we fix any f associated with t by (5), we write $\lambda x \cdot t$ or $\lambda x \cdot t[x]$ for $fy_1 \ldots y_n$. The idea of (5) is that $(\lambda x \cdot t)$ exists because it names a rule, whether or not $t[x]$ is defined. Thus we have

$$(\lambda x \cdot t[x]) \downarrow \wedge (\lambda x \cdot t[x])x \simeq t.$$

As special cases of A5 we get existence of the *identity i* satisfying: $ix \simeq x$; the *combinators k, s* satisfying: $kxy \simeq x$, $sxy \downarrow \wedge sxyz \simeq xz(yz)$; the *successor operation* s_1 satisfying: $s_1 x \simeq x'$; and the *pairing operator p* satisfying $pxy \simeq (x, y)$. A single scheme of explicit definition by certain quantified formulas may also be given which implies A5–A8 and in such a way as to insure the following.

THEOREM 1. $T_{\simeq}$ *is a conservative extension of* T.

A proof of this for the given axioms A5–A8 may be obtained by associating with any model $\mathfrak{M} = (M, \ldots)$ of T a generalization of recursion theory, essentially *prime computability in* $\exists^{(M)}$ defined by Moschovakis [17]. The interpretation of $xy \simeq z$ is $\{x\}(y) \simeq z$ in this generalization.

We can already define partial classifications in T, as those induced by partial characteristic functions: say, $x \rho c \leftrightarrow cx \simeq 0$ and $x \bar{\rho} c \leftrightarrow cx \simeq 1$. Let $cx \simeq 0$ for all x; then $\forall x(x \rho c)$, i.e. c is a universal classification. Write $x: u \rightarrow w$ for $\forall y[y \rho u \rightarrow (xy \downarrow) \wedge (xy) \rho w]$. It is easily proved that

$$\rightarrow \exists d \forall x[x \rho d \leftrightarrow x: c \rightarrow c],$$

i.e. there is no classification which consists exactly of the total operations. For this reason we now pass to a richer theory of classifications.

3.2.   The second step is to expand the language $\mathscr{L}(\simeq)$ by a pair of new binary relation symbols η and $\bar{\eta}$; we further adjoin a new unary propositional operator $\square$; $x \eta c$, $x \bar{\eta} c$ and $\square \varphi$ may be read informally as suggested at the end of Section 2. The extended language is denoted $\mathscr{L}(\simeq, \eta, \bar{\eta})$.

The logic is that of $\mathscr{L}$ augmented by the modal system S4 + BF with rules and axioms as follows (cf. [10] for more details):

B1 from φ infer $\square\varphi$.

B2. $\square\varphi \to \varphi$.

B3. $\square\varphi \to \square\square\varphi$.

B4. $\square(\varphi \to \psi) \to (\square\varphi \to \square\psi)$.

B5. $\forall x \square\varphi(x) \to \square\forall x\varphi(x)$.

We have also the following axioms for atomic formulas special to our situation.

B6. $\varphi \to \square\varphi$ for each atomic formula.

B7. $\to\varphi \to \square\to\varphi$ for each atomic formula of $\mathscr{L}(\simeq)$.

B8. $x\bar{\eta}c \leftrightarrow \square\to(x\eta c)$.

We write $S \vdash \varphi$ if for some $\psi_1, ..., \psi_n \in S$ we can derive $\psi_1 \wedge \cdots \wedge \psi_n \to \varphi$ in this logic. Note that if $\psi \in S$ we cannot in general apply B1 to get $S \vdash \square\psi$. Using B6, B7, we may derive from these axioms:

$$\varphi \leftrightarrow \square\varphi \text{ for each formula } \varphi \text{ of } \mathscr{L}(\simeq).$$

Now the axioms of the main theory $\hat{T}$ are those of $T_\simeq$ together with the following:

D.   (*Disjointness*) $\to (x\eta a \wedge x\bar{\eta}a)$.

C.   (*Comprehension*) For each formula φ of $\mathscr{L}(\simeq, \eta, \bar{\eta})$ with free variables among x, a_i $(i \leqslant n)$,

$$\exists f \forall a_1, ..., a_n \exists c[fa_1 ... a_n \simeq c \wedge \forall x(x\eta c \leftrightarrow \square\varphi)$$
$$\wedge \forall x(x\bar{\eta}c \leftrightarrow \square\to\varphi].$$

THEOREM 2. $\hat{T}$ *is a conservative extension of* $T_\simeq$, *hence of* T.

The idea of the proof is to consider any model $\mathfrak{M} = (M, ..., \simeq)$ of $T_\simeq$ and to apply Kripke semantics to the collection $\mathscr{K}(\mathfrak{M})$ consisting of all $(\mathfrak{M}, R, \bar{R})$ where R, $\bar{R}$ are disjoint binary relations on $\mathfrak{M}$, ordered by $(\mathfrak{M}, R, \bar{R}) \leqslant (\mathfrak{M}, S, \bar{S}) \Leftrightarrow R \subseteq \bar{R}$ and $S \subseteq \bar{S}$. In particular, $(\mathfrak{M}, R, \bar{R}) \vDash \Box \varphi(\mathbf{a}) \Leftrightarrow$ for all $(\mathfrak{M}, S, \bar{S}) \geqslant (\mathfrak{M}, R, \bar{R})$ we have $(\mathfrak{M}, S, \bar{S}) \vDash \varphi(\mathbf{a})$. Next use 0, $'$, and $(\ ,\ )$ to set up a coding in M of arbitrary formulas so that $\ulcorner \varphi \urcorner \in M$ represents φ. Then put $f_\varphi y_1 \cdots y_n \simeq (\ulcorner \varphi \urcorner, y_1, ..., y_n)$. Define η_α, $\bar{\eta}_\alpha$ by transfinite recursion $\eta_0 = \bar{\eta}_0 =$ empty. We take $x \eta_{\alpha+1} c \Leftrightarrow$ for some $\varphi(x, \mathbf{y})$, $c = (\ulcorner \varphi \urcorner, \mathbf{y})$ and $(\mathfrak{M}, \eta_\alpha, \bar{\eta}_\alpha) \vDash \Box \varphi(x, \mathbf{y})$; $x \bar{\eta}_{\alpha+1} c \Leftrightarrow$ for some $\varphi(x, \mathbf{y})$, $c = (\ulcorner \varphi \urcorner, \mathbf{y})$ and $(\mathfrak{M}, \eta_\alpha, \bar{\eta}_\alpha) \vDash \Box \to \varphi(x, \mathbf{y})$. Let $\eta_\lambda = \bigcup_{\alpha < \lambda} \eta_\alpha$, $\bar{\eta}_\alpha = \bigcup_{\alpha < \lambda} \bar{\eta}_\alpha$ for λ a limit number. Finally take $\eta = \eta_\alpha$ and $\bar{\eta} = \bar{\eta}_\alpha$ where $\eta_\alpha = \eta_{\alpha+1}$, $\bar{\eta}_\alpha = \bar{\eta}_{\alpha+1}$. It may be shown that $(\mathfrak{M}, \eta, \bar{\eta})$ is a model of $\hat{T}$ (in the extended logic applied to $\mathfrak{K}(\mathfrak{M})$).

It should be noted that extensionality can actually be disproved for total functions in $T_\simeq$ and for total classifications in $\hat{T}$; this goes by a diagonal argument (cf. [6] 3.6 and 4.7).

3.3. The special case of $T = ZF$ in the language of $\in$, $=$ (with 0, $'$, $(\ ,\ )$ defined as usual) is useful to consider in comparison with present set-theoretical foundations of category theory. Here it is natural to strengthen $\hat{T}$ by the scheme

$$S \ (Separation) \quad \exists b \forall x [x \in b \leftrightarrow x \in a \land \varphi],$$

for each formula φ of $\mathscr{L}(\simeq, \eta, \bar{\eta})$. Let $T^\# = \hat{T} + S$.

THEOREM 3. $T^\#$ is a conservative extension of $T_\simeq$, hence of T.

The proof uses standard models for any fragment of $T^\#$ and the reflection principle for ZF.

3.4. We now draw some basic consequences in these theories. Call φ *persistent* if $(\varphi \to \Box \varphi)$ is provable, and *invariant* if both φ and $\to \varphi$ are persistent. It is easily shown that if both η and $\bar{\eta}$ have only positive occurrences in φ then φ is persistent. With each φ in the language of 1st order predicate calculus is associated a pair of formulas φ^+ and φ^- which are both positive w.r. to η and $\bar{\eta}$ and which are approximations in a certain sense to φ and $\to \varphi$, resp. For example, when φ is in prenex disjunctive normal form, to obtain φ^+ we replace each negated η by $\bar{\eta}$ and each negated $\bar{\eta}$ by η. Then $\vdash (\varphi^+ \to \varphi)$ and $\vdash (\varphi^- \to \to \varphi)$. Under suitable circumstances we actually have $\varphi^+ \leftrightarrow \varphi$ and $\varphi^- \leftrightarrow \rightarrow \varphi$. For example, suppose $\varphi(x, a_1, a_2)$ is $x\eta a_1 \wedge \rightarrow(x\eta a_2)$. Then φ^+ is $x\eta a_1 \wedge x\bar{\eta}a_2$, and φ^- is $x\bar{\eta}a_1 \vee x\eta a_2$. Now if a_1, a_2 are total, i.e. $x\bar{\eta}a_i \leftrightarrow \rightarrow (x\eta a_i)$ we have $\varphi^+ \leftrightarrow \varphi$ and $\varphi^- \leftrightarrow \rightarrow \varphi$. In these cases $(\varphi \leftrightarrow \Box\varphi)$ and $(\rightarrow\varphi \leftrightarrow \Box \rightarrow \varphi)$.

For each $\varphi(x, a_1, ..., a_n)$, take any f satisfying the comprehension scheme C of $\hat{T}$. We write $\hat{x}\varphi(x, a_1, ..., a_n)$ or $\hat{x}\varphi(a, \mathbf{a})$ for $fa_1 \cdots a_n$ and $\lambda \mathbf{a} \cdot \hat{x}\varphi(x, \mathbf{a})$ for f. Hence

$$(1) \qquad [x\eta\hat{x}\varphi(x, \mathbf{a}) \leftrightarrow \Box\varphi(x, \mathbf{a})] \quad \text{and}$$
$$[x\bar{\eta}\hat{x}\varphi(x, \mathbf{a}) \leftrightarrow \Box \rightarrow \varphi(x, \mathbf{a})]$$

are provable in $\hat{T}$. Note we can write $[x\eta\hat{x}\varphi(x, \mathbf{a}) \leftrightarrow \varphi(x, \mathbf{a})]$ only when $\mathbf{a}$ is such that $\varphi(x, \mathbf{a})$ is persistent. One must be cautious to observe whether this is the case.

To begin with, define

$$(2) \qquad CL = \hat{x}\forall z[z\eta x \vee z\bar{\eta}x].$$

Then $x\eta CL$ iff x is a *total classification*. CL is itself not total (by Russell's argument). Next we proceed to define

$(3) \qquad$ (i) $\quad V = \hat{x}(x = x), \quad \Lambda = \hat{x}(x \neq x).$

(ii) $\quad \{a, b\} = \hat{x}(x = a \vee x = b).$

(iii) $\quad a \cup b = \hat{x}(x\eta a \vee x\eta b).$

(iv) $\quad a \cap b = \hat{x}(x\eta a \wedge x\eta b).$

(v) $\quad -a = \hat{x} \rightarrow (x\eta a).$

(vi) $\quad a \times b = \hat{x}\exists x_1, x_2(x = (x_1, x_2) \wedge x_1\eta a \wedge x_2\eta b).$

(vii) $\quad b^a = \hat{x}(x: a \rightarrow b) = \hat{x}\forall u[u\eta a \rightarrow \exists v(xu \simeq v \wedge v\eta b)].$

It may be seen, e.g. that $x\eta(a \cap b) \leftrightarrow x\eta a \wedge x\eta b$ because of the positivity of the defining condition; further if $a\eta CL$, $b\eta CL$ then $(a \cap b)\eta CL$. On the other hand, $x\eta(-a) \leftrightarrow x\bar{\eta}a$ by (1).

In the case of the function classification b^a, it is seen that $x\eta b^a \leftrightarrow (x: a \rightarrow b)$ whenever a is total; moreover, $a\eta CL \wedge b\eta CL \rightarrow (b^a\eta CL).$

Note that for each of these operations we actually have an element that represents it, e.g. $f = \lambda a$, $b(a \cap b)$ gives $fab = a \cap b$.

We may also define extended operations of union, intersection, sum and product as follows:

(4)   (i) $\bigcup a = \hat{x}\exists z(x\eta z \wedge z\eta a).$

         (ii) $\bigcap a = \hat{x}\forall z(z\bar{\eta}a \vee x\eta z).$

         (iii) $\sum_{z\eta a} (gz) = \hat{x}\exists z, w, v[x = (z, w) \wedge z\eta a \wedge gz \simeq v \wedge w\eta v].$

         (iv) $\prod_{z\eta a} (gz) = \hat{x}\forall z[z\bar{\eta}a \vee \exists w, v(xz \simeq w \wedge gz \simeq v \wedge w\eta v)].$

It is clear that $\bigcup$ and $\sum$ behave reasonably in general, that $a\eta CL$ and $a \subseteq CL \rightarrow (\bigcup a)\eta CL$ and that $a\eta CL$ and $h: a \rightarrow CL$ implies $\left(\sum_{z\eta a} (hz)\right)\eta CL$. For $\bigcap$ we get the appropriate defining condition when $a\eta CL$, and we have $(\bigcap a)\eta CL$ when also $\mathbf{a} \subseteq CL$. For $\prod_{z\eta a} (hz)$ we get the appropriate defining condition when $a\eta CL$ and this product is total when $h: a \rightarrow CL$.

It is possible to define power classifications by

(5) $\mathscr{P}(a) = \hat{x}[x\eta CL \wedge \forall u(u\bar{\eta}x \vee u\eta a)].$

Thus for $x\eta CL$, $x\eta\mathscr{P}(a) \leftrightarrow x \subseteq a$. But $\mathscr{P}(a)$ is not in general total even if a is total, e.g. not for $\mathscr{P}(V)$.

If e is an equivalence relation, its equivalence classes may be defined by $[z] = \hat{x}[(x, z)\eta]$. However, since we lack extensionality we cannot conclude that $(x, z)\eta e \leftrightarrow [x] = [z]$, only that $(x, z)\eta e \leftrightarrow [x] \equiv [z]$ where $u \equiv v \leftrightarrow \forall w(w\eta u \leftrightarrow w\eta v)$. But this shows at any rate that all equivalence relations can be reduced to the single one of $\equiv$. Though extensionality is generally thought to be essential for mathematics it is dispensable if we return to an older manner of speaking. When dealing with structures $\mathfrak{A} = (a, \ldots)$ we usually want to consider also some 'equality' relation defined on $\mathbf{a}$, i.e. a congruence relation for the structure $\mathfrak{A}$. This is standard in constructive mathematics (cf. Bishop [1]) and is only slightly complicating; cf. also Section 4 below.

3.5. We have satisfied much of (R3) in $\hat{T}$. However, there remains the question of introducing the natural numbers as a total classification. In other words, we want the existence of an object N satisfying the scheme

(NN) (i) $N\eta CL$.

  (ii) $0\eta N \wedge \forall x(x\eta N \to x'\eta N)$.

  (iii) $\varphi(0) \wedge \forall x[\varphi(x) \to \varphi(x')] \to \forall x(x\eta N \to \varphi(x))$

  for all formulas φ of $\mathscr{L}(\simeq, \eta, \bar{\eta})$.

It is not difficult to modify the proof of Theorem 2 to obtain the following:

THEOREM 4. $\hat{T}+NN$ *is conservative over* $T_{\simeq}$, *hence over* T.

An alternative is to apply Theorem 2 to T regarded as expressed in a logic stronger than 1st order predicate calculus, by adjoining a 'quantifier' which determines inductive generation in general. For this see [6] Section 4c.

4. *Structure of structures and unrestricted category theory in* $\hat{T}$

It is here convenient to use letters of any style as variables ranging over the objects of the theory $\hat{T}$, including Latin, Greek and German letters, as well as the letters $\mathbb{A}, \mathbb{B}, \mathbb{C}, \dots$. We proceed informally but in such a way that all the work can be formalized in $\hat{T}$. For each $k \geq 1$ define a^k by $a^1 = a$, $a^{k+1} = a \times a^k$. Then for each $\tau = ((k_1, \dots, k_n), (l_1, \dots, l_m), p)$ we define $\mathfrak{A}$ to be a *structure of type* τ if it is of the form:

(1) $\mathfrak{A} = (a, r_1, \dots, r_n, f_1, \dots, f_m, 0_1, \dots, 0_p)$

  where $r_i \subseteq a^{k_i}$ $(1 \leq i \leq n)$, $f_i : a^{l_i} \to a$ and $0_i \eta a$.

Here **a** and each r_i are in general only partial classifications. We write $\text{Str}_\tau(\mathfrak{A})$ if (1) holds and $!\text{Str}_\tau(\mathfrak{A})$ if $\mathfrak{A}$ is a *total structure*, i.e. if

(2) $a\eta CL$ and $r_i\eta CL$ $(1 \leq i \leq n)$.

We can form

(3) $\quad !S_\tau = \hat{\mathfrak{A}} \, !Str_\tau \, (\mathfrak{A})$,

so that $!S_\tau$ is the partial classification of all total structures of type τ.

As pointed out in 3.4, since we lack extensionality, in practice each structure $\mathfrak{A}$ must carry along a relation e which acts as an *equality relation for* $\mathfrak{A}$. Thus, for example, a semi-group (associative binary system) is a structure $\mathfrak{A} = (a, e, f)$ where $e \subseteq a^2$, $f: a^2 \to a$, $(f(x, f(y, z)), f(f(x, y)z))\eta e$ and e is an equivalence relation which preserves f. We have a formula Sem Grp $(\mathfrak{A})$ which expresses that $\mathfrak{A}$ is a semi-group and !Sem Grp $(\mathfrak{A})$ which expresses that $\mathfrak{A}$ is a total semi-group. There is then a classification $!SG = \hat{\mathfrak{A}} \, !Sem \, Grp \, (\mathfrak{A})$ *of all total semi-groups.* Write $x \equiv_{\mathfrak{A}} y$ for $(x, y)\eta e$. Given two semi-groups $\mathfrak{A} = (a, \equiv_{\mathfrak{A}}, f)$ and $\mathfrak{B} = (b, \equiv_{\mathfrak{B}}, g)$, a *homomorphism* $\alpha: \mathfrak{A} \to \mathfrak{B}$ is a triple $(h, \mathfrak{A}, \mathfrak{B})$ where h is an operation $h: a \to b$ such that $x \equiv_{\mathfrak{A}} y \to hx \equiv_{\mathfrak{B}} hy$. We shall write αx for hx. An isomorphism is a pair of homomorphisms $\alpha: \mathfrak{A} \to \mathfrak{B}$, $\beta: \mathfrak{B} \to \mathfrak{A}$ such that $x \equiv_{\mathfrak{A}} \beta \alpha x$ for $x\eta a$ and $z \equiv \alpha \beta z$ for $z\eta b$. We write $\mathfrak{A} \cong \mathfrak{B}$ if there exists such an isomorphism. The relation $E = (\mathfrak{A}, \mathfrak{B}) \cdot [\mathfrak{A} \cong \mathfrak{B}]$ is an equivalence relation on SG.

Next, given semi-groups $\mathfrak{A}, \mathfrak{B}$, define $\mathfrak{A} \times \mathfrak{B} = (a \times b, \equiv_{\mathfrak{A} \times \mathfrak{B}}, f \times g)$ as usual. We thus have an operation q such that for any $\mathfrak{A}, \mathfrak{B}$, $q\mathfrak{A}\mathfrak{B} \simeq \mathfrak{A} \times \mathfrak{B}$. By the natural isomorphism $\mathfrak{A} \times (\mathfrak{B} \times \mathfrak{C}) \cong (\mathfrak{A} \times \mathfrak{B}) \times \mathfrak{C}$ we conclude that q is an associative operation up to $\cong$ on semi-groups. Hence we can prove,

(4) $\quad$ Sem Grp $(\mathfrak{S}) \quad$ where $\quad \mathfrak{S} = (!SG, \cong, q)$.

In other words the structure of all total semi-groups with the operation of products forms 'the semi-group of all total semi-groups'.

This is a paradigm for the treatment of other algebraic notions in $\hat{T}$ together with an illustration of the possibilities of self-application. Proceeding similarly we can define the notions: Grp $(\mathfrak{A})$ ($\mathfrak{A}$ is a group), !Grp $(\mathfrak{A})$ ($\mathfrak{A}$ is a total group), $!G = \hat{\mathfrak{A}} \, !Grp \, (\mathfrak{A})$ (the partial classification of all total groups), $\alpha: \mathfrak{A} \to \mathfrak{B}$ (α is a homomorphism from the group $\mathfrak{A}$ to the group $\mathfrak{B}$), Hom $(\mathfrak{A}, \mathfrak{B}) = \hat{\alpha}(\alpha: \mathfrak{A} \to \mathfrak{B})$ for $\mathfrak{A}, \mathfrak{B}\eta!G$, Hom$_{!G} = \hat{\alpha}[\exists\mathfrak{A}, \mathfrak{B}(\mathfrak{A}\eta!G \wedge \mathfrak{B}\eta!G \wedge \alpha: \mathfrak{A} \to \mathfrak{B})]$ and $\alpha \circ \beta \equiv \gamma$ (γ is a composition of α, β). Similarly we can define: Cat ($\mathfrak{A}$) ($\mathfrak{A}$ is a category), !Cat ($\mathfrak{A}$) ($\mathfrak{A}$ is a total category, $!C = \hat{\mathfrak{A}}$!Cat ($\mathfrak{A}$) (the partial classification of all total categories), $\varphi: \mathfrak{A} \to \mathfrak{B}$ (φ is a functor from $\mathfrak{A}$ to $\mathfrak{B}$), $\mathfrak{B}^{\mathfrak{A}} =$ Funct ($\mathfrak{A}, \mathfrak{B}$) $= \hat{\varphi}(\varphi: \mathfrak{A} \to \mathfrak{B})$ for $\mathfrak{A}$, $\mathfrak{B}$ total categories, Funct$_{!C} =$ $\hat{\varphi}[\exists \mathfrak{A}, \mathfrak{B}(\mathfrak{A}\eta!C \wedge \mathfrak{B}\eta!C \wedge \varphi: \mathfrak{A} \to \mathfrak{B})]$ and $\varphi \circ \psi \equiv \vartheta$ (ϑ is a composition of φ, ψ). From $!G$, Hom$_{!G}$ we can assemble a category $\mathbb{G}$ which is the *category of all total groups*. $\mathbb{G}$ itself is not a total structure, but we have

(5)   Cat ($\mathbb{G}$).

Similarly, we can assemble from $!C$, Funct$_{!C}$ a category $\mathbb{C}$ which is the *category of all total categories*, so that

(6)   Cat ($\mathbb{C}$).

In this way we achieve a form of the requirement (R1) in $\hat{\mathrm{T}}$.

Now for (R2), given any *total categories* $\mathbb{A}$, $\mathbb{B}$, we can form as above Funct ($\mathbb{A}, \mathbb{B}$) – which are the objects of $\mathbb{B}^A$ considered as a category; its morphisms are the natural transformations between functors. We obtain

(7)   !Cat ($\mathbb{A}$) $\wedge$!Cat ($\mathbb{B}$) $\to$!Cat ($\mathbb{B}^A$)

as a consequence of the closure of CL under exponentiation. In this way requirement (R2) is met. More generally we can form $\mathbb{B}^A$ with reasonable properties when it is merely assumed that $\mathbb{A}$ is total:

(8)   !Cat ($\mathbb{A}$) $\wedge$ Cat ($\mathbb{B}$) $\to$ Cat ($\mathbb{B}^A$).

A basic structure to consider is $\mathbb{CL}$, the *category of all total classifications*. This has the property that for any objects A, B of $\mathbb{CL}$, Hom $(A, B) = B^A$ is also total. More generally, call a category $\mathbb{A}$ *locally total* if for any objects $\mathfrak{A}$, $\mathfrak{B}$ of $\mathbb{A}$, Hom$_{\mathbb{A}}$ ($\mathfrak{A}, \mathfrak{B}$) is total. Then $\mathbb{CL}$, $\mathbb{G}$, $\mathbb{C}$, etc. are all locally total. With any locally total $\mathbb{A}$ and object $\mathfrak{A}$ of $\mathbb{A}$ is associated the Yoneda functor

(9) $h^{\mathfrak{A}}: \mathbb{A} \to \mathbb{CL}$,

which is defined on objects by $h^{\mathfrak{A}}(\mathfrak{B}) = \mathrm{Hom}_A (\mathfrak{A}, \mathfrak{B})$ and is defined on morphisms in an obvious way. Now *Yoneda's Lemma* (YL) may be expressed in $\hat{T}$ as follows: *If A is locally total and φ is any functor from A to CL then for each object $\mathfrak{A}$ of A, φ is in $1-1$ correspondence with the class of natural transformations from $h^{\mathfrak{A}}$ to φ.*

Call a category *complete* if it is closed under equalizers and products $\prod\limits_{i\eta c} a_i$ where $c\eta CL$. It may be shown that for any total category A, the category CL^A is complete. Further the Yoneda map is an embedding of A into CL^A. Hence every total category can be embedded in a complete (non-total) category.

Suppose we are working over set theory, i.e. $T = ZF$ or an extension of ZF. In this case we work in $T^{\#}$ instead of $\hat{T}$. The objects of principal interest are structures on sets a; these have associated classifications $\bar{a} = \hat{x}(x \in a)$. Denote by Set the classification consisting of all $\bar{a}$; we also call a classification *small* if it belongs to Set. Note that Set is total: $(\text{Set})\eta CL$. We can form a category from Set which we denote CL_s, *the category of all small classifications;* then further we can form G_s, *the category of all small groups,* C_s, *the category of all small categories,* etc. Each of these is a total category and is locally small in the usual sense. In this respect one considers *small-completeness* of a category, i.e. closure under $\prod\limits_{i\eta c} a_i$ for c small. A question of interest would be whether we can associate with any locally small category a small-completion.

These examples relate to several mentioned by MacLane [15] Section 7 as being problematic for the present set-theoretic accounts of category theory. It is evident that some of the hoped-for freedom is gained by passing to theories like $\hat{T}$ or $T^{\#}$. It is true that statements of results must now make distinctions between being *partial* and *total* which previously were made between being *large* and *small*. However, this is no disadvantage if one takes the objects of primary interest to be the categories of small objects such as CL_s, G_s, C_s, etc., all of which are total.

To conclude I wish to emphasize that the kind of expansion of accepted language pursued here is theoretically useful *as a matter of convenience.* That is, the theorems of Section 3 are conservative extension results, which permit us to extend an already accepted theory T to $\hat{T}$ (or $T^{\#}$) without getting new theorems in the language of T. This parallels the classical cases such as introduction of complex numbers. However, there is not yet evidence that expansions such as $\hat{T}$ have any significant mathematical advantage (comparable, say, to the use of complex numbers to obtain results about the real numbers). A further pursuit of the mathematics involved is needed in addition to the logical and aesthetic considerations to guide us to a fully satisfactory foundation of unrestricted structure theory.

Stanford University

NOTES

* Text of a talk for a Symposium on Category Theory in the 5th International Congress of Logic, Methodology and Philosophy of Science (London, Ontario, Aug. 27–Sept. 2, 1975). Research supported by NSF grant MPS74-07505 A01.
[1] This is detailed in [6] and in a succeeding paper which is in preparation.
[2] There are some problems about Lawvere's theory which have been raised by Isbell [11] and others. However, these do not affect its guiding point of view nor the viability of some such theory.
[3] One should mention particularly those raised by Kreisel in his appendix to [3] and in his review [12].
[4] In addition to its appearance in a certain way in the discussions of ftn. 4, I have been told by P. Martin-Löf that he has also raised similar objections.
[5] This is disputed by Myhill [18], which attempts to give an extensional constructive set theory.
[6] Some refinements of this idea should also be mentioned. In [14] MacLane showed that one universe is enough. In [3] I made use of the reflection principle for ZF to show that weaker closure conditions on U suffice for most work.
[7] It is indicated in [12] how to modify Bernays' relation η for the theory of sets and classes (read: properties) so that, e.g., G can also be construed as the category of all large groups. However, this modification does not satisfy (R2). It is also suggested by Kreisel in [12] that one look at 'suitably indexed collections of functors (which can be defined in BG)' rather than 'the' functor category. One development of this idea of getting around (R2) was carried out by the author in some unpublished seminar notes 'Set-theoretical formulation of some notions and theorems of category theory' (Stanford, October 1968). While it is viable and in one way more fitting to a thorough-going algebraic-axiomatic approach to mathematics, in another way it goes against the grain *never* to be able to talk about 'the' category of all functors between any two large categories.
[8] Robinson's non-standard analysis may be viewed as an attempt to justify direct use of infinitesimals in the calculus.

BIBLIOGRAPHY

[1] E. Bishop: 1967, *Foundations of Constructive Analysis*, McGraw-Hill, New York.

[2] A. Chauvin: 1974, *Theorie des objets et théorie des ensembles*, Thèse, Université de Clermont-Ferrand.

[3] S. Feferman: 1967, 'Set-Theoretical Foundations of Category Theory' (with an Appendix by G. Kreisel), in *Reports of the Midwest Category Seminar III*, Lecture Notes in Mathematics 106, Springer, 201–247.

[4] S. Feferman: 1974, 'Some Formal Systems for the Unlimited Theory of Structures and Categories' (abstract), *J. Symbolic Logic* **39**, 374–375.

[5] S. Feferman: 1975, 'A Language and Axioms for Explicit Mathematics', in *Algebra and Logic*, Lecture Notes in Mathematics 450, Springer, 87–139.

[6] S. Feferman: 1975, 'Non-Extensional Type-Free Theories of Partial Operations and Classifications', I, in *Proof Theory Symposion, Kiel 1974*, Lecture Notes in Mathematics 500, Springer, 73–118.

[7] F. B. Fitch: 1963, 'The System CΔ of Combinatory Logic', *J. Symbolic Logic* **28**, 87–97.

[8] F. B. Fitch: 1966, 'A Consistent Modal Set Theory' (abstract), *J. Symbolic Logic* **31**, 701.

[9] P. C. Gilmore: 1974, 'The Consistency of Partial Set Theory Without Extensionality', *Axiomatic set theory* (UCLA 1967), *Proc. Symposia in Pure Math. XIII*, Part II, A.M.S., Providence (1974) 147–153.

[10] G. E. Hughes and M. J. Cresswell: 1968, *An Introduction to Modal Logic*, Methuen, London.

[11] J. Isbell: 1967, Review of Lawvere [13], *Mathematical Reviews* **34**, 7332.

[12] G. Kreisel: 1972, Review of MacLane [15], *Mathematical Reviews* **44**, 25.

[13] F. W. Lawvere: 1966, 'The Category of Categories as a Foundation of Mathematics', *Proc. Conference Categorical Algebra* (LaJolla 1965), Springer, N.Y. 1–20.

[14] S. MacLane: 1969, 'One Universe as a Foundation for Category Theory', in *Reports of the Midwest Category Seminar III*, Lecture Notes in Mathematics '106, Springer, 192–200.

[15] S. MacLane: 1971, 'Categorical Algebra and Set-Theoretical Foundations', in *Axiomatic Set Theory* (UCLA 1967) *Proc. Symposia in Pure Math. XIII*, Part I, A.M.S., Providence, 231–240.

[16] P. Martin-Löf: 1974, 'An Intuitionist Theory of Types: Predicative Part', preprint No. 3, University of Stockholm.

[17] Y. Moschovakis: 1969, 'Abstract First-Order Computability II', *Trans. Am. Math. Soc.* **138**, 465–504.

[18] J. Myhill: 1975, 'Constructive Set Theory', *J. Symbolic Logic* **40**, 347–382.

[19] A. Oberschelp: 1973, 'Set Theory Over Classes', *Dissertationes Mathematicae*, **106**, 1–66.

[20] D. Scott: 1970, 'Constructive Validity', in *Symposium on Automatic Demonstration*, Lecture Notes in Mathematics 125, Springer, 237–275.

[21] S. MacLane: 1975, 'Sets, Topoi and Internal Logic in Categories', in *Logic Colloquium '73*, North-Holland Publ. Co., Amsterdam, 119–133.

[22] G. Osius: 1974, 'Categorical Set Theory: A Characterization of the Category of Sets', *J. Pure Appl. Algebra* **4**, 79–119.

IV

COMPUTABILITY THEORY

GERALD E. SACKS*

RE SETS HIGHER UP

Dedicated to J. B. Rosser

1. THEOREM A

I understand C. G. Jung to say that even the most subjective of all notions, psyche, has objective aspects buried in the collective unconscious [0]. If he is correct, then the notion of RE set, which has at least as many meanings as there are areas of logic, also has objective aspects in the ground common to all logicians. The most inspired account of the notion of RE sets was given by Post [1] in 1944. In his paper an RE set is a set that has an effective enumeration. Thus it is not to be thought of as a set defined by a Σ_1 formula or as a semi-recursive set dependent on computations. Its most apparent feature is positive change. It increases in an effective manner, and is the limit of finite subsets developed along the way. This dynamic view has many consequences. For example RE sets obey an effective selection principle. An element of a non-empty RE set can be selected by simply enumerating the set until its first member appears. In addition complex RE sets can be constructed by enumerating all RE sets simultaneously and diagonalizing.

The dynamic view makes clear the nature of finite sets in recursion theory. G. Kreisel [2] was the first to see that recursion theory higher up would require a broadening of the notion of finite. His earliest example begins by replacing RE sets of numbers by Π_1^1 sets of numbers, and finite by hyperarithmetic. (The same example was discovered independently by H. Rogers, Jr.) A finite set is simply an RE set whose enumeration is eventually complete. I do not mean to imply that a finite set has a last member, but only that its enumeration ends long before that of a universal RE set. To give a finite set in the strong sense is to give a means of enumerating it and a bound on the length of the enumeration. Thus it follows that a finite set is recursive, that is both it and its complement are recursively enumerable. Note

Butts and Hintikka (eds.), Logic, Foundations of Mathematics and Computability Theory, 173–194.
Copyright © 1977 by D. Reidel Publishing Company, Dordrecht-Holland. All Rights Reserved.

that the dynamic view does not imply that the intersection of an RE set and a finite set is finite, a desirable non-theorem in the case of RE sets higher up since otherwise higher recursion theory would have nothing new to say about ways of constructing RE sets. Note further that the dynamic view does not imply that every effective enumeration of a finite set eventually comes to an end. Such preternatural finite sets occur in the setting of β-recursion theory when β is an inadmissible ordinal [3].

If any branch of mathematical logic is in need of hermeneutics, surely recursion theory is. My text for today is Theorem A, and my paper is largely an exegesis of that result. Theorem A is the penultimate version of a positive solution to Post's problem in the setting of finite types. The solution draws on the uniform, hyperregular solution to Post's problem for an admissible ordinal relative to a hyperregular predicate, and on reflection phenomena associated with recursion in objects of finite type. Thus to be guided in and out of Theorem A is to travel through much of the last ten years of higher recursion theory.

The objects of type 0 are the non-negative integers. A typical object of type n $(n>0)$ is a function whose arguments range over all objects of type less than n, and whose values are of type 0. ^nE is the equality predicate for objects of type less $<n$; thus $^nE(x, y)$ is true iff $x=y$, where x and y are of type less than n.

THEOREM A. $(n>0)$ *There exist type n objects G and H, each recursively enumerable in ^nE, such that neither is recursive in the other together with ^nE and any set subrecursive in ^nE, and such that each is a collection of sets subrecursive in ^nE.*

I hope the panic occasioned in some by Theorem A will be allayed by the remark that Theorem A for $n=1$ merely claims the existence of two r.e. sets of numbers whose Turing degrees are incomparable, as was proved by Friedberg [4] and Muchnik [5] two decades ago. A set is subrecursive in ^nE if it is recursive in ^nE, H_σ^{nE} for some ordinal σ subconstructive in ^nE. The sets subrecursive in ^2E are simply the hyperarithmetic reals, a numinous theorem of Kleene [6]. When $n=2$, Theorem A yields to Π_1^1 sets of hyperarithmetic reals such that neither is hyperarithmetic in the other, proved in slightly disguised form in [28, pg. 79].[1]

When $n \geqslant 3$, the proof of Theorem A requires an important lemma due to L. Harrington [7], and referred to as 'further reflection' in Section 3 below. When $n \geqslant 4$, G and H are uncountable and their construction requires the full power of the α-finite injury method [8]. The proof of Theorem A is the same for all $n \geqslant 4$. Consequently some new notions are sorely needed to distinguish between types 4 and 5 in the study of RE sets, notions whose existence is guaranteed by past experience.

Section 2 develops the fundamental definitions of recursion in so-called normal objects via a hierarchial approach, and recalls appropriate bounding and selection principles. Section 3 defines and proves two reflection principles for normal objects of type greater than 2. Section 4 reviews the elements of α-recursion theory as well as its relativization to a hyperregular predicate. Section 5 sketches Shore's uniform solution [9] of Post's problem for $L(\alpha)$ with an additional predicate. Section 6 makes the connection between Sections 4 and 2, thereby leading to a proof of Theorem A for $n \geqslant 3$ in Section 7, as was first done by L. Harrington in his Ph.D thesis [10]. And Section 8 is devoted to further speculations on RE sets based on current work in inadmissible recursion theory.

2. RECURSION IN OBJECTS OF FINITE TYPE

2.1. *Finite Types*

$Tp(m)$ is the set of all objects of type m. $Tp(0) = \omega$, and $Tp(m+1)$ is the set of all functions from $Tp(m)$ into ω. It will prove convenient to identify $Tp(m)$ with its set-theoretic counterpart $R(\omega + m)$. Recall $R(0) = 0$, $R(\delta + 1) = 2^{R(\delta)}$, and $R(\lambda) = \bigcup \{R(\delta) \mid \delta < \lambda\}$ when λ is a limit ordinal. By tricks of coding it is possible to view a type m object as a type q object for any $q > m$, to construe a finite sequence $a_0, ..., a_r$ of objects of type at most m as a single object $\langle a_0, ..., a_r \rangle$ of type m, to regard a function from $Tp(m)$ into $Tp(m+1)$ as an object of type $m+1$, and to equate $Tp(m+1)$ with $2^{Tp(m)}$.

Fix $n > 1$ to establish some conventions useful for discussing objects of finite type without making type levels explicit. The objects of type n are denoted by F, G, ... and are called functionals. The objects of type $n-1$ are denoted by f, g, ... and are called *functions*. The objects of type at most $n-2$ are denoted by a, b, ... and are called *individuals*. The set of all individuals is I. A typical function maps I into ω. Subsets of I, denoted by R, S, ..., are called *sets*. The objects of types less then $n-2$, together with the non-negative integers, are denoted by r, s, ..., and are called *subindividuals;* the set of all such is SI.

F is said to be *normal* if F is of the form $\langle G, E \rangle$, where E is ^nE defined in Section 1. (More generally F is normal if E is recursive in F.) The assumption of normality makes it possible to generate the sets recursive in F by means of a hierarchy.

2.2. The $L(F)$ Hierarchy

Harrington [10] found the hierarchy of sets constructible relative to F useful in the study of recursion in F.

$$L(0, F) = I = Tp(n-2) = R(\omega + n - 2).$$

$X \in L(\sigma+1, F)$ iff $X \subset L(\sigma, F)$ and X is first order definable with parameters over $M(\sigma, F)$.

$$M(\sigma, F) = \langle L(\sigma, F),\ \varepsilon,\ F \restriction (Tp(n-1) \cap L(\sigma, F)) \rangle.$$

$L(\lambda, F) = \bigcup \{L(\sigma, F) \mid \sigma < \lambda\}$ if λ is a limit.

$$L(F) = \bigcup \{L(\sigma, F) \mid \sigma\}.$$

If $n = 2$ and $F = {}^2E$, then $L(F) = L$. Let $L(\mathscr{F})$ be a first order language sharp enough to dissect $L(F)$. $\mathscr{L}(\mathscr{F})$ is the language of ZF augmented by ranked variables x^σ, y^σ, ... for each ordinal σ, set constants to name the elements of $L(F)$, and a function symbol $\mathscr{F}$.

$L(\sigma, F)$ will be of primary interest only if σ is recursive in F in the sense of 2.3. If F is ^2E, then the ordinals recursive in F are the familiar recursive ordinals, but if $n > 2$, then several delightful complexities occur, the simplest of which is the existence of gaps between some of the ordinals recursive in F.

2.3. *Sets and Ordinals Recursive in F*

The use of hierarchies to define Kleene's notion of recursion in a normal F goes back to [11], [12] and [13]. The following version, devised by Harrington [10], stays close to the idea of defining the hyperarithmetic reals by iterating the Turing jump. Let $X \subseteq I$. The F jump of X, denoted by jX, is the effective disjoint union of certain definable subsets of I. Let $\{e\}_{\Pi}^{X}$ be the eth function (eth according to some standard Gödel numbering of formulas) from I into ω whose graph is definable over $\langle I, \varepsilon, X \rangle$ by means of a first order formula whose only parameters are non-negative integers, and whose atomic parts are of the form $y \in z$ and $y \in \chi$, where y and z range over I and χ is a set constant naming X. (Here I is being equated with $R(\omega + n - 2)$.)

Let W_e^X be $\{a \mid a \in I \ \& \ \{e\}_{\Pi}^X(a) = 0\}$.

jX is the set of all $\langle e, a, 0 \rangle$ such that $a \in W_e^X$, and all $\langle e, a, y+1 \rangle$ such that $F(\{e\}_{\Pi}^{\langle x, a \rangle}) = y$. j is the so-called F-jump.

The first part of jX encodes subsets of I first order definable via integer parameters, and the rest of jX encodes the F-images of subsets of I definable via parameters from I.

If F is ^2E, then $j0$ is equivalent to the effective disjoint union of the arithmetic reals.

The sets $\mathbb{O}^F$ and H_σ^F ($\sigma \in$ range $\mid \mid^F$), and the function $\mid \mid^F : \mathbb{O}^F \to$ Ordinals, are defined by a simultaneous recursion. (For notational simplicity, the superscript F is often omitted below.)

(1) $\langle 1, a \rangle \in \mathbb{O}$, $|\langle 1, a \rangle| = 0$ and $H_0 = 0$ $(a \in I)$.

(2) If $\langle e, a \rangle \in \mathbb{O}$ and $|\langle e, a \rangle| = \sigma$, then $\langle 2^e, a \rangle \in \mathbb{O}$, $|\langle 2^e, a \rangle| = \sigma + 1$, and $H_{\sigma+1} = jH_\sigma$.

(3) Suppose $\langle m, a \rangle \in \mathbb{O}$, $|\langle m, a \rangle| = \sigma$, and $W_e^{\langle H_\sigma, a \rangle} \subset \mathbb{O}$. Then $\langle 3^m \cdot 5^e, a \rangle \in \mathbb{O}$ and $|\langle 3^m \cdot 5^e \rangle| = \lambda$, where λ is the least limit ordinal greater than $|b|$ for all $b \in W_e^{\langle H_\sigma, a \rangle} \cup \{\langle m, a \rangle\}$. And H_λ^F is the set of all $\langle b, c \rangle$ such that

$$b \in \mathbb{O} \ \& \ |b| < \lambda \ \& \ c \in H_{|b|}.$$

Let κ^F be the least ordinal not in the range of $|\ |^F$. The hierarchy $\{H_\sigma^F \mid \sigma < \kappa^F\}$ of sets is equivalent to a hierarchy of all the computations needed to compute all functions (from I into ω) recursive in $\langle F, a \rangle$ for each $a \in I$.

$\{e\}^F(a)$ is said to be defined and equal to x if $e = \langle e_0, e_1 \rangle$, $\langle e_0, a \rangle \in \mathbb{O}$, $|\langle e_0, a \rangle| = \sigma$ and $\{e_1\}_{H^\sigma}^{H}(a) = x$.

Thus to compute $\{e\}^F(a)$ is to iterate the F-jump (starting with the empty set of individuals) until e_0 is seen to be a notation for an ordinal σ (if it is) and then to pluck the value x from H_σ in an elementary manner dictated by e_1. Kleene's original definition of recursion in F is based on nine schemes, eight of which resemble those of ordinary recursion theory. The ninth is refractory because of its self-referential nature. Kleene's schemes give rise to infinitely long computation trees which can be linearized and arranged according to ordinal length as above if F is normal.

f is *recursive in* F (in symbols $f \leqslant F$) if there is an e such that $f(a) = \{e\}^F(a)$ for all a.

A set R of individuals is recursive in F ($R \leqslant F$) if the characteristic function of R is recursive in F.

An ordinal σ is *constructive in* F if σ is $|\langle e, 0 \rangle|^F$ for some $\langle e, 0 \rangle \in \mathbb{O}^F$. (In short σ has an integer among its notations.) σ is *recursive in* F if σ is the height of a prewellordering of I recursive in F. The notion of ordinal recursive in F is sometimes said to be *intrinsic*, because it is definable directly in terms of the notion of set recursively enumerable in F. The notion of ordinal constructive in F is not intrinsic if $n > 2$. Each ordinal σ constructive in F is also recursive in F, since σ is the height of the prewellordering of all members of $\mathbb{O}^F$ which are notations for ordinals less than σ, prewellordered by $|\ |^F$. (A set Z is said to be prewellordered if there is a map t from Z into the ordinals.) Each ordinal recursive in F is constructive in F, but the converse is false when $n > 2$. $\aleph_1$ is constructive in ^3E, hence the first countable ordinal not constructive in ^3E is recursive in ^3E, because it can be singled out effectively from the vantage point of $\aleph_1$.

Clause (3) above implies that $\{H_\sigma^F \mid \sigma < \kappa^F\}$ obeys a *bounding principle*: if $f: I \to \mathbb{O}^F$ is recursive in F, then there exists an ordinal σ constructive in F such that $|fa|^F < \sigma$ for all a. It follows from the bounding principle that a set R is recursive in F iff there is an ordinal

σ constructive from F such that R is one-one reducible to $H^F_{\sigma+1}$ as follows. There is an e such that

$$a \in R \leftrightarrow \langle e, a, 0 \rangle \in H^F_{\sigma+1}$$

for all a.

A set R is *recursively enumerable in F* if there is an e such that for all a,

$$a \in R \leftrightarrow \{e\}^F(a) \text{ is defined.}^2$$

Further properties of the hierarchy are based on Grilliot's *selection principle* proved by Harrington and MacQueen [14]: there exists a uniform method for selecting a non-empty recursive (in F) subset of a non-empty recursively enumerable (in F) set of *subindividuals*. To be more precise, there exists a recursive number-theoretic function t such that for all e:

(4)   if $\{e\}^F(r)$ is defined for some subindividual r, then $\{te\}^F(r)$ is the characteristic function of a non-empty subset of SI and (r) $[\{te\}^F(r) = 0 \rightarrow \{e\}^F(r)$ is defined$]$.

One of the reasons (4) is termed a selection principle is that the sets recursive in F behave like generalized finite sets, as is suggested by the bounding principle. In the language of the introduction, a set is generalized finite if it has a recursive enumeration that comes to an end, and the sets recursive in F are just the sets recursively enumerable in F whose enumeration come to end before κ^F_0. Gandy's selection principle [15] antedates Grilliot's. It furnishes a uniform method for selecting an integer from a non-empty recursively enumerable (in F) set of integers. Moschovakis [16] exhibits a non-empty set of reals recursively enumerable in ${}^3E$ with no non-empty recursive subset.

For each $\sigma < \kappa^F$, H^F_σ conveys roughly the same information as the structure $M(\sigma, F)$ of 2.2. See Lowenthal [17] for details and for a proof of the equivalence of Kleene's notion of recursion in F and the above for normal F's.

3. REFLECTION

3.1. *Ordinals that Reflect to* κ_0^F

The notion of reflection is needed to reduce Post's problem for functionals recursively enumerable in F to Post's problem for sets of ordinals α-recursively enumerable relative to an additional predicate.

For each $i < n-1$, let κ_i^F be the supremum of all ordinals with notations in $\mathbb{O}^F$ of type i; that is $\sigma < \kappa_i^F$ if $\sigma = |\langle e, a \rangle|^F$ for some $\langle e, a \rangle \in \mathbb{O}^F$ such that $a \in Tp(i)$. Thus $\kappa^F = \kappa_{n-2}^F$.

An ordinal τ is said to be *reflecting* if for every Σ_1 sentence $\mathscr{S}$ of $\mathscr{L}(\mathscr{F})$ with integer parameters only, it is the case that

$$M(\tau, F) \vDash \mathscr{S} \quad \text{implies} \quad M(\kappa_0^F, F) \vDash \mathscr{S}.$$

The Σ_1 sentence $(E\sigma)[e \in H_\sigma^F]$ is typical, since each $H_{\gamma+1}^F$ encodes all first order facts about $\{H_\sigma^F \mid \sigma < \gamma\}$ in a positive way. To say τ is reflecting is to say: an H_σ^F satisfying a Δ_0 formula[3] (with integer parameters only) is developed before stage τ only if one is developed before stage γ for some γ recursive in F. Note that if γ is recursive in F and $e \in H_\sigma^F$ holds for some $\sigma < \gamma$, then the least such σ is recursive in F because it is first order definable over H_σ^F via integer parameters.

From now on assume $n > 2$ in any statement about reflecting ordinals. Let κ_r^F be the greatest reflecting ordinal. Kechris and Harrington found an intrinsic definition of κ_r^F: call a set R Q-finite if R is corecursively enumerable in F, and if $(Ea)[a \in R \ \& \ P(a, b)]$ is recursively enumerable in F whenever $P(a, b)$ is; κ_r^F is the supremum of $\kappa_0^{\langle F, a \rangle}$ as a ranges over the union of all Q-finite R's. Moschovakis [16] showed $\kappa^F > \kappa_r^F$, and Harrington [7] established that κ_r^F is much larger than κ_{n-3}^F. The next two results suffice for our purposes.

3.2. LEMMA (Simple Reflection [18]). $\kappa_r^F \geqslant \kappa_{n-3}^F$.

Proof. Fix e and assume $e \in H_\sigma^F$ for some $\sigma < \kappa_{n-3}^F$ with the intent of showing the first such σ is less than κ_0^F. Thus there is a subindividual s such that

$$(1) \qquad s \in \mathbb{O}^F \ \& \ (E\sigma)[\sigma < |s|^F \ \& \ e \in H_\sigma^F].$$

Let S be the set of all s that satisfy (1). By Grilliot's selection principle (2.3(4)), there is a non-empty recursive (in F) $T \subset S$. But then the bounding principle (2.3) yields a γ recursive in F such that $|s|^F < \gamma$ for all $s \in T$. Clearly $e \in H_\sigma^F$ for some $\sigma < \gamma$. ■

3.3. LEMMA (Further Reflection, Harrington [7]). *If R is a recursively enumerable (in F) set of subindividuals, then $\kappa_r^F \geqslant \kappa_{n-3}^{\langle F,R \rangle}$.*

Proof. Fix e and assume $e \in H_\sigma^F$ is satisfied by some $\sigma < \kappa_{n-3}^{\langle F,R \rangle}$ with the intent of locating such a σ below κ_0^F. Thus there is a subindividual s such that

$$s \in \mathcal{O}^{\langle F,R \rangle} \ \& \ (E\delta)[\delta < |s|^{\langle F,R \rangle} \ \& \ e \in H_\delta^F].$$

Let W be the set of all individuals c such that c is a wellordering of all subindividuals. W is recursive in F because $n > 2$ and F is normal. For each $c \in W$, let $|c|$ be the ordertype of c. Call an ordinal γ *subconstructive* (in F) if $\gamma = |s|^F$ for some subindividual $s \in \mathcal{O}^F$. Let λ be the ordertype of the set of subconstructive (in F) ordinals. λ splits W in a curious fashion:

(1) $\{c \mid |c| < \lambda\}$ is recursively enumerable in F;

(2)   If $|c| \geqslant \lambda$, then $e \in H_\sigma^F$ for some $\sigma < \kappa_0^{\langle F,c \rangle}$.

For each $s \in \mathcal{O}^F$, let $\|s\|$ be the ordertype of the set of subconstructive (in F) ordinals less than $|s|^F$. To prove (1) observe that

$$(Es)[s \in \mathcal{O}^F \ \& \ |c| < \|s\|]$$

is recursively enumerable in F, since Grilliot's selection principle implies the recursively enumerable (in F) predicates are closed under existential quantification over subindividuals [19].

To prove (2) assume $|c| \geqslant \lambda$. Let c_s be an initial segment of c such that $|c_s| = \lambda$, where c_s is either c or that part of c below some s in the field of c. R is recursive in $\langle F, c, s \rangle$, since the recursive (in F) enumeration of R is interlaced with the development of $\{H_\sigma^F \mid \sigma < \kappa_{n-3}^F\}$, and the latter occurs as soon as $\mathcal{O}^F$ has been enumerated to the point where the set of subconstructive (in F) ordinals has ordertype $|c_s|$. Consequently

(3) $(E\sigma)[\sigma < \kappa_0^{\langle F, c\rangle}$ & $e \in H_\sigma^F].$[4]

The set of all c that satisfy (3) is recursively enumerable in F, since an ordinal is constructible from $\langle F, c \rangle$ iff it has a notation in $\mathcal{O}^F$ of the form $\langle m, c \rangle$ for some integer m. (1) and (2) allow W to be recursively enumerated in F as follows: enumerate c if $|c| < \lambda$ or if c satisfies (3). (Gandy's selection principle implies the union of two sets recursively enumerable in F is recursively enumerable in F.) Since W is recursive in F, the bounding principle implies the enumeration of W is complete by some $\sigma_0 < \kappa_0^F$. By (2) $e \in H_\sigma^F$ for some $\sigma < \sigma_0$. ∎

4. α-RECURSION RELATIVE TO B

Recursion theory on the ordinals was invented by Takeuti [20]. In recent years attention has centered on Σ_1 admissible ordinals. α is such an ordinal if $L(\alpha)$ satisfies Σ_1 replacement with parameters in $L(\alpha)$. A function $f: \alpha \rightarrow \alpha$ is said to be α-recursive if its graph is Σ_1 over $L(\alpha)$. A set x is called α-finite if $x \in L(\alpha)$. Thus $f[x]$ is α-finite if α is Σ_1 admissible, f is α-recursive, and x is α-finite. From now on assume α is Σ_1 admissible. The notion of α-finiteness, and its importance in α-recursion theory, goes back to [21].

4.1. Σ_1 Projecta and Hulls

The Σ_1 projectum of α, denoted by α^* or φ_1^α, is the least γ such that there exists a one-one α-recursive map of α into γ. An equivalent definition is: α^* is the least γ such that some α-recursively enumerable subset of γ is not α-finite. Of course α-recursively enumerable means empty or range of an α-recursive function. The existence of bounded RE sets which are not finite is compatible with the views held in the introduction and is the source of much of the mathematical interest of higher recursion theory.

Z is a Σ_1 substructure of $L(\alpha)$, in symbols $Z <_1 L(\alpha)$, if every Σ_1 sentence (with parameters in Z) true in $L(\alpha)$ is also true in Z. Σ_1 substructures give bounds on partial α-recursive functions. Thus if $f(x)$

is one such, is defined, and x as well as the parameters defining f belong to Z, then $f(x) \in Z$. δ is an α-*cardinal* if $L(\alpha) \models [\delta$ is a cardinal]. Gödel's argument for proving GCH in L is readily adapted to show $L(\delta) <_1 L(\alpha)$ when δ is an α-cardinal: Let h be a master Σ_1 Skolem function. h is partial α-recursive, definable by integer parameters, and for each Σ_1 formula $\mathcal{F}(x, y)$ with no parameters, if $L(\alpha) \models (Ey)\mathcal{F}(b, y)$, then $L(\alpha) \models \mathcal{F}(b, h(b))$, where b is a finite sequence of members of $L(\alpha)$. Clearly $h[X] <_1 L(\alpha)$ for all $X \subset L(\alpha)$. Suppose X is an initial segment of $L(\alpha)$. Then $h[X]$ is isomorphic to an initial segment $\overline{h[X]}$ of $L(\alpha)$ via the 'collapsing' map. But X collapses to X, and so $h[X] = \overline{h[X]}$, since each $v \in h[X]$ is equal to $h(x)$ for some $x \in X$ and so must be its own collapse. Thus the Skolem hull of $L(\delta)$ is $L(\delta)$ when δ is on α-cardinal other than ω. Lemma 4.2 is proved similarly.

4.2. LEMMA (R. B. Jensen). $\alpha^* = \alpha > \omega$ *iff* $L(\alpha)$ *is the limit of* $L(\beta)$'s *which are* Σ_1 *substructures of* $L(\alpha)$.

There is no need for α to be admissible in 4.2.

4.3. *Regularity*

$X \subset \alpha$ is said to be regular [22] if $X \cap y \in L(\alpha)$ whenever $y \in L(\alpha)$. Non-regular α-recursively enumerable sets exist iff $\alpha^* < \alpha$. Their presence is less of a problem than one might think thanks to the regular sets Theorem (4.4). Let $A, B \subset \alpha$. A is α-recursive in B (in symbols $A \leq_\alpha B$) if there exists a partial α-recursive $\emptyset$ such that for all H and I,

(1) $A \in N_{H,I} \leftrightarrow (EJ)(EK)[\emptyset(H, I, J, K) = 0 \,\&\, B \in N_{J,K}]$,

where H, I, J, and K range over the α-finite sets, and $A \in N_{H,I}$ means $H \subset A \,\&\, I \subset \alpha - A$. Clearly $\leq_\alpha$ is transitive. Two sets belong to the same α-*degree* if each is α-recursive in the other.

4.4 THEOREM (Sacks [22]). *Suppose A is α-recursively enumerable. Then there exists a regular, α-recursively enumerable B of the same α-degree as A.*

The best proof of 4.4 is in Simpson [23]. MacIntyre [24] has shown that for some α's, there are A's whose α-degrees contain no regular sets. Let $L(\alpha, A)$ be the result of relativizing $L(\alpha)$ to A: the predicate $x \in A$ is added to the list of atomic formulas and then first order definability is iterated through α. A is regular is equivalent to $L(\alpha) = L(\alpha, A)$.

Let $f: \alpha \to \alpha$. f is said to be weakly α-recursive in B (in symbols $f \leqslant_{w\alpha} B$) if there exists a partial α-recursive $\varnothing$ such that for all γ and δ,

$$f(\gamma) = \delta \leftrightarrow (EJ)(EK)[\varnothing(\gamma, J, K) = \delta \ \& \ B \in N_{J,K}].$$

A is said to be *hyperregular* [22] if $f[H]$ is bounded below α whenever $f \leqslant_{w\alpha} A$ and $H \in L(\alpha)$. A hyperregular A need not be regular, but it is if A is α-recursively enumerable. If A is both regular and hyperregular, then α-recursion theory has a natural relativization to A. The structure $\langle L(\alpha), A, \varepsilon \rangle$ is admissible, that is $f[K]$ is α-finite if f is Σ_1 over $L(\alpha)$ (with $x \in A$ counted as an atomic formula) and $K \in L(\alpha)$. The regularity of A implies computations from A of height less than α are α-finite, and consequently regularity combined with hyperregularity entails admissibility.

4.5. THEOREM ([22]).   *There exists a non-α-recursive, hyperregular, α-recursively enumerable set.*

The proof of 4.5 is a simple priority argument in which certain computations are preserved for the sake of hyperregularity. Its principal virtue is the ease with which it mixes with other priority arguments.

4.6. α-Recursion Relative to B

Let B be a subset of α such that $\langle L(\alpha, B), \varepsilon, B \rangle$ is admissible, that is obeys Σ_1 replacement with $x \in B$ as an atomic formula. It follows that B is hyperregular but not conversely. α-recursion theory relativizes to B as follows. $f: \alpha \to \alpha$ is said to be $\alpha - B$ recursive if its graph is Σ_1 over $\langle L(\alpha, B), \varepsilon, B \rangle$. The $\alpha - B$ finite sets are the elements of $L(\alpha, B)$. The arguments of α-recursion theory tend to extend readily to $\langle L(\alpha, B), \varepsilon, B \rangle$. In particular 4.5 becomes 4.7.

4.7. THEOREM. *There exists a non-α-B recursive, hyperregular, $\alpha - B$ recursively enumerable set.*

To see the point of 4.7, let $\mathscr{B}$ be $\langle L(\alpha, B), \varepsilon, B \rangle$, and let C be a set furnished by 4.7. As noted just before 4.5, a hyperregular, α-recursively enumerable set is regular. That argument extends to $\mathscr{B}$, and so $\mathscr{B}(C) = L(\alpha, \mathscr{B})$, where $\mathscr{B}(C)$ is the set of all sets constructible via ordinals less than α with $x \in B$ and $y \in C$ as additional atomic predicates. Moreover the hyperregularity of C entails the admissibility of $\langle \mathscr{B}, C \rangle$, more precisely the admissibility of $\langle L(\alpha, B), \varepsilon, B, C \rangle$.

4.8. *Projecta and Hulls Relative to B*

The Σ_1 projectum of α relative to B, denoted by $\varphi_1^{\alpha, B}$, is the least γ such that there exists a one-one $\alpha - B$ recursive map of α into γ. (It is assumed that $\langle L(\alpha, B), \varepsilon, B \rangle$ is admissible.) The alternative definition of φ_1^{α} given in 4.1 extends to $\varphi_1^{\alpha, B}$, as is expressed by the next lemma.

4.9. LEMMA. *If $\gamma < \varphi_1^{\alpha, B}$ and Z is an $\alpha - B$ recursively enumerable subset of γ, then $Z \in L(\alpha, B)$.*

The notion of Skolem hull makes sense for $\alpha - B$ recursion theory, but the 'collapsing' argument of 4.1 does not, because the 'collapse' of a subset of B may not be an initial segment of B. Thus the solution of Post's problem for $\langle L(\alpha, B), \varepsilon, B \rangle$ uses Skolem hulls but not 'collapsing' arguments.

5. POST'S PROBLEM RELATIVE TO B

From now on assume $\mathscr{B} = \langle L(\alpha, B), \varepsilon, B \rangle$ is admissible. To give a uniform solution of Post's problem is to give a pair $\langle m, n \rangle$ of integers such that the degrees of the mth and nth sets Σ_1 over $\mathscr{B}$ are incomparable for all α and B. The existence of a solution uniform in $\mathscr{B}$ is implicit in Shore [9], where the assumption of regularity is made for B but never used. An important additional feature of Shore's solution is the hyperregularity (in the sense of $\mathscr{B}$) of the two sets constructed. An outline of the construction follows. It is an instance of the α-finite injury method [8] with the added idea of 'blocking' due to Shore.

$\alpha - B$ recursively enumerable sets are to be constructed so as to satisfy two collections of requirements, $\{P_\beta \mid \beta < \alpha\}$ and $\{Q_\beta \mid \beta < \alpha\}$. A typical P_β requires $\{\beta\}^C$ restricted to β to be an $\alpha - B$ finite object. As the values of $\{\beta\}^C(\gamma)$ $(\gamma < \beta)$ manifest themselves during the enumeration of C, they are preserved in the sense that no new element is added to C if it destroys an existing computation of $\{\beta\}^C(\gamma)$ for some $\gamma < \beta$, unless that new element is demanded by a requirement of higher priority. If the latter occurs, an injury to P_β is said to have occurred.

A typical Q_β requires $C(\gamma) \neq \{\beta\}^D(\gamma)$ for some $\gamma \in Z_\beta$, where $\{Z_\beta \mid \beta < \alpha\}$ is an α-recursive collection of disjoint unbounded α-recursive sets. Q_β is met by waiting for the value of $\{\beta\}^D(\gamma)$ to manifest itself; if that happens with value 1 and there is no current witness to the inequality of C and $\{\beta\}^D$, then γ is added to C unless some requirement of higher priority forbids it. If γ is added to C, then the current computation of $\{\beta\}^D(\gamma)$ is to be preserved.

Conflicts occur because the desire to add an element to C (D respectively) for the sake of some Q_β may run counter to a promise already made to omit that element from C (D respectively) for the sake of some P_β or for the sake of preserving an inequality of the form $D \neq \{\beta\}^C$ ($C \neq \{\beta\}^D$ respectively). Let $\{R_\beta \mid \beta \subset \alpha\}$ be the union of the P_β's and Q_β's. Suppose all conflicts are resolved in favor of the requirement of higher priority. An assignment k of priorities is in essence a one-one, α-recursive map $\alpha \to \alpha$. Thus R_β has higher priority than R_δ if $k\beta < k\delta$; an injury to R_δ (for the sake of R_β) occurs if an element is added to C or D as required by R_β despite a promise made earlier to exclude it as required by R_δ. For each δ, let I_δ be the set of all σ such that R_δ is injured at stage σ.

PRIORITY LEMMA. *For each δ, I_δ is $\alpha - B$ finite.*

The priority lemma implies each R_β is eventually satisfied, since it is eventually free of injuries. If $\alpha = \omega$, then the priority lemma is easily proved by induction on ω (with k equal to the identity). But the proof assumes $\langle L(\alpha, B), \varepsilon, B \rangle$ satisfies Σ_2 replacement, because the function

$$i(\delta) = \text{least } \sigma \, [R_\delta \text{ not injured after stage } \sigma]$$

is Σ_2. Thus Post's problem for $\langle L(\alpha, B), \varepsilon, B \rangle$ is to find a way of doing certain Σ_2 recursions when Σ_2 replacement fails. The trick of [8] is to make the range of k a small initial segment of α in the hope of proving the priority lemma by induction on the ordinals in that small segment. It follows k can no longer be $\alpha - B$-recursive, but it will have $\alpha - B$-recursive approximations, as it must if the enumeration of C and D is to be $\alpha - B$ recursive.

For the moment, let $K: \delta \to \alpha$ be a one-one Σ_2 (over $\langle L(\alpha, B), \varepsilon, B \rangle$) map of δ into α, where $\delta \leqslant \alpha$. The range of k chops α, hence $\{R_\beta \mid \beta \subset \alpha\}$, into δ many blocks. Assume the R_β's have been arranged so that no conflict is possible between any two in the same block. If two requirements do conflict, the resolution is in favor of the one whose block has a higher priority according to k. Thus requirements in block 0, $\{R_\beta \mid \beta < k0\}$, are never injured; those in block 1, $\{R_\beta \mid k0 \leqslant \beta < k1\}$, are injured only by those in block 0 etc.

PRIORITY SUBLEMMA. *Fix σ and γ, and assume no requirement in block γ is injured after stage σ. Then the set of stages after σ at which an element is added to $C \cup D$ for the sake of some requirement in block γ is $\alpha - B$-finite.*

The proof of the Sublemma is based on the opening remarks of 4.1 relativized to B as expressed by 4.8. Let Z be the set of stages after σ at which an element is added for the sake of block γ. Z is in effective $1 - 1$ correspondence with a subset of block γ, because after stage σ there is no reason to make more than one attempt to satisfy a requirement in block γ. Thus Z is essentially an $\alpha - B$-recursively enumerable subset of block γ. Either $\varphi_1^{\alpha, B} = \alpha$ or $\varphi_1^{\alpha, B}$ is the greatest $\alpha - B$ cardinal. In the first case 4.8 implies Z is $\alpha - B$ finite. In the second case assume (for the moment) that k chops α up into blocks of $\alpha - B$ cardinality less than $\varphi_1^{\alpha, B}$ and then apply 4.8.

Define $m(\gamma)$ $(\gamma < \delta)$ to be the least σ such that no injury is inflicted on, and no attempt is made to satisfy, any requirement in block γ after stage σ. The proof of the priority lemma amounts to showing $m(\gamma)$ is defined for all $\gamma < \delta$. Fix γ and suppose $m(\varphi)$ is defined for all $\varphi < \gamma$. If $\{m(\varphi) \mid \varphi < \gamma\}$ is bounded below α, then the Priority Sublemma implies $m(\gamma)$ is defined. Now m is Σ_2, so what is needed is a special instance of Σ_2 replacement, an instance that is always true if k is chosen properly.

Let h be the master Σ_1 Skolem function of 4.1 relativized to B. Define $k(0) = \omega$, $k(\gamma + 1) = h[k(\gamma) + 1]$ and $k(\lambda) = \bigcup \{k(\gamma) \mid \gamma < \lambda\}$.

The domain of k is the least δ such that $\bigcup \{k(\varphi) \mid \varphi < \delta\} = \alpha$. Now it is not difficult to show that $\{m(\varphi) \mid \varphi < \gamma\}$ is bounded, in fact $\alpha - B$-finite, for each $\gamma < \delta$, because the nature of h implies that all attempts to meet requirements in block φ occur at stages less than $h(\varphi + 1)$. Each block contains the Σ_1 hull of all the blocks before it.

All that remains is to develop the $\alpha - B$-recursive approximation of k used in place of k during the enumeration of C and D. Let $h^\sigma(\gamma)$ be the value of $h(\gamma)$ if it is defined by a witness in $L(\sigma, B)$, 0 otherwise. Define $k(\sigma, 0) = 0$, $k(\sigma, \gamma + 1) = \bigcup h^\sigma[k(\sigma, \gamma) + 1]$ and $k(\sigma, \lambda) = \bigcup \{k(\sigma, \gamma) \mid \gamma < \lambda\}$. Clearly $k(\sigma, \gamma)$ is $\alpha - B$-recursive and $k(\sigma, \gamma) \to k(\gamma)$ as $\sigma \to \alpha$.

$k(\sigma, \gamma)$ is tame, a notion due to Lerman [25]. It means that for each γ there is a τ such that

$$(\varphi)_{\varphi < \gamma}(\sigma)_{\sigma > \tau}[k(\sigma, \varphi) = k(\varphi)].$$

Thus any argument above involving $k(\gamma)$ can be replaced by one involving $k(\sigma, \gamma)$ by going to a sufficiently large τ.

One final point claimed in the proof of the Priority Sublemma: if $\varphi_1^{\alpha,B} < \alpha$, then the $\alpha - B$-cardinality of each block is less than $\varphi_1^{\alpha,B}$. Suppose $\varphi_1^{\alpha,B} < \alpha$. Then each block save the last has $\alpha - B$-cardinality less than $\varphi_1^{\alpha,B}$. The problem posed by the last block is overcome by exploiting the Σ_1 injectibility of α into $\varphi_1^{\alpha,B}$. The blocks have to be defined in a slightly more complicated fashion than indicated above. The idea is to make each block save the last do double duty by having it include its image under h. Thus the last block, if there is one, is distributed over the previous blocks, each of which has $\alpha - B$-cardinality less than $\varphi_1^{\alpha,B}$.

An interesting feature of Shore's argument is the definition of a tame Σ_2 injection of α into α^* via an integer parameter, interesting because the Σ_1 injection of α into α^* often requires a parameter greater than α^*.

5.1. THEOREM (R. A. Shore [9]). *There exists a pair of finite Gödel numbers which define hyperregular, $\alpha - B$-recursively enumerable sets such that neither is $\alpha - B$-recursive in the other for all B and α such that $\langle L(\alpha, B), \varepsilon, B \rangle$ is admissible.*

6.   REDUCTION OF FINITE TYPES TO ORDINALS

Recall the notion of subconstructive (in F) ordinal from the proof of 3.3. Let S be the set of all such ordinals, and α its ordertype. τ is the unique orderpreserving map of α onto S. Let B be the set of all $\langle e, \sigma, \beta \rangle$ such that $\sigma < \alpha$ and $M(\tau\beta, F) \vDash \mathcal{G}_e(\tau\sigma)$, where M is as in 2.2 and $\mathcal{G}_e$ is the eth Σ_1 formula of ZF. B is a complete, recursively enumerable (in F) set of subrecursive (in F) ordinals. Let $\mathcal{B}$ be $\langle L(\alpha, B), \varepsilon, B \rangle$.

6.1. LEMMA.   *Let $A \subset \alpha$. Then A is Σ_1 over $\mathcal{B}$ iff τA is recursively enumerable in F.*[5]

Proof.   Suppose τA is r.e. in F. Then there is a Σ_1 formula $\mathcal{G}_e(x)$ such that for all $\sigma < \alpha$,

$$\tau\sigma \in rA \leftrightarrow M(\kappa^F_{n-3}, F) \vDash \mathcal{G}_e(\tau\sigma)$$

$$\leftrightarrow (E\gamma)_{\gamma \in S}(\gamma > \tau\sigma \ \& \ M(\gamma, F) \vDash \mathcal{G}_e(\tau\sigma))$$

$$\leftrightarrow (E\varphi)_{\varphi < \alpha}(\langle e, \sigma, \varphi \rangle \in B).$$

Conversely, the structure $\langle L(\delta, B), \varepsilon, B \cap \omega \times \delta \times \delta \rangle$ is first order definable over $M(\tau^{-1}\delta, F)$ uniformly in δ. Consequently a Σ_1 definition over $\mathcal{B}$ induces one over $M(\kappa^F_{n-3}, F)$. ∎

6.2. LEMMA.   *$\mathcal{B}$ is admissible (in the sense of 4.6).*

Proof.   A Σ_1 (over $\mathcal{B}$) map of some proper initial segment σ of α onto an unbounded subset of α induces (by the concluding argument of 6.1) a map from $\tau\sigma$ onto an unbounded subset of κ^F_{n-3}, a map which is Σ_1 over $M(\kappa^F_{n-3}, F)$ with some subindividual r as parameter. According to the lemma of simple reflection (3.2) such a map is bounded by $\kappa_0^{\langle F, r \rangle} < \kappa^F_{n-3}$. ∎

6.3. *Subgenericity in F*

For each set A of subconstructive (in F) ordinals, let A^H be $\{H^F_\sigma \mid \sigma \in A\}$. A^H is a type n object as is F. A is said to be subgeneric over F if

$$(1) \qquad \kappa_0^{\langle A^H, F, H_\sigma^F \rangle} \leqslant \kappa^F_{n-3}$$

for every subconstructive (in F) ordinal σ. The notion of subgenericity originated in α-recursion theory [22] and was first proposed by Kreisel. The intuitive meaning of (1) is: any computation (encodable by an integer) from $\langle A^H, F \rangle$, together with any fixed subrecursive (in F) set as a parameter, is of subrecursive (in F) ordinal height. Thus the presence of A^H does not lengthen the subrecursive ordinals.

6.4. LEMMA. *Suppose C is regular and hyperregular over $\mathcal{B}$, and τC is subgeneric over F. Then D is Δ_1 over $\langle \mathcal{B}, C \rangle$ iff $(\tau D)^H$ is recursive in $\langle F, (\tau C)^H, H_\sigma^F \rangle$ for some subconstructive (in F) σ.*

Proof. The suppositions concerning C are not needed to prove the left-to-right direction of the lemma; it follows from the reasoning employed in 6.1 and 6.2.

Suppose $(\tau D)^H$ is recursive in $\langle F, (\tau C)^H, H_\sigma^F \rangle$ for some subconstructive (in F) σ. It follows that τD is Δ_1 over

$$M(\beta, \langle F, (\tau C)^H, H_\sigma^F \rangle),$$

where β is $\kappa^{\langle F, (\tau C)^H, H_\sigma^{\ F} \rangle}$. Since τC is subgeneric in F, τD is Δ_1 over

$$M(\kappa_{n-3}^F, \langle F, (\tau C)^H \rangle).$$

It suffices to show the predicate

$$(1) \qquad \beta < \sigma < \alpha \ \& \ M(\tau\sigma, \langle F, (\tau C)^H \rangle) \vDash \mathcal{G}_e(\tau\beta)$$

is Δ_1 over $\langle \mathcal{B}, C \rangle$. Since C is regular and hyperregular (with respect to $\mathcal{B}$), the universe of $\langle \mathcal{B}, C \rangle$ is the same as that of $\mathcal{B}$, and $\langle \mathcal{B}, C \rangle$ is admissible. Thus (1) is equivalent to: there exist $\delta \in S$, $e_0 \in \omega$ and X, $Y \in M(\kappa_{n-3}^F, F)$ such that X is first order definable over $M(\delta, F)$ via the e_0th formula, $X = \tau C \cap \tau\sigma$, $Y = M(\tau\sigma, \langle F, (\tau C)^H \rangle)$ and $Y \vDash \mathcal{G}_e(\tau\beta)$. ∎

6.5. *Existence of Sets Subgeneric over F*

Suppose $C \subset \alpha$ is Σ_1 over $\mathcal{B}$ and has been constructed in the manner described in Section 5 so that it is regular and hyperregular over $\mathcal{B}$. Thus $\langle \mathcal{B}, C \rangle$ is admissible and $L(\alpha, B) = L(\alpha, B, C)$. Some additional requirements, readily added to the construction of Section 5, will insure that τC is subgeneric.[6] Suppose τC is not subgeneric over F in order to see what has to be added. Thus there is an e such that

(1) $\qquad \langle e, 0 \rangle \in \mathbb{O}^{\langle F, (\tau C)^H, H_\sigma^F \rangle}$ & $|\langle e, 0 \rangle| = \kappa_{n-3}^F$

for some subconstructible (in F) ordinal σ. The Lemma of Further Reflection (3.3) implies there is a subconstructive (in F) ordinal γ such that

(2) $\qquad \langle e, 0 \rangle \in \mathbb{O}^{\langle F, (\tau C_{\tau^{-1}\gamma})^H, H_\sigma^F \rangle}$,

where $C_{\tau^{-1}\gamma}$ is that part of C enumerated by stage γ. A typical new requirement consists of preserving situations exemplified by (2); that means $C_{\tau^{-1}\gamma}$ is to equal $C \cap \tau^{-1}\gamma$ if γ satisfies (2). If such a requirement is met, then (1) cannot occur. ∎

7. PROOF OF THEOREM A

The results of Sections 6 and 7 imply Theorem A in short order. Fix $n > 2$ and let F be ^nE. Define $\mathcal{B}$ as in the remarks preceding 6.1. By 5.1 there exist hyperregular sets C and D, each Σ_1 over $\mathcal{B}$, such that neither is $\alpha - B$-recursive in the other. Clearly neither is Δ_1 in the other over $\mathcal{B}$, since both are regular. As indicated in 6.5, the proof of 5.1 can be modified so as to insure that C and D are generic over F.[6]

By 6.1 $(\tau C)^H$ and $(\tau D)^H$ are recursively enumerable in F. By 6.4 neither is recursive in the other together with F and any object of type $n - 3$.

Note that $\langle F, (\tau C)^H \rangle$ and $\langle F, (\tau D)^H \rangle$ have the same extended $(n-2)$-section as F.[7]

8. INADMISSIBLE RECURSION THEORY

Friedman [26] is the first paper on recursion theory over $L(\beta)$, where β is any infinite ordinal. The object of [26] is to show that Σ_1 admissibility is a crude global hypothesis which obscures the finer points of recursion theory. A number of results first proved for Σ_1 admissible β's are sketched in [26] for all β's. The most striking result so far of β-recursion theory is the existence of a pair of sets Σ_1 over $L(\beta)$ such that neither is β-recursive in the other, a theorem of Sy Friedman outlined in [3] and proved in detail in [26] for many β's.

To say β is Σ_1 admissible is to say every β-recursive enumeration (without repetitions) of every β-finite set is eventually complete. The dynamic view of RE sets alluded to in Section 1 does not imply every effective enumeration of every finite set is eventually complete, and hence does not imply that the RE sets satisfy any global hypothesis equivalent to admissibility.

The concept of admissibility retains much of its value, since theorems about β-recursively enumerable sets are proved by classifying the various failures of admissibility. Other ideas are also involved, for example S. Friedman [3, 26] invokes stationary subsets of initial segments of L in much the same way that Jensen [27] does.

It is not yet known if S. Friedman's work on $L(\beta)$ relativizes to $L(\beta, B)$, as it does in Section 5 where $L(\beta, B)$ is admissible. S. Simpson long ago conjectured that for each $n \geqslant 1$, there exist Σ_n subsets of $L(\beta)$ such that neither is Δ_n in the other. Shore's relativization of the solution of Post's problem from an admissible $L(\alpha)$ to an admissible $\langle L(\alpha, B), \varepsilon, B \rangle$ yields Simpson's conjecture when $L(\alpha)$ satisfies Σ_n replacement; simply take B to be the complete Σ_{n-1} set over $L(\alpha)$.

The further development of inadmissible recursion theory may supply the experience, so far lacking, needed to write down the axioms for RE sets.

NOTES

* The author is grateful to the National Science Foundation (P29079) for its support, to L. Harrington and R. Shore for several illuminating conversations, and the International Congress for Logic, Philosophy, and Methodology of Science (1975) for the invitation to give the following lecture.
[1] Suppose J, $K \subset 2^\omega$. J is said to be hyperarithmetic in K if J is recursive in K, ^2E.
[2] A functional G is recursively enumerable in F if there is an e such that all R, $R \in G \leftrightarrow \{e\}^{\langle F, R \rangle}(0)$ is defined.

[3] A formula of ZF is Δ_0 if all its quantifiers are restricted to sets.

[4] Keep in mind that $\{H_\sigma^F \mid \sigma < \kappa^F\}$ can be imbedded effectively in $\{H_\sigma^{\langle F,G \rangle} \mid \sigma < \kappa^{\langle F,G \rangle}\}$.

[5] τA is recursively enumerable in F means $\{H_\sigma^F \mid \sigma \in \tau A\}$ is recursively enumerable in F.

[6] The additional requirements are necessary, since it can be shown by further reflection that hyperregularity over $\mathscr{B}$ does not imply subgenericity over F for $\alpha - B$ recursively enumerable sets.

[7] The set of all type $(n-2)$ objects recursive in F.

BIBLIOGRAPHY

[1] E. Post, 'Recursively Enumerable Sets of Positive Integers and their Decision Problems', *Bull, Am. Math. Soc.* **50** (1944), 284–316.

[2] G. Kreisel, 'Set Theoretic Problems Suggested by the Notion of Potential Totality, *Infinitistic Methods*, Proceedings of the 1959 Warsaw Symposium, Pergamon Press, Oxford, 1961, 103–140.

[3] S. Friedman and G. E. Sacks, 'Inadmissible Recursion Theory', *Bull. Am. Math. Soc.*, to appear.

[4] R. Friedberg, 'Two Recursively Enumerable Sets of Incomparable Degrees of Insolvability', *Proc. Nat. Acad. Sci.* **43** (1957), 236–238.

[5] A. A. Muchnik, 'On the Unsolvability of the Problem of Reducibility in the Theory of Algorithms', *Doklady Akad. Nauk SSSR* **108** (1956), 194–197.

[6] S. C. Kleene, 'Recursive Functionals and Quantifiers of Finite Type', *Trans. Am. Math. Soc.* **91** (1959), 1–52; and **108** (1963), 106–142.

[7] L. Harrington, 'Reflection Principles Associated with Recursion in Objects of Finite Type', to appear.

[8] G. E. Sacks and S. G. Simpson, 'The α-Finite Injury Method', *Ann. Math. Logic*, 1972 (4), 343–367.

[9] R. Shore, 'Σ_n Sets which are Δ_n Incomparable (Uniformly)', *J. Symb. Logic* **39** (1974), 295–304.

[10] L. Harrington, 'Contributions to Recursion Theory in Higher Types', Ph.D. Thesis, M.I.T., 1973.

[11] J. Shoenfield, 'A Hierarchy Based on a Type 2 Object', *Trans. Am. Math. Soc.* **134** (1968), 103–108.

[12] T. Grilliot, 'Hierarchies Based on Objects of Finite Type', *J. Symb. Logic* **34** (1969), 177–182.

[13] G. E. Sacks, *Higher Recursion Theory*, notes by D. MacQueen of lectures given at M.I.T., 1971–72, to be published in revised form by Springer-Verlag.

[14] L. Harrington and D. MacQueen, Grilliot's Selection Principle, *J. Symb. Logic* **41** (1976), 153–158.

[15] R. Gandy, General Recursive Functionals of Finite Type and Hierarchies of Functions, University of Clermont-Ferrand (1962).

[16] Y. N. Moschovakis, 'Hyperanalytic Predicates', *Trans. Am. Math. Soc.* **138** (1967), 249–282.

[17] F. Lowenthal, 'Equivalence of Several Definitions of Recursion in Normal Objects of Finite Type', to appear.

[18] G. E. Sacks, 'The k-Section of a Type n Object', *Amer. J. Math.*, to appear.

[19] T. Grilliot, 'Selection Functions for Recursive Functionals, *Notre Dame J. Form. Logic* **10** (1969), 229–234.

[20] G. Takeuti, 'On the Recursive Functions of Ordinals', *J. Math. Soc. Japan* **12** (1960), 119–128.

[21] G. Kreisel, and G. E. Sacks, 'Metarecursive Sets I, II' (abstracts), *J. Symb. Logic* **28** (1963), 304–305.

[22] G. E. Sacks, 'Post's Problem, Admissible Ordinals and Regularity', *Trans. Am. Math. Soc.* **124** (1966), 1–23.

[23] S. G. Simpson, 'Admissible Ordinals and Recursion Theory', Ph.D. Thesis, M.I.T., 1971.

[24] J. Macintyre, 'Contributions to Metarecursion Theory', Ph.D. Thesis, M.I.T., 1968.

[25] M. Lerman and G. E. Sacks, 'Some Minimal Pairs of α-Recursively Enumerable Degrees', *Ann. Math. Logic* **4** (1972), 415–442.

[26] S. Friedman, 'Inadmissible Recursion Theory', Ph.D. Thesis, MIT, 1976.

[27] R. Jensen, 'The Fine Structure of the Constructible Hierarchy', *Ann. Math. Logic* **4** (1972), 229–308.

[28] G. E. Sacks, 'On the Reducibility of Π_1^1 Sets', *Advances in Math.* **7** (1971), 57–82.

I. A. LAVROV

COMPUTABLE NUMBERINGS

This is a more complete version of the author's invited address at the 5th International Congress of Logic, Methodology and Philosophy of Science held on 27 August to 5 September, 1975, London, Ontario, Canada. This article includes some new results obtained after the Congress. In the first version of the abstract and the address only the authors of respective results or papers were indicated where these results can be found. The present version contains accurate references. Proofs are not presented.

1. Basic definitions

Let $\mathcal{P}$ denote a class of all recursively enumerable sets. Various subclasses of class $\mathcal{P}$ are generally denoted by $\mathcal{S}$. Later on some most important subclasses of class $\mathcal{P}$ will have special notations.

A *numbering* of class $\mathcal{S}$ is a mapping $\nu: \mathcal{N} \to \mathcal{S}$, where $\mathcal{N}$ is a set of natural numbers. The numbering ν of class $\mathcal{S}$ is called *computable*, if the binary relation $x \in \nu y$ is recursively enumerable. Class $\mathcal{S}$ is *computable* if it possesses at least one computable numbering.

Another definition of the computable class equivalent to the previous one can be given. Let π be Post's famous canonical numbering of class $\mathcal{P}$. Class $\mathcal{S}$ is called *computable* iff there is a general recursive function f such that

$$\mathcal{S} = \{\pi_{f(0)}, \pi_{f(1)}, \ldots\}$$

The numbering of class $\mathcal{S}$ can be considered an original way of potentially effective presentation of class $\mathcal{S}$. Complexity of the class's structure is completely encoded in such a way of presentation of class $\mathcal{S}$ and can be often recognized by the properties of the numbering.

Butts and Hintikka (eds.), Logic, Foundations of Mathematics and Computability Theory, 195–206.
Copyright © 1977 by D. Reidel Publishing Company, Dordrecht-Holland. All Rights Reserved.

However there may be cases when class $\mathscr{S}$ is rather simple with respect to its structure, though it possesses complex and 'elegant' numberings. This compels us to compare the numberings of the same class $\mathscr{S}$.

Let ν_1 and ν_2 be two numberings of class $\mathscr{S}$. The numbering ν_1 is *reducible* to ν_2 (denoting $\nu_1 \leqslant \nu_2$) if there exists a general recursive function f such that $\nu_1(x) = \nu_2 f(x)$. If $\nu_1 \leqslant \nu_2$ and $\nu_2 \leqslant \nu_1$ these two numberings are called *equivalent.* Let $\mathscr{L}(\mathscr{S})$ be a partial order set of classes of the equivalent computable numberings of class $\mathscr{S}$. The partial order set $\mathscr{L}(\mathscr{S})$ is some measure of complexity of class $\mathscr{S}$ given in terms of the recursive theory.

Dealing with the elements of the set $\mathscr{L}(\mathscr{S})$ we will speak of numberings and not of the classes of equivalent numberings by choosing to this end any representative of the equivalence class.

The main cases are:

(1) $\mathscr{L}(\mathscr{P})$;
(2) $\mathscr{L}(\Phi)$, where Φ is a class of all unary partial recursive functions. (Speaking of a function as a set we mean the graph of this function which is considered as a linear set);
(3) $\mathscr{L}_m^1$ is a class of m-degrees of recursively enumerable sets; here $\mathscr{S} = \{\varnothing, \{0\}\}$;
(4) $\mathscr{L}_m^2$ is a class of m-degrees of pairs of disjoint recursively enumerable sets $\langle A; B \rangle$; here $\mathscr{S} = \{\varnothing, \{0\}, \{1\}\}$;
(5) $\mathscr{L}(\mathscr{S})$, where $\mathscr{S}$ is a subclass of class $\mathcal{O}$ of all unary general recursive functions.

The main task of the present paper is to inform some recent results and to note the principal ways of investigation into algebraic properties of the set $\mathscr{L}(\mathscr{S})$ depending on how class $\mathscr{S}$ is formed. Most results known in various fields of the recursive theory can be easily reformulated in terms of the set $\mathscr{L}(\mathscr{S})$. On the other hand the statement and the conception of many problems of the recursive theory became possible only after creation of the general theory of numberings.

The fundamental conceptions of the theory of numberings were first stated in [1], [2]. Fruitful development of these ideas, systematization of the cumulative material are in [3]. The first part of this book was translated into German: Ju. L. Eršov, 'Theorie der Numerierungen I', *Zeitschr. math. Logik Grund. Math.* **19,** No. 4 (1973), 289–388.

The author of the present paper gave a course of lectures on the theory of numberings at the Logical semester at S. Banach International Mathematical Center in 1973, (Warsaw, Poland) [4].

Our presentation is divided into some paragraphs, each devoted to a concrete algebraic property of the set $\mathcal{L}(\mathcal{S})$.

2. CARDINALITY OF THE SET $\mathcal{L}(\mathcal{S})$

From the results in [5], [6], [7] and [8] it follows that $\bar{\bar{\mathcal{L}}}(\mathcal{P}) = \bar{\bar{\mathcal{L}}}(\Phi) = \aleph_0$. We have $\bar{\bar{\mathcal{L}}}_m^1 = \bar{\bar{\mathcal{L}}}_m^2 = \aleph_0$ (it follows from R. Friedberg's [9] and A. A. Muchnik's [10] solution of Post's problem of the existence of Turing's incomparable recursively enumerable degrees of unsolvability; the direct proof can be also obtained from [11] and [8].) On the other hand class $\mathcal{O}$ of all unary general recursive functions has no computable numberings and, consequently, $\bar{\bar{\mathcal{L}}}(\mathcal{O}) = 0$. A computable class $\mathcal{S}$ can be constructed so that all computable numberings of this class are equivalent, i.e. $\bar{\bar{\mathcal{L}}}(\mathcal{S}) = 1$. A. B. Khutoretskii's theorem [8] states that $\aleph_0$, 0 and 1 is the whole possible spectrum of cardinalities for $\mathcal{L}(\mathcal{S})$ (see also [41]).

Attempts were made to give a description of classes $\mathcal{S}$ depending on which of the three possible cases for cardinality $\mathcal{L}(\mathcal{S})$ is satisfied. This problem has not yet been solved completely. Only some partial results are considered.

Let $\mathcal{S} \subset \mathcal{O}$. Then $\mathcal{S}$ can be considered as a subset of Baire's topological space. If $\mathcal{S}$ is not a discrete set with respect to the given topology (i.e. $\mathcal{S}$ contains, at least, one limit point) then $\bar{\bar{\mathcal{L}}}(\mathcal{S}) = \aleph_0$ [12]. If $\mathcal{S}$ is an effectively discrete set (at natural effectivization of the notion of discreteness), then $\bar{\bar{\mathcal{L}}}(\mathcal{S}) = 1$ [12]. Unfortunately, examples of discrete but not effectively discrete classes $\mathcal{S}$ can be constructed both with $\bar{\bar{\mathcal{L}}}(\mathcal{S}) = \aleph_0$ [13] and $\bar{\bar{\mathcal{L}}}(\mathcal{S}) = 1$ [14].

In a general case considering class $\mathcal{S}$ of recursively enumerable sets as a subset of the topological space $\mathcal{P}$ (natural topology of $\mathcal{P}$ based on finite sets is presented in [15]) we arrive also at the notions of non-discrete, discrete and effectively discrete sets. Again if $\mathcal{S}$ is an effectively discrete set, then $\bar{\bar{\mathcal{L}}}(\mathcal{S}) = 1$ [16]. A discrete, but not effectively discrete, class $\mathcal{S}$ is constructed so that $\bar{\bar{\mathcal{L}}}(\mathcal{S}) = 1$ [17]. Recently V. L. Selivanov has informed the author that he has an example of a non-discrete class $\mathcal{S}$ in which $\overline{\overline{\mathcal{L}}}(\mathcal{S}) = 1$. Examples when $\overline{\overline{\mathcal{L}}}(\mathcal{S}) = \aleph_0$ were constructed many times.

3. Is $\mathcal{L}(\mathcal{S})$ A LATTICE?

It is not difficult to show that the partial order set $\mathcal{L}(\mathcal{S})$ is an upper semilattice. For two computable numberings ν_1 and ν_2 of class $\mathcal{S}$ the numbering ν_3 defined as

$$\nu_3(x) = \begin{cases} \nu_1(n), & \text{if } \quad x = 2n, \\ \nu_2(n), & \text{if } \quad x = 2n+1, \end{cases}$$

is the least upper bound of ν_1 and ν_2.

The question arises as to whether class $\mathcal{S}$ can be constructed so that $\mathcal{L}(\mathcal{S})$ is not trivial and is a lattice (when $\mathcal{L}(\mathcal{S})$ is trivial, i.e. $\overline{\overline{\mathcal{L}}}(\mathcal{S}) \leqslant 1$, it is evident that $\mathcal{L}(\mathcal{S})$ is a lattice)?

$\mathcal{L}(\mathcal{P})$ and $\mathcal{L}(\Phi)$ are not lattices [5], [6]. Similarly, $\mathcal{L}_m^1$ and $\mathcal{L}_m^2$ are not lattices [18].

It is proved in [19] that if $\mathcal{S} \subset \mathcal{O}$ and $\mathcal{L}(\mathcal{S})$ are not trivial, then $\mathcal{L}(\mathcal{S})$ is not a lattice. Then it has been assumed that it is valid in the general case of class $\mathcal{S}$. Recently V. L. Selivanov has proved the statement: if $\mathcal{L}(\mathcal{S})$ is not trivial, $\mathcal{L}(\mathcal{S})$ is not a lattice.

4. EXISTENCE OF THE GREATEST ELEMENT IN $\mathcal{L}(\mathcal{S})$

It is a well-known fact that Post's numbering π of class $\mathcal{P}$ is the *main computable* numbering, i.e. we have $\nu \leqslant \pi$ for any computable numbering ν of class $\mathcal{P}$. Similarly Kleene's numbering φ of class Φ is the main computable numbering for Φ. If these properties of the numberings π and φ are presented in terms of the sets $\mathcal{L}(\mathcal{S})$ then π and φ are obtained as the greatest elements of the sets $\mathcal{L}(\mathcal{P})$ and $\mathcal{L}(\Phi)$ respectively. The question arises: what properties should class $\mathcal{S}$ possess for $\mathcal{L}(\mathcal{S})$ to have the greatest element?

Thus, if $\mathcal{S} \subset \mathcal{O}$ and $\mathcal{L}(\mathcal{S})$ is not a trivial set, $\mathcal{L}(\mathcal{S})$ has not the greatest element [19].

In spite of numerous attempts, sufficient and necessary conditions for existence of the greatest element in $\mathcal{L}(\mathcal{G})$ have not been found yet.

In many interesting cases Yu. L. Ershov's theorem on the description of families of computable sequences of recursively enumerable sets possessing the creative (m-universal) sequence gives the answer to the question of the existence of the greatest element in $\mathcal{L}(\mathcal{G})$[3]. Let us formulate this theorem and give some examples of its application.

Let $\mathcal{P}'$ denote a class of all computable sequences $(A_0, A_1, ...)$ of recursively enumerable sets, i.e. of such sequences that the binary predicate $x \in A_y$ is recursively enumerable. Class $\mathcal{P}'$ possesses the main computable numbering denoted by ρ (definition of computable and main numberings is naturally generalized to classes of sequences of recursively enumerable sets). Let $\rho(x) = (\rho_0(x), \rho_1(x), ...)$. The sequence $\alpha = (A_0, A_1, ...)$ is *reducible* to the sequence $\beta = (B_0, B_1, ...)$ (we denote $\alpha \leqslant \beta$), if there exists a general recursive function f such that

$$x \in A_y \leftrightarrow f(x) \in B_y.$$

Let $\mathcal{G}_0$ be some subclass of class $\mathcal{P}'$. $\mathcal{G}_0$ is *closed* if for any $\alpha \in \mathcal{G}_0$ and any partial recursive function g the sequence

$$g^{-1}\alpha = (g^{-1}A_0, g^{-1}A_1, ...) \in \mathcal{G}_0.$$

The sequence $\alpha = (A_0, A_1, ...) \in \mathcal{G}_0$ is *creative* for $\mathcal{G}_0$, if there exists a general recursive function f such that

$$(\forall x)\left\{\rho(x) \in \mathcal{G}_0 \rightarrow f(x) \in \bigcap_i [(A_i \cap \rho_i(x)) \cup (\bar{A}_i \cap \bar{\rho}_i(x))]\right\}.$$

The sequence α is *m-universal* for $\mathcal{G}_0$, if $\beta \leqslant \alpha$ for any $\beta \in \mathcal{G}_0$. It is not difficult to show that if $\mathcal{G}_0$ is a closed class and α is a creative sequence for $\mathcal{G}_0$, then α is m-universal for $\mathcal{G}_0$.

Ershov's theorem describes closed classes $\mathcal{G}_0$, possessing creative sequences. We need some more definitions for the exact formulation of this theorem.

Class $\mathcal{G}$ of recursively enumerable sets is *standard* if there exists a general recursive function f such that

(1) $\mathscr{S} = \{\pi_{f(0)},\ \pi_{f(1)},\ ...\}$;

(2) $\pi_x \in \mathscr{S} \to \pi_x = \pi_{f(x)}$.

The class of computable sequences $\alpha = (A_0, A_1, ...)$ of recursively enumerable sets, where all $A_i \in \mathscr{S}$, is denoted by $\mathscr{S}'$. Let for the sequence $\alpha = (A_0, A_1, ...)$, α^* denote the sequence $\beta = (B_0, B_1, ...)$ such that $x \in A_i \leftrightarrow i \in B_x$. Class $\mathscr{S}_0^*$ denotes the class of sequences α^* for $\alpha \in \mathscr{S}_0$.

ERSHOV'S THEOREM.   *Class $\mathscr{S}_0 \subseteq \mathscr{P}'$ is closed and contains a creative sequence iff for some standard class $\mathscr{S} \subseteq \mathscr{P}$ containing ϕ, we have $\mathscr{S}_0 = (\mathscr{S}')^*$.*

EXAMPLES.   (1) If $\mathscr{S} = \{\varnothing, \{0\}\}$, then $(\mathscr{S}')^*$ is a class of sequences of the form $(A, \varnothing, \varnothing, ...)$ where A is a recursively enumerable set. In this case Ershov's theorem is that of Myhill's on creative sets.

(2) If $\mathscr{S} = \{\varnothing, \{0\}, \{1\}\}$, then $(\mathscr{S}')^*$ is a class of sequences of the form $(A, B, \varnothing, \varnothing, ...)$ where A, B are disjoint recursively enumerable sets. In this case Ershov's theorem states the existence of an effectively inseparable pair of recursively enumerable sets.

Many other interesting examples of application of this theorem can be presented.

Another remark should be made. As is known from literature, a few attempts have been made to formulate the notion of creativity for different classes of objects. Some of the attempts have proved unsuccessful. Ershov's theorem gives a correct definition of creative objects and, in a sense, completes Myhill's theory on creative objects.

5.   EXISTENCE OF MINIMAL ELEMENTS IN $\mathscr{L}(\mathscr{S})$

The numbering ν of class $\mathscr{S}$ is *univalent* if the following condition is satisfied:

$x \neq y \to \nu x \neq \nu y$.

If ν is a univalent numbering, then each set of $\mathscr{S}$ has exactly one number. It is clear that any univalent numbering ν is a minimal element in $\mathscr{L}(\mathscr{S})$, i.e. for each computable numbering ν_1 of $\mathscr{S}$ we have

$$\nu_1 \leqslant \nu \rightarrow \nu \leqslant \nu_1.$$

First univalent numberings of classes $\mathscr{P}$ and Φ were constructed in [20]; thus, in $\mathscr{L}(\mathscr{P})$ and $\mathscr{L}(\Phi)$ there exist minimal elements.

Later classes $\mathscr{P}$ and Φ were proved to have infinitely many nonequivalent univalent numberings [5], [6]. Some sufficient conditions connected with the class $\mathscr{S}$ were suggested so that $\mathscr{L}(\mathscr{S})$ should have univalent numberings [6], [21], [22], [23], [24]. In case $\mathscr{S} \subset \mathscr{O}$ and $\mathscr{L}(\mathscr{S}) \neq \varnothing$ all minimal elements of $\mathscr{L}(\mathscr{S})$ are representable by univalent numberings [12]. There may be either 1 or $\aleph_0$ such minimal elements [19], [30]. There are classes $\mathscr{S}$ which have no computable univalent numberings although $\mathscr{L}(\mathscr{S})$ has minimal elements.

The numbering ν of $\mathscr{S}$ is *positive* if the binary relation $\nu x = \nu y$ is recursively enumerable. Positive computable numberings of $\mathscr{S}$ are also minimal elements in $\mathscr{L}(\mathscr{S})$ [16]. For classes $\mathscr{P}$ and Φ infinitely many non-equivalent positive numberings were constructed [6], [25], which were non-equivalent to the univalent ones. There are some sufficient conditions, connected with class $\mathscr{S}$ for $\mathscr{L}(\mathscr{S})$ to have positive numberings [6], [26], [29]. There exist classes $\mathscr{S}$ which have no computable positive numberings although $\mathscr{L}(\mathscr{S})$ has minimal elements [28].

For $\mathscr{P}$ and Φ some other kinds of minimal elements in $\mathscr{L}(\mathscr{P})$ and $\mathscr{L}(\Phi)$ have been constructed. There are infinitely many minimal elements in $\mathscr{L}(\mathscr{P})$ and $\mathscr{L}(\Phi)$ which are non-representable by positive numberings [27].

It appears that in $\mathscr{L}(\mathscr{P})$ there are infinitely many elements under which there are no minimal elements [27].

Finally there exist such computable classes $\mathscr{S}$ that $\mathscr{L}(\mathscr{S})$ has no minimal elements [17], [42].

It should be noted that in $\mathscr{L}_m^1$ (and $\mathscr{L}_m^2$) there is the least element, that is an m-degree of the non-trivial (different from $\varnothing$ and N) recursive set.

6. SEGMENTS IN $\mathscr{L}(\mathscr{S})$

The initial segments $\mathscr{L}_m^1$ are fully described in [31]. Without going into details we will note that such initial segments are direct limits of sequences of finite distributive lattices with 0 and 1, some parameters of which are effectivized by means of general recursive functions. Generalization of these results to arbitrary segments can be found in [32]. Coideals of the above mentioned direct limits of sequences of finite distributive lattices–such is a description of various segments of $\mathscr{L}(\mathscr{P})$ [33]. In [33] one can find a description of (not only initial) segments $\mathscr{L}(\mathscr{P})$ but also of segments $\mathscr{L}(\mathscr{S})$ for a wide range of classes $\mathscr{S}$ (see also [43], [44]).

7. Isomorphisms of $\mathscr{L}(\mathscr{S})$

When $\bar{\bar{\mathscr{L}}}(\mathscr{S}) = \aleph_0$, an important and interesting question arises as to whether semilattices $\mathscr{L}(\mathscr{S}_1)$ and $\mathscr{L}(\mathscr{S}_2)$ at different $\mathscr{S}_1$ and $\mathscr{S}_2$ can be isomorphic. It is very simple to construct examples of $\mathscr{S}_1$ and $\mathscr{S}_2$ when $\mathscr{L}(\mathscr{S}_1)$ and $\mathscr{L}(\mathscr{S}_2)$ are isomorphic. Complicated upper semilattices $\mathscr{L}(\mathscr{S}_1)$ and $\mathscr{L}(\mathscr{S}_2)$ have, as a rule, the same isomorphic types of initial segments. Therefore to construct classes $\mathscr{S}_1$ and $\mathscr{S}_2$ such that $\mathscr{L}(\mathscr{S}_1)$ and $\mathscr{L}(\mathscr{S}_2)$ are non-isomorphic, is a very difficult task. First these examples were constructed in [34]. Other examples of non-isomorphic $\mathscr{L}(\mathscr{S})$ can be found in [17].

At present it is not known whether $\mathscr{L}_m^1$ and $\mathscr{L}_m^2$ are isomorphic. Some hints at the positive answer can be found in [35] and [36].

8. Subobjects of Post's numbering

The theory of numberings distinguishes different kinds of subobjects of Post's numbering π [3]. Some kinds of subobjects have been systematically studied and described in detail. It would be of interest to obtain some additional properties of the set $\mathscr{L}(\mathscr{S})$ for the cases when $\mathscr{S}$ is one kind of subobjects of Post's numbering. Let us present definitions of some of these kinds of subobjects and the basic results connected with their description.

Let $\mathscr{D}_0$, $\mathscr{D}_1$, ... be a canonic numbering of all finite sets. Any sequence $\mathscr{T} = \{\mathscr{D}_{f(x)}\}$ where f is some general recursive function is a *strong array* of finite sets.

(1) Class $\mathscr{S}$ of recursively enumerable sets is a *completely recursive enumerable class* if $\{x/\pi_x \in \mathscr{S}\}$–a set of numbers of members of this class in Post's numbering–is recursively enumerable.

THEOREM [37] *Class $\mathscr{S}$ is a completely recursive enumerable class iff there is a strong array $\mathfrak{T} = \{\mathfrak{D}_{f(x)}\}$ such that*

$$\mathscr{S} = \{\pi_x/(\exists y)[\mathfrak{D}_{f(y)} \subseteq \pi_x]\}.$$

(2) Class $\mathscr{S}$ of recursively enumerable sets is an *effectively main class* if there exists a partial recursive function h such that

>   (a) $\mathscr{S} = \{\pi_{h(a_0)}, \pi_{h(a_1)}, ...\}$ where $\{a_0, a_1, ...\}$ is domain h;
>
>   (b) $\pi_x \in \mathscr{S} \rightarrow h(x)$ is defined & $\pi_x = \pi_{h(x)}$.

At present there is no theorem describing satisfactorily the effectively main classes.

(3) The effectively main class $\mathscr{S}$ is a *weak special standard class* if the h presented in the above definition satisfies the following property

>   (c) $h(x)$ is   defined $\rightarrow \pi_{h(x)} \subseteq \pi_x$.

THEOREM [38] *Class $\mathscr{S}$ is a weak special standard class iff there exists a strong array $\mathfrak{T} = \{\mathfrak{D}_{f(x)}\}$ such that each set of $\mathscr{S}$ is a union of the increasing chain of sets of $\mathfrak{T}$.*

(4) Class $\mathscr{S}$ of recursively enumerable sets is *standard* if there exists a general recursive function h such that

>   (a) $\mathscr{S} = \{\pi_{h(0)}, \pi_{h(1)}, ...\}$;
>
>   (b) $\pi_x \in \mathscr{S} \rightarrow \pi_x = \pi_{h(x)}$.

At present there is no theorem describing satisfactorily standard classes.

(5) A standard class $\mathscr{S}$ is a *special standard class* if the h given in the above definition satisfies the following property:

>   (c) $\pi_{h(x)} \subseteq \pi_x$.

THEOREM [39]   *Only weak special standard classes containing the empty set $\varnothing$ are special standard classes.*

(6)   Class $\mathscr{S}$ of recursively enumerable sets is *factorization* if there exists a general recursive function h such that

(a) $\mathscr{S} = \{\pi_{h(0)}, \pi_{h(1)}, \ldots\}$;

(b) $\pi_x = \pi_y \to \pi_{h(x)} = \pi_{h(y)}$.

THEOREM [40]   *Class $\mathscr{S}$ is a factorization iff there exist two strong arrays $\mathscr{T}_1 = \{\mathscr{D}_{\alpha(x)}\}$ and $\mathscr{T}_2 = \{\mathscr{D}_{\beta(x)}\}$ where $\varnothing \in \mathscr{T}_1$ are such that*

$$\pi_{h(x)} = \bigcup_{\mathscr{D}_{\alpha(i)} \subseteq \pi_x} \mathscr{D}_{\beta(i)}.$$

(7) Class $\mathscr{S}$ of recursively enumerable sets is *retract* if is $\mathscr{S}$ factorization and a standard class.

THEOREM [40]   *Factorization $\mathscr{S}$ for the strong arrays $\mathscr{T}_1 = \{\mathscr{D}_{\alpha(x)}\}$ and $\mathscr{T}_2 = \{\mathscr{D}_{\beta(x)}\}$ is retract iff*

(a) $(\forall i j_1 \cdots j_\zeta)[\mathscr{D}_{\alpha(i)} \subseteq \mathscr{D}_{\beta(j_1)} \cup \cdots \cup \mathscr{D}_{\beta(j_\zeta)} \to (\exists \tau_1 \cdots \tau_\kappa)$

$(\mathscr{D}_{\alpha(\tau_1)} \cup \cdots \cup \mathscr{D}_{\alpha(\tau_\kappa)} \subseteq \mathscr{D}_{\alpha(j_1)} \cup \cdots \cup \mathscr{D}_{\alpha(j_\zeta)} \&$

$\mathscr{D}_{\beta(i)} \subseteq \mathscr{D}_{\beta(\tau_1)} \cup \cdots \cup \mathscr{D}_{\beta(\tau_\kappa)})]$;

(b) $(\forall i)(\exists j_1 \cdots j_\zeta \tau_1 \cdots \tau_\kappa)[\mathscr{D}_{\alpha(\tau_1)} \cup \cdots \cup \mathscr{D}_{\alpha(\tau_\kappa)} \subseteq \mathscr{D}_{\alpha(i)} \&$

$\mathscr{D}_{\beta(i)} \subseteq \mathscr{D}_{\beta(j_1)} \cup \cdots \cup \mathscr{D}_{\beta(j_\zeta)} \&$

$\mathscr{D}_{\alpha(j_1)} \cup \cdots \cup \mathscr{D}_{\alpha(j_\zeta)} \subseteq \mathscr{D}_{\beta(\tau_1)} \cup \cdots \cup \mathscr{D}_{\beta(\tau_\kappa)}]$

Siberian Branch of the USSR Academy of Sciences, Novosibirsk

BIBLIOGRAPHY

[1] В. А. Успенский, *Лекции о вычислимых функциях*, Физматгиз, 1960.
[2] А. И. Мальцев, 'Конструктивные алгебры,' *Успехи мат. наук* **16**, No. 3 (1961), 3–60.
[3] Ю. Л. Ершов, *Теория нумераций*, ч. 1 (1969), ч. 2 (1973), ч. 3 (1974), Новосибирск, НГУ.

[4] I. A. Lavrov, 'Theory of Numberings', Logical semester, 1973, January-June, Stefan Banach International Mathematical Center for Raising Research Qualifications, Warsaw, Poland.

[5] M. B. Pour-El, 'Gödel Numberings versus Friedberg Numberings', *Proc. Am. Math. Soc.* **15**, No. 2 (1964), 252–255.

[6] А. Б. Хуторецкий, 'О сводимости вычислимых нумераций,' *Алгебра и логика* **8**, No. 2 (1969), 251–264.

[7] А. Б. Хуторецкий, 'О неглавных нумерациях,' *Алгебра и логика*, **8**, No. 6 (1969), 726–732.

[8] А. Б. Хуторецкий, 'О мощности верхней полурешётки вычислимых нумераций,' *Алгебра и логика* **10**, No. 5 (1971), 561–569.

[9] R. M. Freidberg, 'Two Recursively Enumerable Sets of Incomparable Degrees of Unsolvability (Solution of Post's Problem 1944)', *Proc. Nat. Acad. Sci.*, *USA* **43**, No. 2 (1957), 236–238.

[10] А. А. Мучник, 'Неразрешимостъ проблемы сводимости теории алгоритмов,' *Доклады АН СССР* **108**, No. 2 (1956), 194–197.

[11] С. Д. Денисов, 'Об *m*-степенях рекурсивно перечислимых множеств' *Алгебра и логика* **9**, No. 4 (1970), 422–427.

[12] Ю. Л. Ершов, 'Нумерации семейств общерекурсивных функций,' *Сибир. мат. журнал* **8**, No. 5 (1967), 1015–1025.

[13] Ю. Л. Ершов, И. А. Лавров, 'О вычислимых нумерациях II,' *Алгебра и логика* **8**, No. 1 (1969), 65–71.

[14] В. Л. Селиванов, 'О нумерациях семейств общерекурсивных функций,' готовится к печати.

[15] В. А. Успенский, 'Системы перечислимых множеств и их нумерации,' *Доклады АН СССР* **105**, No. 6 (1955), 1155–1158.

[16] А. И. Мальцев, 'Позитивные и негативные нумерации,' *Доклады АН СССР* **160**, No. 2 (1965), 278–280.

[17] В. В. Вьюгин, 'О некоторых примерах верхних полурешеток вычислимых нумераций,' *Алгебра и логика* **12**, No. 5 (1973), 512–529.

[18] Ю. Л. Ершов, 'Гипергиперпростые *m*-степени,' *Алгебра и логика* **8**, No. 5 (1969), 523–552.

[19] С. С. Марченков, 'О вычислимых нумерациях семейств общерекурсивных функций,' *Алгебра и логика* **11**, No. 5 (1972), 588–607.

[20] R. M. Freidberg, 'Three Theorems on Recursive Enumeration', *J. Symb. Logic* **23**, No. 3 (1958), 309–316.

[21] A. H. Lachlan, 'On Recursive Enumeration without Repetition', *Zeit. math. Logik und Grund. Math.* **11**, No. 3 (1965), 209–220; **13**, No. 2 (1967), 99–100.

[22] M. B. Pour-E. and W. A. Howard, 'A Structural Criterion for Recursive Enumeration without Repetition', *Zeit. math. Logik und Grund. Math.* **10**, No. 2 (1964), 105–114.

[23] M. B. Pour-El and H. Putnam, 'Recursively Enumerable Classes and their Application to Recursive Sequences of Formal Theories', *Arch. Math. Log. Grund. Math.* **8**, No. 3/4 (1965), 104–121.

[24] С. С. Марченков, 'О минимальных нумерациях систем рекурсивно перечислимых множеств,' *Доклады АН СССР* **198**, No. 3 (1971), 530–532.

[25] Ю. Л. Ершов, 'О вычислимых нумерациях,' *Алгебра и логика* **7**, No. 5 (1968), 71–99.

[26] В. В. Вьюгин, 'О дискретных классах рекурсивно перечислимых множеств,' *Алгебра и логика* **11**, No. 3 (1972), 243–256.

[27] А. Б. Хуторецкий, 'Две теоремы существования для вычислимых нумераций,' *Алгебра и логика* **8,** No. 4 (1969), 483–492.

[28] С. С. Марченков, 'О существовании семейств без позитивных нумераций,' *Матем. заметки* **13,** No. 4 (1973), 597–604.

[29] С. А. Бадаев, *Одно простое условие существования позитивных нумераций*, Третья Всесоюзняя конференция по матем. логике, Новосибирск, 1974, pp. 8–9.

[30] Ф. И. Валидов, *О минимальных нумерациях семейств общерекурсивных функций*, Третья Всесоюная конференция по матем. логике, Новосибирск, 1974, p. 26.

[31] A. H. Lachlan, 'Recursively Enumerable Many-One Degrees', *Алгебра и логика* **11,** No. 3(1972), 326–358.

[32] В. В. Вьюгин, 'Сегменты рекурсивно перечислимых *m*-степеней,' *Алгебра и логика* **13,** No. 6 (1974), 635–654.

[33] В. В. Вьюгин, *Структура начальных сегментов верхней полурешетки вычислимых нумераций класса всех р.п. множеств*, Третья Всесоюзная конференция по матем. логике, Новосибирск, 1974, 36–37.

[34] Ю. Л. Ершов и И. А. Ловров, 'Верхняя полурешетка $L(\gamma)$' *Алгебра и логика* **12,** No. 2 (1973), 167–189.

[35] Ю. Л. Ершов, 'Верхняя полурешетка нумераций конечного множества,' *Алгебра и логика* **14,** No. 3 (1975), 258–283.

[36] Е. А. Палютин, 'Дополнение к статье Ершова Ю.Л. "Верхняя полурешетка нумераций конечного множества",' *Алгебра и логика* **14,** No. 3 (1975), 284–287.

[37] H. G. Rice, 'On Completely Recursively Enumerable Classes and their Key Arrays', *J. Symb. Logic* **21,** No. 3 (1956), 304–308.

[38] А. И. Мальцев, 'К теории вычислимых семейств объектов,' *Алгебра и логика* **3,** No. 4 (1964), 5–31.

[39] A. H. Lachlan, 'Standard Classes of Recursively Enumerable Sets', *Zeit. Math. Logik und Grund Math.* **10,** 1 (1964), 23–42.

[40] И. А. Лавров, 'Некоторые свойства ретрактов нумерации Поста,' *Алгебра и логика* **13,** No. 6 (1974), 662–675.

[41] С. А. Бадаев, 'О несравнимых нумерациях,' *Сиб. мат. журнал* **15,** No. 4 (1974), 730–738.

[42] В. В. Вьюгин, 'О минимальных нумерациях вычислимых классов рекурсивно перечислимых множеств,' *Доклады АН СССР* **212,** No. 2 (1973), 273–275.

[43] В. В. Вьюгин, 'О верхних полурешетках нумераций,' *Доклады АН СССР* **217,** No. 4 (1974), 749–751.

[44] С. С. Марченкjв, 'О полуструктуре вычислимых нумераций,' *Доклады АН СССР* **198,** No. 4 (1971), 766–768.

YIANNIS N. MOSCHOVAKIS*

ON THE BASIC NOTIONS IN
THE THEORY OF INDUCTION

My main purpose here is to describe an abstract, algebraic context in which one can develop at least the rudiments of the theory of inductive definability. It will be obvious that all the standard examples of induction fit the abstract model in a natural way: these include ordinary recursion theory on the integers and its generalizations to various kinds of abstract structures as well as the theories of positive and non-monotone elementary induction, e.g. see Moschovakis (1974a, b). Despite this generality, the model is very simple and one can establish for it easily the basic results which are common to all these theories. My hope is that future monographs on inductive definability will start with a brief chapter on 'algebraic preliminaries' which will contain some version of this material.

The last section deals with Kleene's beautiful theory of recursion in higher types which also fits naturally in this abstract approach. In the forthcoming Kechris–Moschovakis [a] we will give an elementary exposition of recursion in higher types from this point of view, as a chapter in the theory of inductive definability.

1. INDUCTION ALGEBRAS

In some theories of induction we deal with partial functions, in others we deal with relations. Also the basic operations that we study vary widely from one theory to the next. Thus in an abstract theory it pays to take the basic objects and operations as (almost) completely arbitrary.

1.1. DEFINITION. An *induction algebra* is a structure

$$\mathbb{F} = \langle \{X^\alpha\}_{\alpha \in I}, \{\leq^\alpha\}_{\alpha \in I}, \{\vee^\alpha\}_{\alpha \in I}, \mathfrak{F} \rangle,$$

Butts and Hintikka (eds.), Logic, Foundations of Mathematics, and Computability Theory, 207–236.

where the following conditions hold:

(1)   Each X^α is a *non-empty set.*
(2)   For each index $\alpha \in I$, $\leq^\alpha$ is a *partial ordering* on X^α such that chains have sups (least upper bounds).

In particular, each X^α has a *minimum,* the supremum of the empty chain,

$$0 = 0^\alpha = \sup(\varnothing).$$

(3)   For each index $\alpha \in I$, $\vee^\alpha$ is a *binary operation (adjunction)* on X^α which satisfies the following two properties:

$$x \leq x \vee y,$$

$$x \leq y \Rightarrow x \vee y = y \quad (x, y \in X^\alpha).$$

(We write $x \leq y$ and $x \vee y$ for $x \leq^\alpha y$, $x \vee^\alpha y$ when the index is clear from the context or irrelevant.)

(4)   Each f in the set $\mathfrak{F}$ is an *operation* with domain some $X^{\alpha_1} \times \cdots \times X^{\alpha_n}$ and range some X^α,

$$f \colon X^{\alpha_1} \times \cdots \times X^{\alpha_n} \to X^\alpha;$$

moreover, the identity maps

$$f(x) = x \qquad (x \in X^\alpha)$$

are all in $\mathfrak{F}$ and $\mathfrak{F}$ is closed under composition with projections.

Here a projection is a map of the form

$$(x_1, \ldots, x_n) \mapsto x_i \qquad (1 \leq i \leq n)$$

and we are assuming that if g is in $\mathfrak{F}$ and f satisfies

$$f(x_1, \ldots, x_n) = g(\pi_1(x_1, \ldots, x_n), \ldots, \pi_m(x_1, \ldots, x_n))$$

with $\pi_1, ..., \pi_m$ given projections, then f is also in $\mathfrak{F}$. This implies immediately that we can permute, identify and add to the variables of some function in $\mathfrak{F}$ and stay in $\mathfrak{F}$.

A look at the two main classes of examples that we have in mind will illuminate the definition.

(I) *Induction Algebras of Relations*

Let A be a fixed non-empty set and for each $n = 0, 1, 2, ...$ take

$$X^n = \text{all } n\text{-ary relations on } A$$
$$= \{R: R \subseteq A^n\}.$$

It is convenient here to identify X^0 with the set of two constant relations of no arguments, *truth* and *falsity*,

$$X^0 = \{T, F\}.$$

In this kind of model, we naturally take *inclusion* for our partial ordering and *union* for adjunction,

$$R \leqslant S \Leftrightarrow R \subseteq S$$
$$R \vee S = R \cup S \qquad (R, S \text{ } n\text{-ary relations on } A).$$

A *second order relation* on A is a relation

$$\varphi(\bar{R}, \bar{s}) \Leftrightarrow \varphi(R_1, ..., R_n, s_1, ..., s_m)$$

with arguments relations on A and members of A – e.g. R_i here may vary over the k_i-ary relations on A, $i = 1, ..., n$. Each such second order relation determines an operation

$$f: X^{k_1} \times \cdots \times X^{k_n} \to X^m$$

in the obvious way,

$$f(\bar{R}) = \{\bar{s}: \varphi(\bar{R}, \bar{s})\}.$$

Conversely, each operation f determines a second order relation φ by

$$\varphi(\bar{R}, \bar{s}) \Leftrightarrow \bar{s} \in f(\bar{R}).$$

Given a set Φ of second order relations on A, let $\mathfrak{F}$ be the set of associated operations and put

$$\mathbb{R}[A, \Phi] = \langle \{X^n\}_{n \in \omega}, \{\subseteq\}_{n \in \omega}, \{\cup\}_{n \in \omega}, \mathfrak{F} \rangle.$$

Of course Φ must satisfy certain reasonable conditions if this structure is to be an induction algebra.

(II) *Induction Algebras of Partial Functions*

Let A, B be fixed non-empty sets and for each $n = 0, 1, 2, \ldots$ take

$$X^n = \text{all } n\text{-ary partial functions on } A \text{ to } B$$
$$= \{p : p \subseteq A^n \times B, p \text{ a function}\}.$$

Here X^0 consists of all the constants in B and the totally undefined partial function (of no arguments).

Again we take *inclusion* for our partial ordering,

$$p \leq q \Leftrightarrow p \subseteq q \quad (p, q, n\text{-ary partial functions on } A \text{ to } B).$$

For adjunction we take the less familiar operation of *consistent union*, where we adjoin to p the part of q which is consistent with p,

$$p \vee q = p \cup \{(x_1, \ldots, x_n, q(x_1, \ldots, x_n)) : p(x_1, \ldots, x_n)$$
$$\text{is undefined and } q(x_1, \ldots, x_n) \text{ is defined}\};$$

this operation clearly satisfies the conditions in (3) above.

In these algebras we typically have $B \subseteq A$, usually $B = A$ or $B = \omega \subseteq A$. As in (1), we often start with a collection Φ of *partial second order operations* (or *functionals*) on A to B, i.e. partial functions

$$\varphi(\bar{p}, \bar{s}) = \varphi(p_1, \ldots, p_n, s_1, \ldots, s_m)$$

with arguments partial functions on A to B and members of A and values in B. With each such φ we associate the obvious operation on partial functions

$$f(\bar{p}) = \lambda \bar{s} \varphi(\bar{p}, \bar{s});$$

conversely with each operation f we associate the functional

$$\varphi(\bar{p}, \bar{s}) = f(\bar{p})(\bar{s}).$$

Now let $\mathfrak{F}$ be the set of all operations associated with functionals in Φ and let

$$\mathbb{P}[A, B, \Phi] = \langle \{X^n\}_{n \in \omega}, \{\subseteq\}_{n \in \omega}, \{\vee\}_{n \in \omega}, \mathfrak{F} \rangle.$$

We will come back to a more detailed examination of examples of both types in Section 3 after we define the basic notions of the abstract theory in the next section.

It is often convenient to think of an induction algebra

$$\mathbb{F} = \langle \{X^\alpha\}_{\alpha \in I}, \{\leq^\alpha\}_{\alpha \in I}, \{\vee^\alpha\}_{\alpha \in I}, \mathfrak{F} \rangle$$

as a pair

$$\mathbb{F} = (\mathbb{X}, \mathfrak{F}),$$

where

$$\mathbb{X} = \langle \{X^\alpha\}_{\alpha \in I}, \{\leq^\alpha\}_{\alpha \in I}, \{\vee^\alpha\}_{\alpha \in I} \rangle$$

is *the domain* of $\mathbb{F}$. Sometimes we consider several algebras over the same fixed domain $\mathbb{X}$ – in which case it is natural to identify $\mathbb{F} = (\mathbb{X}, \mathfrak{F})$ with the set of operations $\mathfrak{F}$. For the particular types of algebras of relations and partial functions that we described above we use the notation

$$\mathbb{R}[A, \Phi] = (\mathbb{R}[A], \Phi),$$
$$\mathbb{P}[A, B, \Phi] = (\mathbb{P}[A, B], \Phi),$$

i.e. we denote by $\mathbb{R}[A]$ and $\mathbb{P}[A, B]$ the appropriate sets of relations and partial functions together with $\subseteq$ and $\cup$ or consistent union.

2.   INDUCTIVE POINTS AND OPERATIONS

Suppose

$$\mathbb{F} = \langle \{X^\alpha\}_{\alpha \in I}, \{\leq^\alpha\}_{\alpha \in I}, \{\vee^\alpha\}_{\alpha \in I}, \mathfrak{F} \rangle$$

is an induction algebra and

$$f: X^\alpha \to X^\alpha$$

is some operation in $\mathfrak{F}$. For each ordinal ξ, let f^ξ be defined by the recursion

$$(1) \qquad f^\xi = f^{<\xi} \vee f(f^{<\xi}),$$

where

$$(2) \qquad f^{<\xi} = \sup \{f^\eta : \eta < \xi\}$$

and let

$$f^\infty = \sup \{f^\xi : \xi \text{ is an ordinal}\};$$

we call f^∞ the *inductive point* in X^α determined by the operation f.

To justify this definition, we must prove that (1) and (2) determine uniquely an element f^ξ of X^α for each ordinal ξ. This is very easy – we will do it presently in the more general context of *simultaneous inductive definitions*.

Suppose $f_1, ..., f_n$ are n given operations in $\mathfrak{F}$ such that

$$f_i: X_1 \times \cdots \times X_n \to X_i \qquad (1 \leq i \leq n),$$

where of course each $X_i = X^{\alpha_i}$ is one of the basic domains of $\mathbb{F}$. This *system of operations* determines a mapping of $X_1 \times \cdots \times X_n$ into itself by

$$\bar{x} \mapsto (f_1(\bar{x}), f_2(\bar{x}), ..., f_n(\bar{x})),$$

where $\bar{x} = (x_1, ..., x_n)$. We want to obtain inductive points by iterating this mapping as we iterated f above.

First the lemma that justifies this recursion.

2.1. LEMMA. *Suppose $f_1, ..., f_n$ are operations in $\mathfrak{F}$ such that $f_i: X_1 \times \cdots \times X_n \to X_i$. For each ordinal ξ, there are unique points $f_1^{\xi}, ..., f_n^{\xi}$ in $X_1, ..., X_n$ respectively which satisfy the conditions*

$$(*) \qquad f_i^{\xi} = f_i^{<\xi} \vee f_i(f_1^{<\xi}, f_2^{<\xi}, ..., f_n^{<\xi}) \qquad (1 \le i \le n)$$

with

$$f_j^{<\xi} = \sup\{f_j^{\eta} : \eta < \xi\} \qquad (1 \le j \le n).$$

Proof. We prove simultaneously by induction on ξ that $\{f_1^{\xi}, ..., f_n^{\xi}\}$ are uniquely determined by (*) and that $\{f_j^{\eta} : \eta < \xi\}$ is a chain.

Assuming the result for $\eta < \xi$, it is enough to verify that $\{f_j^{\eta} : \eta < \xi\}$ is a chain, since then (*) determines f_i^{ξ} uniquely. This is trivial if ξ is limit (or 0) since an increasing union of chains is a chain. If $\xi = \zeta + 1$, then

$$\{f_j^{\eta} : \eta < \xi\} = \{f_j^{\eta} : \eta < \zeta\} \cup \{f_j^{<\zeta}\}$$

and by induction hypothesis,

$$f_j^{\zeta} = f_j^{<\zeta} \vee f_j(f_1^{<\zeta}, f_2^{<\zeta}, ..., f_n^{<\zeta});$$

now by the properties of adjunction

$$\sup\{f_j^{\eta} : \eta < \zeta\} = f_j^{<\zeta} \le f_j^{\zeta},$$

hence $\{f_j^{\eta} : \eta < \zeta + 1\}$ is obviously a chain. ∎

2.2. DEFINITION. For each system of operations

$$f_i\colon X_1 \times \cdots \times X_n \to X_i \qquad (1 \leq i \leq n)$$

in an induction algebra $\mathbb{F}$, put

(*) $f_i^\xi = f_i^{<\xi} \vee f_i(f_1^{<\xi}, f_2^{<\xi}, \ldots, f_n^{<\xi}) \qquad (1 \leq i \leq n)$

with

$$f_j^{<\xi} = \sup\{f_j^\eta\colon \eta < \xi\} \qquad (1 \leq j \leq n)$$

and let

$$f_i^\infty = \sup\{f_i^\xi\colon \xi \text{ is an ordinal}\}.$$

A point x in some X^α is $\mathbb{F}$-*inductive* if

$$x = f_i^\infty$$

for some system $f_1, \ldots, f_n$ in $\mathfrak{F}$ with $f_i\colon X_1 \times \cdots \times X_n \to X_i$ and $X_i = X^{\alpha_i}$.

We denote the collection of $\mathbb{F}$-inductive points in all the domains of $\mathbb{F}$ by $\mathbb{F}$-IND, or $\mathfrak{F}$-IND when $\mathbb{F} = (\mathbb{X}, \mathfrak{F})$ with $\mathbb{X}$ understood or held fixed for the discussion.

It is immediate from cardinality consideration that for each system

$$\bar{f} = (f_1, \ldots, f_n)$$

there is some ordinal κ such that

$$f_i^\infty = f_i^\kappa = f_i^{<\kappa} \qquad (1 \leq i \leq n);$$

the least such κ is *the closure ordinal of* $\bar{f}$, $\kappa(\bar{f})$. The *closure ordinal of the algebra* $\mathbb{F}$ is the supremum of these closure ordinals of the systems in $\mathbb{F}$,

$$\kappa(\mathbb{F}) = \sup\{\kappa(\bar{f})\colon \bar{f} = (f_1, \ldots, f_n) \text{ with}$$
$$f_i\colon X_1 \times \cdots \times X_n \to X_i \text{ in } \mathfrak{F}, \qquad 1 \leq i \leq n\}.$$

The basic definition can be easily *relativized* to an arbitrary sequence of parameters in the following way.

2.3. DEFINITION.   Suppose $f_1, ..., f_n$ are operations in some induction algebra $\mathbb{F}$, where

$$f_i\colon X_1 \times \cdots \times X_n \times Y_1 \times \cdots \times Y_m \to X_i.$$

For each ordinal ξ and each tuple

$$\bar{y} = (y_1, ..., y_m),$$

define the operation

$$f_i^\xi\colon Y_1 \times \cdots \times Y_m \to X_i$$

by the recursion

(*) $f_i^\xi(\bar{y}) = f_i^{<\xi} \vee f_i(f_1^{<\xi}(\bar{y}), f_2^{<\xi}(\bar{y}), ..., f_n^{<\xi}(\bar{y}), \bar{y}),$

where

$$f_j^{<\xi}(\bar{y}) = \sup \{f_j^\eta(\bar{y})\colon \eta < \xi\}$$

and let

$$f_i^\infty(\bar{y}) = \sup \{f_i^\xi(\bar{y})\colon \xi \text{ is an ordinal}\}.$$

An *operation*

$$f\colon Y_1 \times \cdots \times Y_m \to X_i$$

is $\mathbb{F}$-*inductive* if

$$f = f_i^\infty$$

for some such system $f_1, ..., f_n$ in $\mathbb{F}$ and some i.

The basic lemma that justifies (*) is just like Lemma 2.1.

We end this section with some remarks about the simplified case of *monotone* induction.

An operation $f: X_1 \times \cdots \times X_n \to X_{n+1}$ is *monotone* if

$$x_1 \leqslant x_1' \,\&\, x_2 \leqslant x_2' \,\&\, \cdots \,\&\, x_n \leqslant x_n' \Rightarrow f(x_1, ..., x_n)$$
$$\leqslant f(x_1', ..., x_n').$$

More generally, $f: X_1 \times \cdots \times X_n \times Y_1 \times \cdots \times Y_m \to X_{n+1}$ is *monotone in* $x_1, ..., x_n$ if for each $\bar{y} = (y_1, ..., y_n)$,

$$x_1 \leqslant x_1' \,\&\, x_2 \leqslant x_2' \,\&\, \cdots \,\&\, x_n \leqslant x_n' \Rightarrow f(x_1, ..., x_n, \bar{y})$$
$$\leqslant f(x_1', ..., x_n', \bar{y}).$$

2.4. LEMMA. *Let $\bar{f} = (f_1, ..., f_n)$ be a system of operations in some induction algebra $\mathbb{F}$ and assume that for some fixed i, the operation f_i is monotone. Then for every ordinal ξ,*

$$f_i^\xi = f_i(f_1^{<\xi}, f_2^{<\xi}, ..., f_n^{<\xi}).$$

Similarly, given a system $\bar{f} = (f_1, ..., f_n)$ with parameters,

$$f_i: X_1 \times \cdots \times X_n \times Y_1 \times \cdots \times Y_m \to X_i,$$

if f_i is monotone in $x_1, ..., x_n$, then for each ξ and $\bar{y}$,

$$f_i^\xi(\bar{y}) = f_i(f_1^{<\xi}(\bar{y}), f_2^{<\xi}(\bar{y}), ..., f_n^{<\xi}(\bar{y}), \bar{y}).$$

Proof is by induction on ξ. Granting the result for $\eta < \xi$, we have for each such η

$$f_i^\eta = f_i(f_1^{<\eta}, f_2^{<\eta}, ..., f_n^{<\eta})$$
$$\leqslant f_i(f_1^{<\xi}, f_2^{<\xi}, ..., f_n^{<\xi})$$

by monotonicity, since evidently $f_j^{<\eta} \leqslant f_j^{<\xi}$, when $\eta < \xi$. Taking the sup on the left we have

$$f_i^{<\xi} \leqslant f_i(f_1^{<\xi}, f_2^{<\xi}, ..., f_n^{<\xi})$$

so that

$$f_i^\xi = f_i^{<\xi} \vee f_i(f_1^{<\xi}, f_2^{<\xi}, ..., f_n^{<\xi})$$
$$= f_i(f_1^{<\xi}, f_2^{<\xi}, ..., f_n^{<\xi})$$

by the second basic property of adjunction.

The proof with parameters is identical. ∎

It follows from this simple result that if all the operations in $\mathfrak{F}$ are monotone, then the adjunction operation does not come into the definition of $\mathbb{F}$-IND and we might as well identify $\mathbb{F}$ with the simpler reduct $\langle \{X^\alpha\}_{\alpha \in I}, \{\leqslant^\alpha\}_{\alpha \in I}, \mathfrak{F} \rangle$.

In the case of monotone induction we can also give a simple ordinal-free characterization of the inductive points.

2.5. LEMMA. *Let $f_1, ..., f_n$ be monotone operations in some induction algebra $\mathbb{F}$, where*

$$f_i: X_1 \times \cdots \times X_n \to X_i,$$

put

$$\bar{X} = X_1 \times \cdots \times X_n$$

and define

$$\bar{f}: \bar{X} \to \bar{X}$$

by

$$\bar{f}(\bar{x}) = (f_1(\bar{x}), ..., f_n(\bar{x})).$$

On $\bar{X}$ define the partial ordering

$$(x_1, ..., x_n) \leqslant (y_1, ..., y_n) \Leftrightarrow x_1 \leqslant y_1 \& \cdots \& x_n \leqslant y_n.$$

Then $\bar{f}$ is monotone in this partial ordering and

$$\bar{f}^\infty = (f_1^\infty, ..., f_n^\infty)$$

is the least fixed point of $\bar{f}$. Spelled out, this means that

$$f_i(f_1^\infty, ..., f_n^\infty) = f_i^\infty \qquad (i = 1, ..., n)$$

and if for some $x_1, ..., x_n$,

$$f_i(x_1, ..., x_n) \leqslant x_i \qquad (i = 1, ..., n),$$

then

$$f_1^\infty \leqslant x_1, ..., f_n^\infty \leqslant x_n.$$

Proof is very easy using Lemma 2.4. ■

3. EXAMPLES

The results in this section are either immediate or follow very easily from well-known facts and the simple theorems to come in Section 4.

(I) *Induction Algebras of Relations*

Recall the notation $\mathbb{R}[A, \Phi]$ of Section 1, where A is some non-empty set and Φ is a collection of relations of the form $\varphi(R_1, ..., R_n, s_1, ..., s_m)$, where the s_i vary over A and each R_i ranges over the n_i-ary relations on A.

   A system of operations in an algebra $\mathbb{R}[A, \Phi]$ of this type is determined by n relations

$$\varphi_i(R_1, ..., R_n, s_1, ..., s_m) \Leftrightarrow \varphi_i(\bar{R}^i, \bar{s}^i) \qquad (1 \leqslant i \leqslant n)$$

in Φ, with the corresponding operations in the algebra defined by

$$f_i(\bar{R}) = \{\bar{s}^i : \varphi_i(\bar{R}^i, \bar{s}^i)\}.$$

The basic inductive definitions of the relations φ_i^ξ corresponding to f_i^ξ take the form

$$\varphi_i^\xi(\bar{s}^i) \Leftrightarrow \varphi_i^{<\xi}(\bar{s}^i) \vee \varphi_i(\varphi_1^{<\xi}, \varphi_2^{<\xi}, ..., \varphi_n^{<\xi}, \bar{s}^i),$$
$$\varphi_j^{<\xi}(\bar{s}^i) \Leftrightarrow (\exists \eta < \xi) \varphi_j^\eta(\bar{s}^i),$$

where '$\vee$' is the logical 'or'. If the relations $\varphi_i(\bar{R}^i, \bar{s}^i)$ are monotone, then we have the simpler equivalence

$$\varphi_i^\xi(\bar{s}^i) \Leftrightarrow \varphi_i(\varphi_1^{<\xi}, \varphi_2^{<\xi}, ..., \varphi_n^{<\xi}, \bar{s}^i).$$

Similarly, for the case of operations with parameters, we are given second order relations

$$\varphi_i(\bar{R}^i, \bar{U}, \bar{s}^i) \Leftrightarrow \varphi_i(R_1, ..., U_1, ..., U_n, s_1, ...) \qquad (1 \leq i \leq n)$$

and define by induction on ξ the relations

$$\varphi_i^\xi(\bar{U}, \bar{s}^i) \Leftrightarrow \varphi_i^{<\xi}(\bar{U}, \bar{s}^i)$$
$$\vee \varphi_i(\{\bar{s}^1 : \varphi_1^{<\xi}(\bar{U}, \bar{s}^1)\}, ..., \{\bar{s}^n : \varphi_n^{<\xi}(\bar{U}, \bar{s}^n)\}, \bar{s}^i),$$

where

$$\varphi_j^{<\xi}(\bar{U}, \bar{s}^i) \Leftrightarrow (\exists \eta < \xi)\varphi_j^\eta(\bar{U}, \bar{s}^i).$$

To illustrate the definitions, suppose $\omega \subseteq A$ and the following two relations are in Φ:

$$\varphi_1(R_1, R_2, U, \bar{s}, n) \Leftrightarrow n \in \omega \ \& \ [U(\bar{s}, n) \vee R_1(s, n+1)]$$
$$\varphi_2(R_1, R_2, U, \bar{s}) \Leftrightarrow R_1(\bar{s}, 0).$$

Then

$$\varphi_1^\xi(U, \bar{s}, n) \Leftrightarrow n \in \omega \ \& \ [U(\bar{s}, n) \vee \varphi_1^{<\xi}(U, \bar{s}, n+1)]$$
$$\varphi_2^\xi(U, \bar{s}) \Leftrightarrow \varphi_1^{<\xi}(U, \bar{s}, 0),$$

from which follows easily that

$$\varphi_2^\infty(U, s) \Leftrightarrow (\exists n)U(\bar{s}, n).$$

Thus the operation of *existential quantification* on ω

$$U \mapsto \exists^\omega U = \{\bar{s} : (\exists n)U(\bar{s}, n)\}$$

is inductive in this algebra; equivalently, the second order relation

$$\varphi(U, \bar{s}) \Leftrightarrow (\exists n)U(\bar{s}, n)$$

is inductive.

Examples of this type arise most often by fixing a structure $\mathfrak{U}$ with domain A and putting in Φ all relations definable by *formulas* of a certain form in some language appropriate for $\mathfrak{U}$. Of course we must choose a language with both individual and relation free variables.

(Ia) *Semi-recursive relations on ω.* Let

$$\mathbb{N} = \langle \omega, ', 0 \rangle$$

be the natural structure on the integers with the successor operation and the designated constant 0, let $\mathfrak{L}$ be the first-order language of this structure with equality and with free variables for relations (of any number of variables) added. Take

$$\Phi = \operatorname{Pos} \Pi_0^0$$

> $=$ all relations definable by formulas
> $\varphi(R_1, ..., R_n, s_1, ..., s_m)$ which are quantifier-free
> and in which $R_1, ..., R_n$ have only
> positive occurrences.

It is simple to check that

$$\mathbb{R}[\omega, \operatorname{Pos} \Pi_0^0]\text{-IND} = \textit{all semi-recursive (recursively}$$
$$\textit{enumerable) relations on } \omega.$$

In fact there is considerable flexibility in this basic example. We can take

> $\Phi = \operatorname{Pos} \Sigma_1^0 = $ all relations definable by positive formulas
> which are existential except for bounded
> quantifiers,

and still the inductive points are exactly the semirecursive relations on ω. In another direction, we can take

$$\Phi = \Pi_0^0 = \text{all relations definable by quantifier-free formulas}$$

and the inductive relations again are the semirecursive relations on ω.

This example generalizes directly to arbitrary sets and yields simple inductive characterizations of the *semi-prime computable* and the *semi-search computable* relations of Moschovakis (1969); see Grilliot (1971). We will omit a detailed discussion of these generalizations here.

(Ib) *Positive elementary induction.* Let $\mathfrak{U} = \langle A, - \rangle$ be a structure with domain A and take

$$\Phi = \mathrm{Pos}\, \Pi^0_\infty = \text{all relations definable by formulas}$$

$\varphi(R_1, ..., R_n, s_1, ..., s_m)$ of the first-order language of A allowing parameters from A and with only positive occurrences of $R_1, ..., R_n$.

If A is infinite, then in the terminology of Moschovakis [1974a]

$$\mathbb{R}[A, \mathrm{Pos}\, \Pi^0_\infty]\text{-IND} = \text{all inductive relations on } A.$$

If $\mathfrak{U} = \langle \omega, ', 0 \rangle$ is the natural structure on the integers, then the inductive relations of this example are precisely the Π^1_1 relations on ω. For arbitrary structures, this induction algebra is studied in detail in Moschovakis (1974a).

Both Ia and Ib are examples of *monotone* induction – all the operations in these algebras are monotone.

(Ic) *Non-monotone induction.* Consider $\mathbb{R}[A, \Phi]$ where Φ is a *reasonable, non-monotone* class of (second order) relations on some infinite set A in the sense of Moschovakis (1974b).

These examples were introduced in Richter (1971) (at least some of them, on ω) and were studied in Aczel–Richter (1972), (1974) as well as Moschovakis (1974b). They include the very interesting cases of Σ^0_n-induction ($n \geq 2$).

It is worth collecting in a definition some simple properties of collections of relations Φ which insure that $\mathbb{R}[A, \Phi]$ is an induction algebra and in fact yield several useful properties of $\mathbb{R}[A, \Phi]$-induction.

3.1. DEFINITION. Consider the following conditions on a collection Φ of second order relations on some set A.

(1)   For each n, the relation

$$\varphi(R, s_1, ..., s_n) \Leftrightarrow R(s_1, ..., s_n)$$

is in Φ (where R varies over the n-ary relations).

(2) Φ is closed under & and $\vee$; i.e. if $\varphi(\bar{R}, \bar{s})$, $\psi(\bar{R}, \bar{s})$ are in Φ, so are $\chi_1(\bar{R}, \bar{s})$, $\chi_2(\bar{R}, \bar{s})$ defined by

$$\chi_1(\bar{R}, \bar{s}) \Leftrightarrow \varphi(\bar{R}, \bar{s}) \,\&\, \psi(\bar{R}, \bar{s}),$$

$$\chi_2(\bar{R}, \bar{s}) \Leftrightarrow \varphi(\bar{R}, \bar{s}) \vee \psi(\bar{R}, \bar{s}).$$

(3) $\bar{\Phi}$ is closed under addition of variables of either type i.e. if $\varphi(\bar{R}, \bar{s})$ is in Φ (where either $\bar{R}$ or $\bar{s}$ may be an empty list of variables) so are the relations

$$\chi_1(\bar{R}, \bar{U}, \bar{s}) \Leftrightarrow \varphi(\bar{R}, \bar{s}),$$

$$\chi_2(\bar{R}, \bar{s}, t) \Leftrightarrow \varphi(\bar{R}, \bar{s}).$$

A *projection into A* is a map $\pi: A^n \to A$, where for some i, $1 \leqslant i \leqslant n$,

$$\pi(s_1, ..., s_n) = s_i.$$

(4) Φ is closed under substitution of projections into A; i.e. if $\pi_1(\bar{s}), ..., \pi_m(\bar{s})$ are projections into A, $\varphi(\bar{R}, \bar{t})$ is in Φ and

$$\chi(\bar{R}, \bar{s}) \Leftrightarrow \varphi(R, \pi_1(\bar{s}), \pi_2(\bar{s}), ..., \pi_m(\bar{s})),$$

then $\chi(\bar{R}, \bar{s})$ is also in Φ.

A *projection into the n-ary relations* is a map

$$(\bar{R}, \bar{s}) \to \pi(\bar{R}, \bar{s}) \subseteq A^n,$$

where for suitable projections $\pi_1, ..., \pi_m$ into A and some i,

$$\pi(\bar{R}, \bar{s}) = \{\bar{t}: R_i(\pi_1(\bar{t}, \bar{s}), \pi_2(\bar{t}, \bar{s}), ..., \pi_m(\bar{t}, \bar{s}))\};$$

e.g. the map

$$\pi(R_1, R_2, s_1, s_2, s_3) = \{t: R_2(s_1, s_1, s_2, t)\}$$

is a projection.

(5) Φ is closed under substitutions of projections into the n-ary relations for each n; i.e. if $\pi_1(\bar{R}, \bar{s}), ..., \pi_m(\bar{R}, \bar{s})$ are such projections, $\varphi(U_1, ..., U_m, \bar{s})$ is in Φ and

$$\chi(\bar{R}, \bar{s}) \Leftrightarrow \varphi(\pi_1(\bar{R}, \bar{s}), ..., \pi_m(\bar{R}, \bar{s}), \bar{s}),$$

then $x(\bar{R}, \bar{s})$ is also in Φ.

If Φ satisfies (1)–(5) we call it *barely adequate* and we call $\mathbb{R}[A, \Phi]$ an *induction algebra of relations* – it is immediate that $\mathbb{R}[A, \Phi]$ is in fact an induction algebra.

Clearly the Examples Ia–Ic are induction algebras of relations in this sense. We will not pursue here the theory of these algebras beyond the next lemma which illustrates the kind of simple result that we can prove.

3.2. LEMMA. *Let $\mathbb{R}[A, \Phi]$ be an induction algebra of relations and assume that $\omega \subseteq A$ and Φ contains the set $\{(m, n): m = n \ \& \ n \in \omega\}$ (as a binary relation on A) and is closed under substitution of 0 and the successor operation. Prove that the second order relations*

$$\varphi(U, \bar{s}) \Leftrightarrow (\exists n) U(\bar{s}, n)$$
$$\varphi(U, \bar{s}, n) \Leftrightarrow (\forall m \leqslant n) U(\bar{s}, m)$$

are both F-*inductive.*

Proof. The first result follows from the remarks in Section 2. To prove the second, verify easily from the hypothesis that the relations

$$\varphi_1(R_1, R_2, U, \bar{s}, m, n)$$
$$\Leftrightarrow U(\bar{s}, m) \ \& \ \{[m = n \in \omega] \lor R_1(\bar{s}, m+1, n)\}$$
$$\varphi_2(R_1, R_2, U, \bar{s}, n) \Leftrightarrow R_1(\bar{s}, 0, n)$$

are in $\mathfrak{F}$ and then check

$$\varphi_2^\infty(U, \bar{s}, n) \Leftrightarrow (\forall m \leq n)U(\bar{s}, m). \quad \blacksquare$$

(II) *Induction Algebras of Partial Functions*

Recall again the notation $\mathbb{P}[A, B, \Phi]$ of Section 1, where Φ is a set of (partial) functionals on A to B. As with relations, we can obtain examples of this type by fixing a language and putting in Φ all functionals definable by *terms* of a certain form. Perhaps it is simpler here to give first a definition of an *induction algebra of partial functions* and discuss the examples within that context.

3.3. DEFINITION. Consider the following conditions on a collection Φ of (partial) functionals on some A to B, where

$$\varnothing \neq B \subseteq A.$$

(1)   For each n, the functional

$$\varphi(p, s_1, ..., s_n) = p(s_1, ..., s_n)$$

is in Φ (where p varies over n-ary partial functions on A to B).

(2) Φ is closed under composition and definition by cases relative to some fixed $b_0 \in B$; i.e. if

$$\chi(\bar{p}, \bar{s}) = \varphi(\psi(p, \bar{s}), p, \bar{s})$$

and φ, ψ are in Φ, then so is χ and if

$$\chi(u, \bar{p}, \bar{s}) = \begin{cases} \varphi(\bar{p}, \bar{s}) & \text{if} \quad u = b_0 \\ \psi(\bar{p}, \bar{s}) & \text{if} \quad u \neq b_0 \end{cases}$$

with φ, ψ in Φ, then again χ is in Φ.

(3) Φ is closed under addition of variables of either kind.

(4) Φ is closed under substitution of projections into A; i.e. if $\pi_1(\bar{s}), ..., \pi_m(\bar{s})$ are projections into A, $\varphi(\bar{p}, \bar{t})$ is in Φ and

$$\chi(p, s) = \varphi(\bar{p}, \pi_1(\bar{s}), ..., \pi_m(\bar{s})),$$

then $\chi(\bar{p}, \bar{s})$ is also in Φ.

A *projection into the n-ary partial functions* on A to B is a map

$$(\bar{p}, \bar{s}) \mapsto \pi(\bar{p}, \bar{s}): A \to B,$$

where for suitable projections $\pi_1, ..., \pi_m$ into A and some i,

$$\pi(\bar{p}, \bar{s}) = \lambda \bar{t} p_i(\pi_1(\bar{t}, \bar{s}), ..., \pi_m(\bar{t}, \bar{s})).$$

(5) Φ is closed under substitutions of projections into the n-ary partial functions on A to B; i.e. if $\pi_1(\bar{p}, \bar{s}), ..., \pi_m(\bar{p}, \bar{s})$ are such projections, $\varphi(q_1, ..., q_m, \bar{s})$ is in Φ and

$$\chi(\bar{p}, \bar{s}) = \varphi(\pi_1(\bar{p}, \bar{s}), ..., \pi_m(\bar{p}, \bar{s}), \bar{s}),$$

then $\chi(\bar{p}, \bar{s})$ is also in Φ.

If Φ satisfies (1)–(5) we call it a *barely adequate* collection of functionals and we call $\mathbb{P}[A, B, \Phi]$ an *induction algebra of partial functions*.

(IIa) *Recursive partial functions on ω.* Take $A = B = \omega$ and let Φ be the least barely adequate collection of functionals which contains the successor operation $s \mapsto s+1$, the constant function $s \mapsto 0$ and the characteristic function of equality,

$$\chi_=(m, n) = \begin{cases} 0 & \text{if} \quad m = n, \\ 1 & \text{if} \quad m \neq n. \end{cases}$$

One can verify quite easily that the inductive points in this algebra are precisely the recursive partial functions.

As with relations, this example too generalizes easily to arbitrary sets and yields a characterization of *prime computability*.

We leave for Section 6 the very important example of recursion in higher types.

(III) *Induction Algebras of Mixed Type*

A natural example here is *search computability* as developed in Moschovakis (1969). One has an algebra

$$\mathbb{F} = \langle \{X^n\}_{n\in\omega}, \{\subseteq\}_{n\in\omega}, \{\cup\}_{n\in\omega}, \mathfrak{F} \rangle,$$

where the basic objects in each X^n are n-ary *partial multiple-valued functions* on some set A, i.e. $(n+1)$-*ary relations;* the operations in $\mathfrak{F}$ on the other hand reflect the fact that we think of our objects as functions – they include definition by cases, (multiple-valued) composition, etc.

(IV) *Shrinking Induction*

Take again some $A \neq \varnothing$,

$$X^n = \text{all } n\text{-ary relations on } A$$

but this time put

$$R \leqslant S \Leftrightarrow R \supseteq S,$$
$$R \vee S = R \cap S.$$

Now the minimum point in X^n is A^n and the inductive relations are obtained by shrinking down from A^n to some first, largest fixed point. Examples of such inductive definitions occur quite often, both with relations and with partial functions.

(V) *Direct Products – Global Theories*

Suppose

$$\mathbb{X}_j^\infty = \langle \{X^{\alpha,j}\}_{\alpha\in I}, \{\leqslant^{\alpha,j}\}_{\alpha\in I}, \{\vee^{\alpha,j}\}_{\alpha\in I} \rangle$$

is a structure for each $j \in J$ and put

$$\mathbb{X} = \prod_{j \in J} \mathbb{X}_j = \left\langle \left\{ \prod_{j \in J} X^{\alpha, j} \right\}_{\alpha \in I}, \{\leqslant^\alpha\}_{\alpha \in I}, \{\vee^\alpha\}_{\alpha \in I} \right\rangle,$$

where for arbitrary x, y in the product $\prod\limits_{j \in J} X^{\alpha, j}$,

$$x \leqslant^\alpha y \Leftrightarrow \quad \text{for each} \quad j, x(j) \leqslant^{\alpha, j} y(j),$$
$$(x \vee^\alpha y)(j) = x(j) \vee^{\alpha, j} y(j).$$

We can define interesting induction algebras by taking

$$\mathbb{F} = \langle \mathbb{X}, \mathfrak{F} \rangle,$$

where the operations in $\mathfrak{F}$ act componentwise on $\mathbb{X}$, in the obvious way.

For example, the basic members in $\mathbb{X}$ might be pairs of relations and partial functions.

For a more interesting example let

$$\mathfrak{U}_j = \langle A_j, - \rangle \qquad (j \in J)$$

be an indexed set of structures of the same similarity type and take $\mathbb{X}$ as above, with

$$\mathbb{X}_j = \langle \{X^{n, j}\}_{n \in \omega}, \{\subseteq\}_{n \in \omega}, \{\cup\}_{n \in \omega} \rangle,$$

where

$$X^{n, j} = \text{all } n\text{-ary relations on } A_j.$$

For $\mathfrak{F}$, take all operations on the relations in the A_j's which are defined *uniformly* by fixed positive formulas in the common elementary language of the structures $\mathfrak{U}_j$. This gives the theory of *global* or *uniform*, positive elementary induction on $\{\mathfrak{U}_j : j \in J\}$. One can formulate similarly global theories for various non-monotone inductions, prime computability, etc.

Undoubtedly there are many other interesting examples of induction algebras that the reader will think up, but these exhibit the range of the theory. We should point out that induction on relations (or partial functions) *on a finite set* is also covered naturally, so there may be some interesting algebras that arise in *computer science*.

4. RESULTS ABOUT INDUCTION ALGEBRAS

The great diversity of examples in Section 3 may suggest that nothing of any importance can be proved about all induction algebras. In fact the general results are simple but not entirely uninteresting.

Take first an easy but typical fact.

4.1. THEOREM. *Let* $\mathbb{F} = \langle \{X^\alpha\}_{\alpha \in I}, ... \rangle$ *be an induction algebra and suppose that*

$$f: X_1 \times \cdots \times X_n \to X$$

is a monotone operation in $\mathbb{F}$. *If* $x_1, ..., x_n$ *are* $\mathbb{F}$-*inductive, then* $f(x_1, ..., x_n)$ *is also* $\mathbb{F}$-*inductive.*

Proof. Suppose for simplicity of notation that f is unary and $x = f_1^\infty$, with $f_1, ..., f_n$ a system in $\mathbb{F}$. Consider the system

$$g_1(x_1, ..., x_n, y) = f_1(x_1, ..., x_n),$$
$$\cdots \qquad\qquad \cdots$$
$$g_n(x_1, ..., x_n, y) = f_n(x_1, ..., x_n)$$
$$g_{n+1}(x_1, ..., x_n, y) = f(x_1);$$

this is clearly in $\mathbb{F}$ and obviously

$$g_i^\xi = f_i^\xi \qquad (i = 1, ..., n)$$

so in particular

$$g_1^\infty = f_1^\infty = x.$$

Also, for any ξ,

$$g_{n+1}^\xi = f(g_1^{<\xi})$$

since f is monotone and hence taking ξ large enough,

$$g_{n+1}^\xi = f(g_1^\infty) = f(x),$$

so $f(x)$ is $\mathbb{F}$-inductive. ∎

This result already shows that in an induction algebra of relations the set of all inductive relations is closed under & and $\vee$. Similarly, in an induction algebra of partial functions, the set of inductive partial functions is closed under composition and definition by cases.

Call an operation

$$f\colon Y_1 \times \cdots \times Y_n \to X$$

naturally monotone $\mathbb{F}$-*inductive* if

$$f(\bar{y}) = f_i^\infty(\bar{y})$$

for some i and some system $f_1(\bar{x}, \bar{y}), ..., f_n(\bar{x}, \bar{y})$ in $\mathbb{F}$, where all these operations are monotone.

The following two results can be proved easily by slight extensions of the argument in 4.1.

4.2. THEOREM.   *Let* $\mathbb{F} = \langle \{X^\alpha\}_{\alpha \in I}, ... \rangle$ *be an induction algebra and suppose that*

$$f\colon X_1 \times \cdots \times X_n \to X$$

is naturally monotone $\mathbb{F}$-*inductive. If* $x_1, ..., x_n$ *are* $\mathbb{F}$-*inductive, then so is* $f(x_1, ..., x_n)$. ∎

4.3. THEOREM.   *Let* $\mathbb{F} = \langle \{X^\alpha\}_{\alpha \in I}, ... \rangle$ *be an induction algebra and suppose that* $g_1(\bar{x}), ..., g_m(\bar{x})$ *are* $\mathbb{F}$-*inductive operations and* $f(\bar{y})$ *is naturally monotone* $\mathbb{F}$-*inductive; then the composition*

$$h(\bar{x}) = f(g_1(\bar{x}), ..., g_m(\bar{x}))$$

is $\mathbb{F}$-*inductive.*

In particular, the set of naturally monotone $\mathbb{F}$-*inductive operations is closed under composition.* ∎

The first of these results combines with Lemma 3.1 (for example) to show that in $\mathbb{R}[A, \Phi]$ of the appropriate kind, the set of inductive relations is closed under unbounded existential and bounded universal quantification over ω. The second result implies that there are lots of naturally monotone inductive operations – which is important in view of the following basic fact.

4.4. THE COMPLETENESS THEOREM FOR MONOTONE INDUCTION. *Let* $\mathbb{F} = \langle\{X^\alpha\}_{\alpha \in I}, ...\rangle$ *be an induction algebra and suppose* $f_1(\bar{x}), ..., f_n(\bar{x})$ *is a system of operations*

$$f_i : X_1 \times \cdots \times X_n \to X_i,$$

where each f_i *is naturally monotone* $\mathbb{F}$-*inductive. If* f_i^ξ, f_i^∞ *are defined in the usual way, then* $f_1^\infty, ..., f_n^\infty$ *are all* $\mathbb{F}$-*inductive.*

Proof is just like that of the Positive Induction Completeness Theorem 6B.4 of Moschovakis (1974a) and we will omit it. ∎

In Example IIa, where the inductive points are the recursive partial functions on ω, the inductive operations are precisely Kleene's *recursive functionals* – this is easy to check. Thus Theorem 4.4 is a direct generalization of Kleene's *First Recursion Theorem*.

One can also prove various extensions of 4.4 to inductive operations, the simplest being the following:

4.5. THEOREM. *Let* $\mathbb{F} = \langle\{X^\alpha\}_{\alpha \in I}, ..., \mathfrak{F}\rangle$ *be an induction algebra where all the operations in* $\mathfrak{F}$ *are monotone, put*

$$\bar{\mathbb{F}} = \langle\{X^\alpha\}_{\alpha \in I}, ..., \bar{\mathfrak{F}}\rangle,$$

where $\bar{\mathfrak{F}}$ *consists of all* $\mathbb{F}$-*inductive operations* $(\mathfrak{F} \subseteq \bar{\mathfrak{F}})$; *then every* $\bar{\mathbb{F}}$-*inductive operation is in fact* $\mathbb{F}$-*inductive.* ∎

5. An abstract notion of finiteness

As evidence that our choice of primitive notions is rich enough, we give here an abstract definition of $\mathbb{F}$-*finiteness* which yields the standard objects in the usual examples. The results of this section are tentative and incomplete – we are not entirely confident that the present notion of finiteness will prove the best choice in the development of the theory.

Let $\mathbb{F} = \langle \{X^\alpha\}, ..., \mathfrak{F} \rangle$ be an induction algebra and let

$$Q \subseteq X_1 \times \cdots \times X_n = \bar{X}$$

be an relation with arguments in some of the domains of $\mathbb{F}$. If $\bar{Y} = Y_1 \times \cdots \times Y_m$ and

$$g : \bar{X} \times \bar{Y} \to Z$$

is any operation, define $f : \bar{X} \times \bar{Y} \to Z$ by

$$f(\bar{x}, \bar{y}) = \begin{cases} g(\bar{x}, \bar{y}) & \text{if} \quad Q(\bar{x}) \\ 0 & \text{if} \quad -Q(\bar{x}), \end{cases}$$

where 0 is the least element of Z; we call f *the restriction of* g *to* Q.

We say that Q is an *admissible restriction* for $\mathbb{F}$ if whenever g is in $\mathfrak{F}$, so is f.

Suppose for example that $\mathbb{F} = \mathbb{R}[A, \Phi]$ is an induction algebra of relations. Then $Q(R_1, ..., R_n)$ is an admissible restriction for $\mathbb{F}$ if whenever $\psi(\bar{R}, \bar{T}, \bar{s})$ is a second order relation in Φ, so is $\varphi(\bar{R}, \bar{T}, \bar{s})$ defined by

$$\varphi(\bar{R}, \bar{T}, \bar{s}) \Leftrightarrow \psi(\bar{R}, \bar{T}, \bar{s}) \,\&\, Q(\bar{R}).$$

Similarly, if $\mathbb{F} = \mathbb{P}[A, B, \Phi]$ is an induction algebra of partial functions, then $Q(p_1, ..., p_n)$ is an admissible restriction for $\mathbb{F}$ if whenever $\psi(\bar{p}, \bar{q}, \bar{s})$ is in Φ, so is the functional $\varphi(\bar{p}, \bar{q}, \bar{s})$ defined by

$$\varphi(\bar{p}, \bar{q}, \bar{s}) = \begin{cases} \psi(\bar{p}, \bar{q}, \bar{s}) & \text{if} \quad Q(\bar{p}), \\ \text{undefined} & \text{if} \quad -Q(\bar{p}). \end{cases}$$

These admissible restrictions play an important part in the theories of the standard algebras.

5.1. DEFINITION. Let $\mathbb{F} = \langle \{X^\alpha\}, ... \rangle$ be an induction algebra. A point x in some X^α is $\mathbb{F}$-*finite* if there is a system of operations $\bar{f} = (f_1, ..., f_n)$ in $\mathbb{F}$ and an admissible restriction Q for $\mathbb{F}$ such that the following hold:

(1) $x = f_i^\infty$ for some i,

(2)   for some ordinal λ such that $x = f_i^\lambda$, we have

$$-Q(f_1^{<\eta}, ..., f_n^{<\eta}) \quad \text{for all} \quad \eta \leqslant \lambda$$

$$Q(f_1^{<\eta}, ..., f_n^{<\eta}) \quad \text{for all} \quad \eta > \lambda.$$

Intuitively, x is $\mathbb{F}$-finite if there is an induction which generates x in such a way that we can tell when x is completed.

The next result is fairly easy by standard methods and we will not prove it here.

5.2. THEOREM.   *If* $\mathbb{F} = \mathbb{R}[\omega, \text{Pos } \Pi_0^0]$ *is the induction algebra of example* Ia *which generates the semirecursive relations on* ω, *then the* $\mathbb{F}$-*finite points are precisely the finite subsets of some* ω^n.

In the induction algebras of examples Ib *and* Ic, *the* $\mathbb{F}$-*finite points are the* $\mathbb{F}$-*hyperdefinable relations – the relations which are* $\mathbb{F}$-*inductive and have* $\mathbb{F}$-*inductive complements.*

In the induction algebra of example IIa *which generates the recursive partial functions on* ω, *the* $\mathbb{F}$-*finite points are precisely the partial functions with finite domain.* ∎

There is only one result about $\mathbb{F}$-finiteness which is worth putting down at this point. It is an obvious extension of Theorem 4 and it can be proved by the same method.

5.3. THEOREM.   *Let* $\mathbb{F} = \langle \{X^\alpha\}_{\alpha \in I}, ... \rangle$ *be an induction algebra and suppose that*

$$f: X_1 \times \cdots \times X_n \to X$$

is an operation in $\mathbb{F}$, *not necessarily monotone. If* $x_1, ..., x_n$ *are* $\mathbb{F}$-*finite, then* $f(x_1, ..., x_n)$ *is* $\mathbb{F}$-*inductive.* ■

This result is significant only in the case of non-monotone induction, when $\mathbb{F}$-IND is not (in general) closed under all the operations in $\mathbb{F}$.

6. RECURSION IN HIGHER TYPES

The starting point for the research on which we are reporting here was an attempt to develop Kleene's theory of *recursion in higher types* [introduced in Kleene (1959)] within the context of a general theory of induction. The details of this program are carried out in the forthcoming expository paper Kechris–Moschovakis [a]. Here we confine ourselves to a brief discussion of some of the ideas involved.

Consider a set A such that $\omega \subseteq A$ and a collection Φ of functionals on A into ω such that the following conditions hold:

(1) $\mathbb{P}[A, \omega, \Phi]$ is an induction algebra of partial functions, in the sense of II of Section 3.

(2) All functionals $\varphi(\bar{p}, \bar{s})$ in Φ are monotone.

(3) In Φ we have the following functions, all $= 0$ for arguments not in ω;

$$\chi_\omega(n) = 1 \quad \text{if} \quad n \in \omega, \quad \text{(characteristic function of } \omega\text{)},$$

$$i(n) = n \qquad\qquad \text{(identity)},$$

$$s(n) = n + 1 \qquad\quad \text{(successor)},$$

$$\chi_=(n, m) = \begin{cases} 1 & \text{if} \quad n = m \\ 0 & \text{if} \quad n \neq m \end{cases}.$$

$$\text{(characteristic function of} = \text{on } \omega\text{)}$$

(4) Φ is closed under functional substitution in the following sense; if $\psi(p_1, ..., p_n, \bar{s})$ and $\psi_1(\bar{q}, \bar{t}^1, \bar{s}), ..., \psi_n(\bar{q}, \bar{t}^n, \bar{s})$ are all in Φ, so is $\varphi(\bar{q}, \bar{s})$ defined by

$$\varphi(\bar{q}, \bar{s}) = \psi(\lambda \bar{t}^1 \psi_1(\bar{q}, \bar{t}^1, \bar{s}), ..., \lambda \bar{t}^n \psi_m(\bar{q}, \bar{t}^n, \bar{s})).$$

If (1)–(4) hold, call $\mathbb{P}[A, \omega, \Phi]$ a *monotone recursion algebra into* ω.

The induction algebra of Example IIa which leads to ordinary recursion theory clearly satisfies (1)–(4). *So does recursion in higher types, with a suitable (natural) choice of A and Φ.* Moreover, all the *basic* properties of recursion in higher types are easy to formulate in this more general context – and much easier to prove.

To take a single but important example, suppose $\varphi(p, \bar{s})$ is a functional in some monotone recursion algebra into ω which determines *the fixed point $\varphi^\infty(\bar{s})$* by

$$\varphi^\xi(\bar{s}) = \varphi(\varphi^{<\xi}, \bar{s}).$$

For each $\bar{s}$, put

$$|\bar{s}|_\varphi = \begin{cases} \text{least } \xi \text{ such that } \varphi^\xi(\bar{s}) \text{ is defined,} \\ \text{if } \varphi^\infty(\bar{s}) \text{ is defined,} \\ \infty, \text{ if } \varphi^\infty(\bar{s}) \text{ is undefined,} \end{cases}$$

and let $\psi(\bar{s}, \bar{t})$ be the *stage comparison partial function* defined by

$$\psi(\bar{s}, \bar{t}) = \begin{cases} 0, \text{ if } \varphi^\infty(\bar{s}) \text{ is defined and } |\bar{s}|_\varphi \leq |\bar{t}|_\varphi, \\ 1, \text{ if } \varphi^\infty(\bar{t}) \text{ is defined and } |\bar{t}|_\varphi < |\bar{s}|_\varphi, \\ \text{undefined otherwise.} \end{cases}$$

A natural problem is to determine conditions on φ which insure that ψ is Φ-inductive.

Call $\varphi(p, \bar{s})$ *normal (in Φ, i.e. in the algebra $\mathbb{P}[A, \omega, \Phi]$)* if there is some Φ-inductive functional $\Delta_\varphi(p, q, \bar{s})$ such that the following conditions hold:

(1) if $\varphi(p \upharpoonright \{\bar{t}:q(\bar{t}) = 0\}, \bar{s})$ is defined, then $\Delta_\varphi(p, q, \bar{s}) = 0$,

(2) if q is a total function, $\{\bar{t}: q(\bar{t}) = 0\} \subseteq \text{domain}(p)$ and $\varphi(p \upharpoonright \{\bar{t}: q(\bar{t}) = 0\}, \bar{s})$ is not defined, then $\Delta_\varphi(p, q, \bar{s}) = 1$.

6.1. THE STAGE COMPARISON THEOREM FOR NORMAL FUNCTIONALS.
Let $\mathbb{P}[A, \omega, \Phi]$ be a monotone recursion algebra into ω and let $\varphi(p, \bar{s})$ be a normal (in Φ) functional in Φ. Then the stage comparison partial function ψ associated with φ^∞ is Φ-inductive. ■

The details of this proof are given in Kechris-Moschovakis [a], but it is really just a simple variation of the proof of the Stage Comparison Theorem 2A.2 in Moschovakis [1974a]. On the other hand, when we apply this result to the algebra associated with recursion in higher types, we obtain the so-called selection theorems which are basic to that theory – see Gandy [1962], Platek [1966], Moschovakis [1967] Grilliot [1969], and Hinman [1969]. The original proofs of these results were quite complicated and did not bring out the natural relationship of the selection theorems with the stage comparison results of inductive definability.

University of California, Los Angeles

BIBLIOGRAPHY

Aczel, P. and Richter, W.: 1972, 'Inductive Definitions and Analogues of Large Cardinals', Conf. in Mathematical Logic, London 1970, *Springer Lecture Notes in Math.* **255**, 1–9.

Aczel, P. and Richter, W.: 1974 'Inductive Definitions and Reflecting Properties of Admissible Ordinals', in J. E. Fenstad and P. Hinman (eds.), *Generalized Recursion Theory*, North Holland, Amsterdam.

Gandy, R. O.: 1962, 'General Recursive Functionals of Finite Type and Hierarchies of Functions', in *Proc. Logic Colloq. Clermont Ferrand*, pp. 5–24.

Grilliot, T. J.: 1969, 'Selection Functions for Recursive Functionals', *Notre Dame Journal of Formal Logic* **10**, 225–234.

Grilliot, T. J.: 1971, 'Inductive Definitions and Computability', *Trans. Am. Math. Soc.* **158**, 309–317.

Hinman, P. G.: 1969, 'Hierarchies of Effective Descriptive Set Theory', *Trans. Am. Math. Soc.* **142**, 111–140.

Kechris, A. S. and Moschovakis, Y. N.: a, 'Recursion in Higher Types', to appear in K. J. Barwise (ed.), *Handbook of Mathematical Logic*.

Moschovakis, Y. N.: 1967, 'Hyperanalytic Predicates', *Trans. Am. Math. Soc.* **129**, 249–282.

Moschovakis, Y. N.: 1969, 'Abstract First Order Computability I and II', *Trans. Am. Math. Soc.* **138**, 427–504.

Moschovakis, Y. N.: 1974a, *Elementary Induction on Abstract Structures*, North Holland, Amsterdam, 1974.

Moschovakis, Y. N.: 1974b, 'On Nonmonotone Inductive Definability', *Fund. Math.* **82**, 39–83.

Platek, R.: 1966, 'Foundations of Recursion Theory', Doctoral Dissertation, Stanford University, Stanford, California.

Richter, W: 1971, 'Recursively Mahlo Ordinals and Inductive Definitions', in: R. O. Gandy and C. E. M. Yates (eds.), *Logic Colloquium '69,* North Holland, Amsterdam.

NOTE

* During the preparation of this paper the author was partially supported by NSF, Grant MPS74-06732.

KAREL ČULÍK

BASIC CONCEPTS OF COMPUTER
SCIENCE AND LOGIC

0. PREFACE

Some clarifications and mathematical definitions which are concerned
with the many fundamental concepts of computability theory based
directly on the computer field, are presented. The sole aim of this
paper is to attempt to bridge the gulf between the computer field on
the one hand, and mathematics and logic on the other. The particular
concepts are presented here for justification only: and therefore almost
no theorems are included. (Fortunately there is no employer who could
press me to prove the theorems!). The intention is to provide such
mathematical definitions as would be very relevant to computers and
to their languages, and even more, to their development and design.

The crucial concept of language is discussed in Sections 1–6 because
its relevance in the computer field cannot be overestimated, and only
in the remaining Sections 7–12 are the more complex computer
concepts treated.

1. LANGUAGE

We may recognize a language anywhere where we are able to differentiate three main components: a set Symb of *symbols*, a set Mean of
meanings and a binary *understanding relation* (or semantics) $\triangleright$ such
that, if $x \in$ Symb, $m \in$ Mean and $x \triangleright m$ holds, then 'm is the meaning
(or understanding) of x'. Thus a language is a structure $\langle$Symb,
Mean, $\triangleright \rangle$ such that

$$(1.1) \quad \varnothing \neq \triangleright \in \text{Symb} \times \text{Mean};$$

$$(1.2) \quad \text{Symb} \cap \text{Mean} = \varnothing,$$

is a simplified concept, sufficient (after further clarifications) for natural
languages as well as for the usual mathematics, but not for computer
storage.

We introduce a language in order to *use* it; this is characterized by

Butts and Hintikka (eds.), Logic, Foundations of Mathematics and Computability Theory, 237–266.
Copyright © 1977 by D. Reidel Publishing Company, Dordrecht-Holland. All Rights Reserved.

the following basic language convention:

(1.3)   one writes a symbol $x \in$ Symb, but one thinks of a meaning
          of x, i.e. of an $m \in$ Mean, where $x \triangleright m$.

Therefore when a language $\langle$Symb, Mean, $\triangleright\rangle$ is used, only the symbols from Symb do actually occur; this is the reason why often only the set Symb itself is considered as the language. This concept of a formal language (namely a set of strings of some basic symbols) has its origin in logic and is studied rather intensively by pure mathematicians. Although the formal language is concerned with the syntax of the 'actual' language only–the heart of the 'actual' language is the understanding relation. It is an awkward confusion not to differentiate between formal and 'actual' languages.

The lack of a clear concept of an 'actual' language, with all further specifications, is one of the reasons for many confusions concerning programming and natural languages, and for conceptual difficulties with their translations. Here the independent basic concept of the meaning seems to be of the same explanatory importance that Newton's concept of force is in physics.

With each language $\langle$Symb, Mean, $\triangleright\rangle$ an equality '=' (i.e. the synonymy relation) is associated according to the following natural definition:

(1.4) $x = y$, where x, $y \in$ Symb, iff $(x \triangleright m \Leftrightarrow y \triangleright m)$ holds for each
          $m \in$ Mean.

Obviously the equality = is an equivalence relation in Symb.

Further we shall restrict ourselves only to the *understanding functions*, i.e. to the case when $\triangleright$ satisfies the condition of *unambiguity*:

(1.5)   to each symbol $x \in$ Symb there exists at most one meaning
          $m \in$ Mean such that $x \triangleright m$,

and then instead of $\triangleright$ we shall often use the functional symbol S. Note that in the general case it is obviously incorrect to write $Sx = m$ for $x \triangleright m$.

All programming languages, the language of mathematics–but not necessarily natural languages–are unambiguous.

2. Variables and constants

The need for clarification of the concept of a variable was stressed by my friend and teacher L. Rieger many times. According to our intuition a variable is a symbol, the different occurrences of which may have different meanings. This can occur for unambiguous languages only with respect to several understanding functions and under the condition

(2.1) $\text{Mean} \geqslant 2$.

Therefore let a *language system* be a set $LS =_{df} \{\langle \text{Symb}, \text{Mean}, S_i \rangle;\ i \in \mathfrak{I}\}$ of unambiguous languages, all of which have the same sets of symbols and meanings. Then with respect to LS

(2.2)   a symbol $x \in \text{Symb}$ is called a variable or a constant, if there exist or do not exist two indices $i, j \in \mathfrak{I}$ such that $S_i x \neq S_j x$ respectively.

If Var or Const is the set of all variables or constants then

(2.3) $\text{Symb} = \text{Var} \cup \text{Const}$ and $\text{Var} \cap \text{Const} = \varnothing$.

With regard to (2.3) each understanding function S_i can be determined by two of its partial functions: *variable understanding function* $S_i |_{\text{Var}}$ and *constant understanding function* $S_i |_{\text{Const}}$. Obviously from (2.2) it follows that

(2.4) $S_i |_{\text{Const}} = S_j |_{\text{Const}}$ for all $i, j \in \mathfrak{I}$, i.e. there is exactly one constant understanding function.

Therefore using these new specifications we can say that a simplified language is the following structure:

(2.5) $\langle \text{Var}, \text{Const}, \text{Mean}, \text{Mean}^{\text{Var}}, CS \rangle$ where Mean^{Var} is the set of all possible (total) variable understanding functions, while CS is one (total) constant understanding function, i.e. $CS \in \text{Mean}^{\text{Const}}$, and obviously (1.1), (1.2), (2.3) and (2.1) are satisfied.

Let us point out that the introduction of a huge set Mean^{Var} of understanding functions is of crucial importance. (In [6] only one understanding function is considered.)

For example, the language Intg, for speaking about integers (natural numbers), is as follows: $\langle$Var Intg, Const Intg, Intg, $\text{Intg}^{\text{Var Intg}}$, $CS \in \text{Intg}^{\text{Const Int}}\rangle$. In this case if $S_1 \in \text{Intg}^{\text{Var Intg}}$ and $x \in$ Var Intg, $4 \in$ Const Intg, we can write $x = 4$ (when $S_1(x) = 4$).

3. DENOTATION AND INTERPRETATION.
COMPOSED MEANINGS

It is a simplification of the concept of language to omit properties of the set Mean, but in fact a *categorization of meanings*, i.e. a decomposition $\overline{\text{Mean}}$ of the set Mean is always assumed. This implies a corresponding decomposition of symbols (in general of both variables and constants) $\overline{\text{Symb}}$ such that if $\text{Symb}^* \in \overline{\text{Symb}}$ corresponds to $\text{Mean}^* \in \overline{\text{Mean}}$, then the understanding functions S_i^* are differentiated such that Domain $S_i^* = \text{Symb}^*$ and Range $S_i^* \subset \text{Mean}^*$.

In mathematics a set $\text{Obj} \subset$ Mean of 'individual' objects, a set $\text{Rel} \subset$ Mean of *relations* in Obj, and a set Opr of *operations* in Obj, are usually differentiated; this constitutes a *'relational' structure* $\langle$Obj, Rel, Opr$\rangle$, and therefore the following subsets of symbols are differentiated: $\text{SymbObj} \cup \text{SymbRel} \cup \text{SymbOpr} \subset \text{Symb}$ generally $\text{SymbObj} = \text{Var}$ Symb $\text{Obj} \cup \text{Const Symb Obj}$, etc.

In logic each understanding function concerning objects or relations and operations, is called a *denotation* or an *interpretation* (*connotation*) respectively.

In a formal language of a first order logic no individual constants and no variables for relations and operations are admitted (i.e. $\text{ConstSymbObj} = \text{VarSymbRel} = \text{VarSymbOpr} = \varnothing$). Further an unrestricted number of individual variables is assumed in VarSymbObj, which is common to all theories, and the same is true for all logical and other auxiliary symbols. Therefore they are not specified separately in each particular case. Thus the formal language is specified only by two sets ConstSymbRel and ConstSymbOpr.

By an interpretation int of this formal language in a model $\langle$Obj, Rel, Opr$\rangle$ (i.e. Domain int $|_{\text{ConstSymbRel}} = $ Rel and Domain int $|_{\text{ConstSymbOpr}} = $ Opr) the actual alnguage:

(3.1) $\langle$VarSymbObj,   ConstSymbRel$\cup$ConstSymbOpr,   Obj$\cup$ Rel$\cup$Opr, Obj^{VarSymbObj}, int$\rangle$ is determined in full accordance with (2.5).

The objects are atomic meanings, i.e. they do not depend on any other meaning category, and therefore in a denotation there is no dependence between a possible structure of symbols and the structure of objects, denoted by them. The same is true about names of relations and operations (when their arity is not specified). Therefore, according to [7], arbitrary atomic symbols can be used as names for atomic meanings.

On the other hand formulas and terms (or programs) are not atomic symbols, because they are composed of atomic ones, and actually they express composed meanings of situations and activities, respectively, because they depend on other categories of meanings, i.e. on objects and relations and operations. The corresponding understanding function cannot be chosen freely, but with careful regard to the structures of symbols and meanings. Only here the dependancy between syntax and semantics of a language starts.

It is important to remember, that within the usual mathematical theory of the arithmetics of natural numbers the formula $x = 4$ cannot occur (as 4 is a constant), although such a formula is used by a numerical mathematician very often.

4.   DETERMINATION OF A LANGUAGE. EXHIBITIONS. METALANGUAGE

A language can be sometimes determined by another language (i.e. by usual definitions, See Section 6), but if the regressus ad infinitum is to be avoided, there must be a first language, given independently of any language. Often there is no difficulty with Symb, as it is sufficient to put down a list of all particular symbols from Symb, but obviously then the basic language convention (1.3) is violated, namely we do not use the occurrences of symbols in that list as expressing anything, we are exhibiting (or showing) them only. In fact there is no other way of determining the concrete alphabet of basic symbols of any formal language.

Probably in order to save the basic language convention R. Carnap [1] introduced his 'Autonyme Redeweise', i.e. he assumed that the exhibited symbol expresses itself. Thus one must admit

$$(4.1) \qquad x \triangleright x.$$

The possible confusions between a use of a symbol and its exhibition can be avoided by parenthesis according to A. Tarski's metalanguage [2], i.e. no exhibition of symbol $x \in$ Symb is allowed, instead its metasymbol 'x' must be used. Then the validity of basic language convention is saved again. Thus we may write 'x'$\triangleright x$ for each $x \in$ Symb (or for each 'x' $\in$ Symb?).

Unfortunately the use of parenthesis everywhere is not convenient and therefore in the theories of free semigroups, or monoids, grammars and automata the exhibition of symbols and strings of symbols is accepted in general, although some confusion e.g. between basic symbols and variables for them can occur easily.

It remains to determine the meanings of Mean and the understanding relation $\triangleright$. In general there is no other way than the ostensive definitions of B. Russell [3] e.g. we are showing (or observing) the city of Prague and giving it its name Prague. This is easy to understand for objects, but is less clear for relations or operations.

Thus the second part of the famous thesis of L. Wittgenstein [4] "... and what cannot be said, we must be silent about" may be supplemented by saying "what cannot be said, can be exhibited or observed sometimes", when we understand that the meaning cannot be said very often.

In the case of numerical mathematics (or information processing in general) again the exhibition of certain meanings is used. In actual computation (either by a mathematician or by a computer) it is useless and superflous to ask questions of the type "what is the meaning of the numerical constant 4?" because the only aim of the computation is to start and finish with the numerical constants themselves (unless the integers concerned are used within a program, e.g. for the determination of the number of iterations: these are in [5] clearly distinguished from those which are stored in memory cells).

Therefore instead of the language $\langle$VarIntg, ConstIntg, Intg, $\text{Intg}^{\text{VarIntg}}$, $CS\rangle$, according to (2.5), the language

(4.2) $\langle$VarIntg, Mean = ConstIntg, $\text{ConstIntg}^{\text{VarIntg}}\rangle$

is used, where again no individual constants appear, as in the languages of first order mathematical theories (See Section 3).

5. STATE OF STORAGE

The crucial correspondence between a language and a computer storage consists in the fact that the understanding function is realized by a storing function, i.e. the assertion 'a variable x expresses the meaning m' corresponds to the assertion 'at a location x the content m is stored', where the term location is a convenient simplification of the pair (memory cell and its address).

In contradistinction to the usual languages satisfying (1.1) the computer languages satisfy a weaker requirement:

(5.1) $\emptyset \neq \triangleright \subset \text{Symb} \times (\text{Symb} \cup \text{Mean})$, i.e. a symbol does not necessarily express, a meaning but may express a symbol itself. This is a consequence of the use of computer storage.

The symbols which express symbols are called *pointers*, or general *addresses* (locations) *of higher degree*, etc.

Assuming the unambiguity (1.5) of $\triangleright$ from (5.1) we have

(5.2)   (antitransitivity) $x_1 \triangleright x_2 \triangleright \cdots \triangleright x_n$, where $n \geqslant 3$ and x_1 and x_{i+1} are different symbols for each $i = 1, 2, ..., n-1 \Rightarrow x_1 \ntriangleright x_n$.

This does not exclude (4.1), but it definitely proves that the relation $\triangleright$ must not be confused with the relation $=$, as would be possible were (1.1) satisfied.

Allowing (5.1) and using (4.2) with $\triangleright \in$ (ConstIntg $\cup$ VarIntg) $\uparrow$ VarIntg it can happen that $x \triangleright 4$, where $x \in$ VarIntg and $4 \in$ ConstIntg $=$ Mean, and also $y \triangleright x$. Then by (5.2) $y \not\triangleright 4$, but after replacing '$\triangleright$' by '$=$', $y = 4$ would follow from $x = 4$ and $y = x$.

Using the functional notation S instead of $\triangleright$ no inconsistency occurs by writing $Sx = 4$ and $Sy = x$, which obviously means $SSy = Sx = 4$ (thus y is a pointer), from which it does not follow that $Sy = 4$.

The storage of a computer is characterized by the set Loc of available *locations* (memory cells), by the set Cont of their possible *contents* (which can be specified further) and by all possible *states of storage* $S \in \text{Cont}^{\text{Loc}}$ which are nothing else but the understanding functions of a computer storage language. Obviously (5.1) is modified to the requirement

(5.3) $\text{Loc} \subset \text{Cont}$.

6. Definitions and exhibitions

I would like to note that M. I. Malcev proposed, in Prague 1968, a further development of the theory of definition in logic.

Each defining is an *activity* with respect to an (unambiguous) language $\langle \text{Symb}, \text{Mean}, \triangleright \rangle$, by which a meaning $m \in$ Mean of a symbol $x \in$ Symb (thus $x \triangleright m$ is assumed) is *assigned* to another symbol y. The *definition*:

(6.1) $y =_{\text{df}} x$, where x, y are called *definiens, definiendum*, respectively, is a *command* for the activity of defining. The *execution* of (6.1) leads to the new language $\langle \text{Symb}^*, \text{Mean}^*, \triangleright^* \rangle$ where either

(6.2) $\text{Symb}^* = \text{Symb} \cup \{y\}$, $\text{Mean}^* = \text{Mean}$ and $\triangleright^* = \triangleright \cup \{(y, m)\}$, if $y \notin$ Symb (which is the usual case in logic where y is a new symbol, introduced e.g. as an abbreviation of x) and

(6.3) $\text{Symb}^* = \text{Symb}$, $\text{Mean}^* = \text{Mean}$ and $\triangleright^* = (\triangleright - \{(y, m^*)\}) \cup \{(y, m)\}$, if $y \in$ Symb and $y \triangleright m^*$ (which is the usual case in the computer storage where no further symbols are available besides those in Symb). Within the new language it holds

$$(6.4) \qquad y = x$$

where the same symbol $=$ for equality is used in all the languages (see [8], [9]).

In programming languages instead of the definition symbol '$=_{df}$' the *assignment symbol* '$=$' is used, and the entry

$$(6.5) \qquad y := x, \quad \text{where} \quad x, y \in \text{Loc}$$

is called the *assignment statement*, the execution of which causes the change of a current state of storage S to the new state S^* in accordance with (6.3).

Each exhibiting is also an activity, which can, but need not have any influence on a language. The command for this activity is the *exhibition*

$$(6.6) \qquad y =_{exh} x,$$

the execution of which leads to a new language, which differs from the old one at most in the assertion

$$(6.7) \qquad y \triangleright x$$

where the same symbol $\triangleright$ for the understanding function is used in all the languages. We use this notation to clarify the essential difference between (6.4) and (6.7).

In programming languages, exhibitions are not allowed, which is one of the causes of difficulties with inputs. The corresponding commands (such as *LOAD*) are again of the definition type.

If Symb and Mean contain a large number of elements, the usual definitions provide insufficient tools, and therefore recursive definitions are used. From the language point of view *syntactical* (concerning only Symb) and *semantical* (concerning also Mean) definitions are distinguished (see [9]), and they are considered as *prescriptions for activities*, i.e. as programs or algorithms. Thus the recursive definitions are computations.

7. COGNITIVE PROCESS (CHANGE), KNOWLEDGE AND SITUATION. PARALLELITY

There is no other way of human cognition of the real world in time, thus the history of the world, than to observe the world in many consecutive instants. In this sentence the term world is used ambiguously: once as 'the history of the world' and secondly as 'the world in an instant'. Therefore in general these two concepts must be distinguished: a *historical object* and an *instantaneous object*. Usually the name W of a historical object is used as a *historical variable*, all the possible meanings of which are the names W_i of the corresponding instantaneous objects occurring in the prescribed *time sequence* $(W_1, W_2, ..., W_r)$, where W_i appeared in time sooner than W_j for $i < j$.

Similarly some 'dynamic concepts' such as state of storage, stack, queue, etc. are used ambiguously, and therefore special terms such as history, *process*, *activity*, etc. are used in order to remind us that a historical object is considered.

Each historical object $(W_1, W_2, ..., W_r)$ or an arbitrary length $r \geqslant 2$ can be determined by historical objects of the length 2, only when their time sequence $(Ch_1, Ch_2, ..., Ch_{r-1})$ is considered, where $Ch_i = (W_i, W_{i+1})$ for $i = 1, 2, ..., r-1$. Therefore the historical object of the length 2 (W, W^*) *is* called a *change* (or a transition in [16]), although the proper change requires $W \neq W^*$.

An *observational formula* is any one of the following strings:

(7.1) (i) $r^{(n)} (x_1, x_2, ..., x_n)$;

(ii) $r^{(n)} (x_1, x_2, ..., x_n)$;

(iii) $f^{(n)} (x_1, x_2, ..., x_n) = x_0$ where 'x_0' $\neq$ 'x_i' for $i = 1, 2, ..., n$,

where $r^{(n)} \in \text{SymbRel}$, $f^{(n)} \in \text{SymbOpr}$ and $x_i \in \text{Loc}$ for $i = 0, 1, 2, ..., n$.

The functional unit of a computer is a relational structure $\langle \text{Obj}, \text{Rel}, \text{Opr} \rangle$ such that the relations and operations from $\text{Rel} \cup \text{Opr}$

must be understood *intensionally* (and not extensionally) as *cognitive procedures* which can be applied to some objects from Obj, i.e. they are certain algorithms performable by the unit.

With respect to the pair (S, int), where int is an interpretation into $\langle \text{Obj, Rel, Opr} \rangle$ (determined by the choice of the computer, and therefore fixed) and S is a state of storage $S \in \text{Obj}^{\text{Loc}}$, the meaning of the observational formula (7.1) (i), (ii), (iii) is the *true assertion* (or the 'holding condition' in [16]), called *observation*, determined by (7.2) (i), (ii), (iii), respectively, where

(7.2) (i) the cognitive procedure int $r^{(n)}$ applied to the objects Sx_1, Sx_2, ..., Sx_n (on the places x_1, x_2, ..., x_n, respectively) gave the result true, which is abbreviated as $(\text{int} r^{(n)})$ $(Sx_1, \ldots Sx_n) = \text{true}$;

(ii) int $r^{(n)}$ applied to S_{x1}, ..., Sx_n gave result false, abreviated as $(\text{int} r^{(n)})$ $(Sx_1, \ldots, Sx_n) = \text{false}$;

(iii) the cognitive procedure int $f^{(n)}$ applied to the objects Sx_1, ..., Sx_n gave the result $(\text{int} f^{(n)})$ $(Sx_1, \ldots, Sx_n)$ and this result is stored at x_0. This is again abbreviated as $(\text{int} f^{(n)})$ $(Sx_1, \ldots, Sx_n) = Sx_0$.

Usually it is assumed that there is a set Axi of axioms which are satisfied by the structure $\langle \text{Obj, Rel, Opr} \rangle$ (with respect to the interpretation int).

A set Sit of observational formulas (7.1) is called a *situation* if the following set (of first order logic formulas) Sit $\cup$ Axi is *consistent* (where the locations which occur in observational formulas from Sit are considered as object ($=$ individual) constants. The term situation is rather special but is in accordance with its use in the *situation calculus* [19, 18]. Obviously, the index i of the member W_i in a time sequence $(W_1, W_2, \ldots, W_r)$ is the value of the (discrete) *situation parameter*, etc.

The meaning of a situation Sit with respect to a state S (and to the fixed interpretation int) is the set Know of all observations which are meanings of all observational formulas from Sit. It is natural to say that Know is a *knowledge*.

A *cognitive process* is a time sequence $(\text{Know}_1, \text{Know}_2, \ldots, \text{Know}_r)$ of knowledges and it can be determined by two time sequences of situations $(Sit_1, Sit_2, ..., Sit_r)$ and of states of storage $(S_1, S_2, ..., S_r)$. The *cognitive changes* (called events in [16]) (Know, Know*) can be classified with respect to the classification of changes (S, S^*) of states of storage of computers as follows

(7.3)   the change (S, S^*) is called *m-parallel* if there exist at most m different locations $x_1, x_2, ..., x_m \in$ Loc such that $S^*z = Sz$ for each $z \in$ Loc $-\{x_1, x_2, ..., x_m\}$, i.e. S and S^* differ in the contents of at most m locations.

A computer in which each change of state of its storage is m-parallel and m is as small as possible, is itself called *m-parallel*. For a long time only 1-parallel, i.e. *serial computers* were considered (but the starting point in [20] is a general parallel computer). Each serial change of the state of storage is determined by the application of an operation from Opr to the contents of some locations and by storing the result at a new location [17]. And this is only the basic change of a state, to which all other changes can be reduced, and which is uniquely determined by the operational command (8.1) (i).

8. THE SYNTACTICAL AND SEMANTICAL DEFINITION OF A PROGRAM

With respect to the storage of Section 5 and to the functional unit of Section 7 the following three sorts of *commands* (instructions) are used:

(8.1) (i) *'operational'* $f^{(n)}(x_1, x_2, ..., x_n) = : x_0;$

(ii) *'relational'* $r^{(n)}_{[a_1,a_2]}(x_1, x_2, ..., x_n);$

(iii) *'stopping'*   STOP,

where $x_1 \in$ Loc for $i = 0, 1, ..., n$; $a_1, a_2 \in$ Loc; $f^{(n)} \in$ SymbOpr$^{(n)}$ and $r^{(n)} \in$ SymbRel$^{(n)}$ $\left(n \text{ is the } arity \text{ and SymbOpr} = \bigcup_{n=1}^{\infty} \text{SymbOpr}^{(n)}, \right.$ SymbRel $= \bigcup_{n=1}^{\infty}$ SymbRel$^{(n)} \Big)$.

Obviously the commands are only strings over a certain alphabet. Let

Com be the set of all commands over $SymbOpr \cup SymbRel \cup Loc$ (and over some additional auxiliary symbols such as $=:$ STOP etc.).

With respect to Section 5 the possibility of storing the commands (and then of modifying them, by which John v. Neumann [10] was fascinated so much) is expressed by the requirement

(8.2) $Com \subset Cont$.

Each ordered pair $\langle b^{(i)}, C^{(i)} \rangle$ where $b^{(i)} \in Loc$ and $C^{(i)} \in Com$ is called a *stored* (or *labelled*) *command*.

The *syntactical definition of a program over Com* is very simple in our lower level programming language (but is extremely complicated in higher level programming languages): each finite (or infinite) sequence of stored commands $P =_{df} (K^{(1)}; K^{(2)}; ..., K^{(N)})$ where $K^{(i)} = \langle b^{(i)}, C^{(i)} \rangle$ for each $i = 1, 2, ..., N$ (such that '$b^{(i)}$' $\neq$ '$b^{(j)}$' if $i \neq j$), is called *stored program* (see [21] where additional natural and obvious restrictions concerning stored programs are introduced and studied.

The *semantical definition of the stored program P* assumes an interpretation int of operation and relation symbols which occur in P in a computer (i.e. in its functional unit $\langle Obj, Rel, Opr \rangle$), and determines the meaning of P as the set of all possible *activities* (of the computer under consideration) which are defined for an arbitrary *initial state of storage* S_0 recursively as follows:

(8.3) $Act_{int} (P, S_0) =_{df} ((K_1, S_0), (K_2, S_1), ..., (K_i, S_{i-1}), ...)$ where
$K_i =_{df} \langle b_i, C_i \rangle$, $b_i \in Loc$ and $C_i \in Com$ for $i = 1, 2, ...$ *and*

(i) $K_1 =_{df} K^{(1)}$;

(ii) if $i \geqslant 1$ and (K_i, S_{i-1}) has already been defined such that
$K_i =_{df} K^{(h)}$ where $1 \leqslant h \leqslant N$, and $S_{i-1} \in Obj^{Loc}$ then three
possibilities (which correspond to the three sorts of commands) (8.1) (i)–(iii) can appear:
(1) $C_i =_{df} f^{(n)} (x_1, x_2, ..., x_n) =: x_0$;
if $h < N$ and $(S_{i-1} x_1, S_{i-1} x_2, ..., S_{i-1} x_n) \in Domain (int f^{(n)})$
then we say that the *activity continues* and we define
(a) $K_{i+1} =_{df} K^{(h+1)}$;
(b) $\begin{cases} S_i x_0 =_{df} (int\ f^{(n)}) (S_{i-1} x_1, S_{i-1} x_2, ..., S_{i-1} x_n) \\ S_i z =_{df} S_{i-1} z \quad \text{for each} \quad z \in Loc -\{x_0\}, \end{cases}$

otherwise we say that the *activity finishes* and nothing more is defined;

(2) $C_i =_{df} r^{(n)}_{[a_1,a_2]} (x_1, x_2, ..., x_n)$;

if either $(S_{i-1}x_1, ..., S_{i-1}x_n) \in$ Field (int $r^{(n)}$) and $a_1 = b^{(j)}$ for some j such that $1 \leq j \leq N$ and j is minimal, or $(S_{i-1}x_1, ..., S_{i-1}x_n) \in$ Field (int $r^{(n)}$) and $a_2 = b^{(j)}$ for some j such that $1 \leq j \leq N$ and j is minimal; then we say that the *activity continues* and we define

(a) $K_{i+1} =_{df} K^{(j)}$;

(b) $S_i =_{df} S_{i-1}$,

otherwise we say that the *activity finishes* and nothing more is defined;

(3) $C_i =_{df}$ STOP: then we always say that the *activity finishes* and it is called *stopped*; nothing more is defined but the last state of storage S_{i-1} is called the *resulting state* and is denoted by $Res_{int}(P, S_0)$.

Let us recall that the phrase 'j is minimal' is concerned with the general case where the requirement '$b^{(i)}$' $\neq$ '$b^{(j)}$' need not be satisfied. Furthermore it is assumed that the cognitive procedure int $r^{(n)}$ gives a result when applied to any n-tuple from Obj^n (as it is assumed in logic but need not be true in general).

The activity $Act_{int}(P, S_0)$ can be split into two parts:

(8.4) (i) the *stored branch* $StBr_{int}(P, S_0) =_{df} (K_1, K_2, ..., K_i, ...)$ which is a time sequence of actually executed commands, and

(ii) the *computation* $Comput_{int}(P, S_0) =_{df} (S_0, S_1, ..., S_{i-1}, ...)$ which is a time sequence of states that actually occur.

Obviously we are interested only in the stopped activities and in the stopped stored branches and the stopped computations. It is easy to see ([21]) that the union of all sets of all stopped stored branches of P in all interpretations int, which can be defined without respect to semantics, is a regular event in Kleene's sense (Chomsky's type 3 language). However it is probably undecidable whether or not the set of all stopped stored branches of P for an arbitrary interpretation is a regular event.

There are rather natural reasons for the choice of two sets of locations which occur in the program P: the set Imp_p of *input locations* and the set Outp_p of *output locations* of P, in such a way that these sets are independent of any interpretation of P. If $\text{Inp}_p = \{x_1, x_2, ..., x_p\}$ and $\text{Outp}_p = \{y_1, y_2, ..., y_q\}$ and int is an interpretation of P then the system of qp-ary functions $\lambda[[x_1, x_2, ..., x_p; y_1, y_2, ..., y_q], P_{int}] =_{df} (\lambda_1[[x_1, ..., x_p], P_{int}], \lambda_2[[x_1, ..., x_p], P_{int}], ..., \lambda_q[[x_1, ..., x_p], P_{int}])$ is called the *functional meaning of P in int* where

$$(8.5) \qquad \lambda_i[[x_1, ..., x_p], P_{int}] \ S_0 x_1, ..., S_0 x_p) =_{df} (\text{Res}_{int}(P, S_0)) \ (y_i)$$
$$\text{if } \text{StBr}_{int}(P, S_0) \text{ is stopped and } 1 \le i \le q,$$

where Church's λ-notation is used in a slightly modified way (concerning the output locations) and the index refers to the ordering in the system. It can happen that $\lambda_i[[x_1, ..., x_p], P_{int}] = \lambda_j[[x_1, ..., x_p], P_{int}]$ even if $i \ne j$.

Each function of the system is called *evaluable* (or computable) by P in int within the system (because, in fact, very often the particular functions of the system cannot be evaluated without evaluating all of them).

9.   PROGRAMMING ALGORITHMIC THEORIES AND FLOW DIAGRAMS

Using *program variables* $P(x_1, ..., x_p; y_1, ..., y_q)$ as abbreviations for programs P in the prescribed order of their input and output locations (thus λ-notation is no more necessary) we say that $P(x_1, ..., x_p; y_1, ..., y_q)$ *is functionally weaker or equal than* $P^*(x_1^*, ..., x_{p*}^*; y_1^*, ..., y_{q*}^*)$ *in the interpretation int*, and we write $P(\cdots) \le_{int} P^*(\cdots)$ if

(9.1) (i) $p \le p^*$ and $q \le q^*$;
     (ii) $\lambda_i[[x_1, ..., x_p]; P_{int}] = \lambda_i[[x_1^*, ..., x_{p*}^*]; P_{int}^*]$ for each $i = 1, ..., q$;

and if $P(\cdots) \le_{int} P^*(\cdots)$ holds for each int then we write $P(\cdots) \le P^*(\cdots)$. Further if $P \le_{int} P^*$ and $P^* \le_{int} P$ then we write $P =_{int} P^*$, and if $P \le P^*$ and $P^* \le P$ then we write $P = P^*$.

One can easily prove the Skolem-Löwenheim theorem for programming theories: if $P(x_1, ..., x_p; y_1, ..., y_q) =_{int} P^*(x_1, ..., x_p; y_1, ..., y_q)$ holds in each countable interpretation int them $P(\cdots) = P^*(\cdots)$.

In programming theories the results of the following sort are required (e.g. for any optimization, etc.): from certain structural properties of programs (i.e. from syntax) some properties of meanings (i.e. semantics) should be derived. E.g. if two programs P and P^* have the same set of all stopped branches (they are called *branch equivalent* in [21]) $P^*(x_1, ..., x_p; y_1, ..., y_q)$.

The concept of *programming* (algorithmic) *theory* is in some sense similar to the concept of the usual *algebraic theory*, but in fact it is a refinement of it and simultaneously an essential generalization.

The similarity is concerned with the atomic formulas: in programming theories they have the form either $P(x_1, ..., x_p; y_1, ..., y_q) \leqslant P^*(x_1^*, ..., x_{p*}^*; y_1^*, ..., y_{q*}^*)$ or $P(\cdots) = P^*(\cdots)$, where P and P^* are arbitrary programs, but in algebraic theories they have the form $T(x_1, ..., x_p) = T^*(x_1^*, ..., x_{p*}^*)$, where T and T^* are just terms.

The refinement is concerned with the crucial fact that the function $\lambda[[x_1, ..., x_p]; T_{int}]$ is really defined only by a choice of a *course* [23] of evaluation of all components of T. E.g. having in mind the term $((x_1 + x_2) * (x_1 - x_2))$, abbreviated as $T(x_1, x_2)$,

$$P(x_1, x_2; y_1) =_{df} (\langle 1, x_1 + x_2 =: z \rangle;$$
$$\langle 2, x_1 - x_2 =: w \rangle; \langle 3, z * w =: y_1 \rangle; \langle 4, STOP \rangle)$$

and

$$P^*(x_1, x_2; y_1) =_{df} (\langle 1^*, x_1 - x_2 =: w \rangle;$$
$$\langle 2^*, x_1 + x_2 =: z \rangle; \langle 3^*, z * w =: y_1 \rangle; \langle 4^*, STOP \rangle)$$

are programs which correspond to the only two different (serial) courses of T, and one sees that $T(x_1, x_2) =_{int} P(x_1, x_2; y_1) =_{int} P^*(x_1, x_2; y_1)$ holds in all interpretations int of T, P and P^*.

The generalization is concerned with the following facts: (1) each interpreted term expresses only one function, while interpreted programs can express arbitrary many functions; (2) $T(x_1, ..., x_p) =_{int} T^*(x_1, ..., x_p)$ holds in all interpretations int iff T and T^* are identical terms. This is not true for programs (see the example above), and moreover, it is probably undecidable whether or not $P(x_1, ..., x_p; y_1, ..., y_q) =_{int} P^*(x_1, ..., x_p; y_1, ..., y_q)$ holds for arbitrary P, P^* and arbitrary int; (3) each model of an algebraic theory is an algebra $\langle Obj, \{=\}, Opr \rangle$ while models of programming theories are general relational structures $\langle Obj, Rel, Opr \rangle$ (or algorithmic algebras according to [22]).

The usual algebraic axioms which express the associativity of '+' $(x_1 + x_2) + x_3 = x_1 + (x_2 + x_3)$, *or the distributivity of* '+' *and* '*' $(x_1 + x_2) *$ $x_3 = (x_1 * x_3) + (x_2 * x_3)$ have the following programming forms:

$$(9.2) \quad \text{(associativity)} \quad (\langle 1, x_1 + x_2 = :z \rangle; \langle 2, z + x_3 = :y_1 \rangle; \langle 3, \text{STOP} \rangle)$$
$$= (\langle 4, x_2 + x_3 = :z \rangle; \langle 5, x_1 + z = :y_1 \rangle; \langle 6, \text{STOP} \rangle)$$
$$\text{(distributivity)} \quad (\langle 7, x_1 + x_2 = :z \rangle; \langle 8, z * x_3 = :y_1 \rangle; \langle 9, \text{STOP} \rangle)$$
$$= (\langle 10, x_1 * x_3 = :z \rangle; \langle 11, x_2 * x_3 = :t \rangle;$$
$$\langle 12, z + t = :y_1 \rangle; \langle 13, \text{STOP} \rangle).$$

If we are not interested in the locations at which the commands of a stored program are stored, then we can abstract from them by passing on from stored programs to flow diagrams, which are directed graphs with labelled vertices (by commands) and edges (by truth values), and therefore the well known plane representation can be used. (The inventor of interpreted flow diagrams was J. v. Neumann [10]. For their simplicity and transparency flow diagrams are taken as a base for any theoretical study (the uninterpreted flow diagrams are called *program schemes*). The clear connection between programs and flow diagrams is established in [21]. By transferring the definitions from programs to flow diagrams the programming axioms (9.2) can be expressed in Figure 1 and in Figure 2.

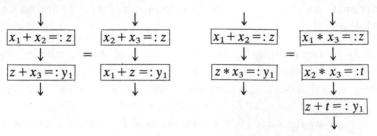

Fig. 1.   Fig. 2.

254   K. ČULÍK

A flow diagram is called *serial* if each vertex is labelled by exactly one command (see e.g. [10, 27, 28, 21]), otherwise it is called *parallel*. In [25] only the operational commands can occur in parallel, but in [26, 11] the general case is allowed, and moreover the '*flow relation*' is replaced by the '*permit relation*' in *permit diagrams* (defined by using the concept of *scope* which is rather complicated for a presentation here).

Using the corresponding definitions and notations the following functional equalities hold in all interpretations:

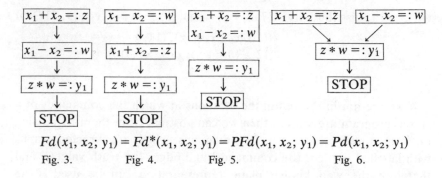

$$Fd(x_1, x_2; y_1) = Fd^*(x_1, x_2; y_1) = PFd(x_1, x_2; y_1) = Pd(x_1, x_2; y_1)$$

Fig. 3.   Fig. 4.   Fig. 5.   Fig. 6.

where Fd and Fd^* are serial flow diagrams, which correspond to the stored programs P and P^* mentioned above, but PFd is a parallel flow diagram (a program scheme according to [25]) and Pd is a permit diagram [11].

In an analogy with logic the functional equality $Fd(x_1, ..., x_p; y_1, ..., y_2) = Fd^*(x_1, ..., x_p; y_1, ..., y_2)$ is called a *programming (algorithmic) tautology*. In order to axiomatize the set of all programming tautologies a series of *elementary transformations* of flow diagrams (or programs), which preserve the functional equality, are introduced in [24]. E.g.

(9.3)   (reordering) if the vertex v, v^* of Fd is labelled by the operational command $f^{(n)}(x_1, ..., x_n =: x_0$ and $g^{(m)}(z_1, ..., z_m) =: z_0$ such that 'x_0' $\neq$ 'z_j' for $j = 1, 2, ..., m$, and 'z_0' $\neq$ 'x_i' for $i = 1, 2, ..., n$, and if v is connected with v^* by the edge (v, v^*) then the new Fd^* arises from Fd by the exchange of labels of the vertices v and v^*, or

(9.4)   (splitting and joining) if 'x_0'$\neq$'y_j' for $j = 1, 2, ..., m$ then
           from *Fd*, which contains the diagram from Figure 7, the
           new *Fd*** arises by replacement of the diagram from Figure
           7 by that of Figure 8 (and conversely).

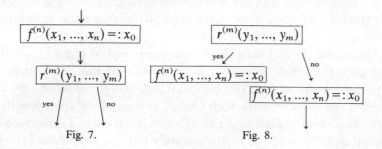

Fig. 7.   Fig. 8.

Other sorts of elementary transformations are concerned with *re-naming* of locations (the concept of scope is necessary in their defini-tion). Futhermore the *developing* (and *wrapping up* of cycles is an elementary transformation related to *homomorphisms* of flow diagrams [30].

Using these homomorphisms Engeler's *normal forms* [27, 28, 30] can be obtained (each two homomorphic images or patterns of a flow diagram are branch equivalent).

The set of all branch equivalent flow diagrams can be considered as a (rather strong) *algorithmic method*, because the same way (= method of evaluation) is prescribed for all the members of the set. Some other (weaker) concepts of algorithmic methods are proposed in [24].

10. STRUCTURED OBJECTS AND LOGICAL TYPES

The locations of storage are characterized by all their possible con-tents. Usually a string of zeros and ones–thus a binary numeral–can be stored at one location and a fixed length, usually 16–64 bits, is prescribed for all of them. These numerals are only *basic objects* for a computer. If more *complex objects* are needed, e.g. vectors or matrices of basic objects, or some graph structures, the vertices and edges of which are labelled by basic objects, etc., then two things are necessary:

(A) to store their basic objects at several locations, and
(B) to store their 'structure' in some way.

There is a great variety of *data structures* used in computing and it is often unclear what they are when considered independently of computer storage. We shall show that the 'structures' of data structures are local types of finite objects over certain basic ones.

Unfortunately and surprisingly the usual logical types [13, 14] are not general enough, because the restrictions of the simple theory of types cannot be accepted in computers. Although we need only finite objects, we need to work with sets of vectors of different lengths, or with matrices, the dimensions of which may vary, etc. Furthermore the logical types from [13] concern sets only but not their elements unless they are sets themselves. For example, each binary relation R, which is not empty, has the type $(*, *)$ but its element $(a, b) \in R$ has no type, although it seems natural to use the string $*, *$ for its type. Of course it is necessary to accept the additional basic concept of an element (or structured individual [12]) beside the concept of a set.

Using braces for sets and parentheses for elements the syntactical definition of [13] can be extended as follows: two sorts of *logical types* LT are distinguished: *set logical type* SLT and *element logical type* ELT, which are strings of braces, parentheses, commas and asterisk symbols *

(10.1) (i) $(\underbrace{*, *, ..., *}_{n \text{ times}})$ is an ELT and $AR(\underbrace{*, *, ..., *}_{n \text{ times}}) = n$ for each

$n = 1, 2, ...$;

(ii) $n \geqslant 1$, t_i is either an ELT or an SLT for $i = 1, 2, ..., n$ then

$(t_1, t_2, ..., t_n)$ is an ELT and $AR(t_1, t_2, ..., t_n) = \sum_{i=1}^{n} AR t_i$;

(iii) t is an ELT then $\{t\}$ is an SLT.

Thus, we can have SLT $R = \{(*, *)\}$ and ELT $(a, b) = (*, *)$, where $(a, b) \in R$ and R is a binary relation (among basic objects).

Let Elem_{Obj} or Set_{Obj} be the set of all finite *elements over Obj* or *sets over Obj* respectively. These are defined recursively together with *equality* =, *length LG* (of element) and *cardinality CR* (of set) as follows:

(10.2) (i) $n \geqslant 1$, $a_i \in \text{Obj}$ for $i = 1, 2, ..., n \Rightarrow (a_1, a_2, ..., a_n) \in \text{Elem}_{\text{Obj}}$ and $LG(a_1, a_2, ..., a_n) = n$;

(ii) $a_1, ..., a_n \in \text{Obj}$ and $b_1, ..., b_m \in \text{Obj} \Rightarrow (a_1, ..., a_n) = (b_1, ..., b_m)$ iff $n = m$ and a_i, b_i are identical objects for each $i = 1, 2, ..., n$;

(iii) $n \geqslant 1$, $a_i \in \text{Elem}_{\text{Obj}} \cup \text{Set}_{\text{Obj}}$ for $i = 1, 2, ..., n \Rightarrow (a_1, ..., a_n) \in \text{Elem}_{\text{Obj}}$ and $LG(a_1, ..., a_n) = n$;

(iv) $(a_1, ..., a_n) \in \text{Elem}_{\text{Obj}}$, $(b_1, ..., b_m) \in \text{Elem}_{\text{Obj}}$, $n \geqslant 1$ and $m \geqslant 1 \Rightarrow (a_1, ..., a_n) = (b_1, ..., b_m)$ iff $n = m$ and $a_i = b_i$ for $i = 1, 2, ..., n$.

(v) $n \geqslant 1$, $a_i \in \text{Elem}_{\text{Obj}}$ and $a_i \neq a_j$, where $i \neq j$, for each $j, i = 1, 2, ..., n \Rightarrow \{a_1, ..., a_n\} \in \text{Set}_{\text{Obj}}$ and $CR\{a_1, ..., a_n\} = n$;

(vi) $\{a_1, ..., a_n\} \in \text{Set}_{\text{Obj}}$, $\{b_1, ..., b_m\} \in \text{Set}_{\text{Obj}}$, $n \geqslant 1$ and $m \geqslant 1 \Rightarrow \{a_1, ..., a_n\} = \{b_1, ..., b_m\}$ iff $n = m$ and there exists a permutation π of integers $1, 2, ..., n$ such that $a_i = b_{\pi i}$ for $i = 1, 2, ..., n$.

Let us note that the construction of finite objects (see [15, 31, 9]) is split into two parts here: the construction of finite elements and of finite sets.

Now some sets over Obj (but also some elements over Obj) will have their logical types determined according to the following definition (which is missing in [13]):

(10.3) (i) $n \geqslant 1$, $a_i \in \text{Obj}$ for $i = 1, 2, ..., n \Rightarrow LT(a_1, ..., a_n) = (\underbrace{*, ..., *}_{n \text{ times}})$;

(ii) $n \geqslant 1$, $(a_i, ..., a_n) \in \text{Elem}_{\text{Obj}}$ and LTa_i is defined for $i = 1, 2, ..., n \Rightarrow LT(a_1, ..., a_n) = (LTa_1, LTa_2, ..., LTa_n)$;

(iii) $n \geqslant 1$, $\{a_1, ..., a_n\} \in \text{Set}_{\text{Obj}}$, LTa_1 is defined and $LTa_i = LTa_j$ for all $i, j = 1, 2, ..., n \Rightarrow LT\{a_1, ..., a_n\} = \{LTa_i\}$ where $1 \leqslant i \leqslant n$.

The restrictive requirement (10.3) (v) and the corresponding requirement (10.1) (iii) are accepted in [13] (the principle of purity of logical types), but they can be omitted and replaced by

(10.1) (iii*) $n \geqslant 1$, t_i is an ELT and $t_1 \neq t_j$, where $i \neq j$, for each $i, j = 1, 2, ..., n \Rightarrow \{t_1, t_2, ..., t_n\}$ is an SLT,

where the equality $=$ (and the inequality $\neq$) is defined by (10.2), when there exists just one basic object $*$, i.e. $\text{Obj} = \{*\}$, and

(10.3) (v*) $n \geqslant 1$, $\{a_1, ..., a_n\} \in \text{Set}_{\text{Obj}} \Rightarrow LT\{a_1, ..., a_n\} = \{LTa_i; \ i = 1, 2, ..., n\}$.

Now an element over Obj which does not contain any set over Obj, i.e. which is constructed independently on any construction of sets over Obj, is called a *structured object*. Let StrObj be the set of all structured objects over Obj (thus $\text{StrObj} \cap \text{Obj} = \varnothing$, although there is a one-to one correspondence between basic objects a and the simplest structured objects (a)).

B. Russel's concept of reduction can be used for structural objects as follows: the *reduction* $\text{RED}a$ *of the structured object* $a \in \text{StrObj}$ is a structured object again:

(10.4) (i) $a \in \text{Obj} \Rightarrow \text{red } a = a$;

(ii) $n \geqslant 1$, $(a_1, ..., a_n) \in \text{StrObj} \Rightarrow \text{red } (a_1, ..., a_n) = \text{red } a_1, ..., \text{red } a_n$;

(iii) $n \geqslant 1$, $(a_1, ..., a_n) \in \text{StrObj} \Rightarrow \text{RED } (a_1, ..., a_n) = (\text{red}(a_1, ..., a_2))$

Obviously

(10.5) $a \in \text{StrObj} \Rightarrow AR \ LT \ a = LG \ \text{RED } a$.

According to (B) the concept of *structured location* (here we shall restrict ourselves to the storing of structured objects only) must be similar in some sense to the concept of structured object over Obj. Its definition includes the definition of the set of all (basic) *locations BL*, of all *free* (basic) *locations FL* (of a structured location):

(10.6) (i) $n \geqslant 1$, $x_i \in \text{Loc}$ and $x_i \neq x_j$, where $i \neq j$, for each $i, j = 0, 1, ..., n \Rightarrow x_0(x_1, x_2, ..., x_n) \in \text{StrLoc}$;

(ii) $n \geqslant 1$, (a) $x_i \in \mathrm{StrLoc}$ for $i = 1, 2, ..., n$;

(b) $BLx_i \cap BLx_j = \varnothing$, where $i \neq j$, for each $i, j = 1, 2, ..., n$

(c) $x_o \in \mathrm{Loc}$ and $x_0 \notin \bigcup_{i=1}^{n} BLx_i$ $x_0(x_1, ..., x_n) \in \mathrm{StrLoc}$,

$$BLx_0(x_1, ..., x_n) = \{x_0\} \cup \bigcup_{i=1}^{n} BLx_i \quad \text{and} \quad FLx_0(x_1,$$

$$..., x_n) = \bigcup_{i=1}^{n} FLx_i,$$

where StrLoc denotes the set of all structured locations over Loc.

One says that a structured location $x_0(x_1, ..., x_n) \in \mathrm{StrLoc}$ *belongs to the state of storage* S if

(10.7) (i) $Sx_0 = (x_1, ..., x_n)$ (where the parentheses are not stored 1);

(ii) $x_i \in \mathrm{StrLoc} \Rightarrow x_i$ belongs to S for each $i = 1, 2, ..., n$.

This assumes the following property of storage

(10.8) (i) $\mathrm{Loc}^n \subset \mathrm{Cont}$ for each $n = 1, 2, ...$, i.e. $S \in$

$$(\mathrm{Obj} \cup \bigcup_{n=1}^{\infty} \mathrm{Loc}^n)^{\mathrm{Loc}}.$$

Obviously, (10.8) (i) is a theoretical requirement, which is convenient for program-writers (because the same assumption is accepted in mathematics), but it must be replaced by a more restrictive one together with a certain convention. E.g. in J. McCarthy's programming language LISP [7] lists are replaced by S-expressions, which means that the following stronger requirement is accepted

(10.8) (ii) $\mathrm{Loc}^2 \subset \mathrm{Cont}$

with the convention that (10.7) (i) is replaced by

(10.7) (i*) $Sx_0 = (x_1, ?), Sx_1 = (x_2, ?), ..., Sx_{n-1} = (x_n, ?), Sx_n = \mathrm{NIL}$,

where NIL is the termination symbol from LISP and the question marks stand for anything.

By omitting all basic locations which belong to BLx, but which are not free, i.e. locations from $BLx - FLx$, we are getting 'a structured red object over Loc' which is called the *structure of x*, and is denoted by str x, where $x \in$ StrLoc. The logical type LT str x is then determined for all structured locations x.

A *structured object $a \in$ StrObj is stored at the structured location x at the state $S \in (\text{Obj} \cup \bigcup_{n=1}^{\infty} \text{Loc}^n)^{\text{Loc}}$ if*

(10.9) (i) LT str $x = $ LT a;
      (ii) x belongs to S;
      (iii) RED $a = (a_1, ..., a_k)$ and   RED   str $x = (x_1, x_2, ..., x_k) \Rightarrow Sx_i = a_i$ for each $i = 1, 2, ..., k$.

It is clear that the free locations of a structured location and nothing else are used for storing the basic objects of a structured object in full accordance with (A).

Let us note that by accepting two basic concepts: sets and elements, no antinomy of the set theory can occur even without applying any restriction on logical types.

11. Situation proof for programmes which are correct with respect to an evaluation problem

One can speak about correctness of a program only with respect to a goal which should be reached, and one can prove the correctness only if the reaching of the goal is expressed formally. Therefore let us consider first-order formulas over VarObj, VarRel, VarOpr and interpretations int into relational structures ⟨Obj, Rel, Opr⟩ which satisfy axioms from the set Axi. Let us abreviate an open formula by its name followed by all its free (individual) variables in some order.

An open interpreted formula Req $(x_1, ..., x_p; y_1, ..., y_q)$, called the *requirements*, determines a usual mathematical *evaluation problem* if its free variables are divided in two subsets: *input variables* from $\text{Inp}_{\text{Req}} = \{x_1, ..., x_p\}$ and *output variables* from $\text{Outp}_{\text{Req}} = \{y_1, ..., y_q\}$. If a_i and b_j are individual constants (with respect to int) such that $\text{Req}_{\text{int}}(a_i, ..., a_p; b_1, ..., b_q)$ is true, then $(b_1, ..., b_q)$ is called a *solution* for the *outset*

$(a_1, ..., a_p)$. The evaluation problem itself consists of finding solutions for prescribed outsets, and must be specified either as *particular* (if just one outset is prescribed) or *universal* (if all outsets are prescribed), and can yield *one*, or *several*, or *all* solutions. The evaluation problem must not be confused with the *existence problem*.

A flow diagram *Fd* (and similarly a program) is called *correct with respect to the particular* (one-solution) *problem* $\text{Req}_{\text{int}}(a_1, ..., a_p; y_1, ..., y_q)$ if

(11.1) (i) $\text{Inp}_{Fd} = \text{Inp}_{\text{Req}}$; $\text{Outp}_{Fd} = \text{Out}_{\text{Req}}$;

(ii) if S_0 is an initial state such that $S_0 x_i = a_i$ for $i = 1, 2, ..., p$, then the computation $\text{Comput}_{\text{int}}(Fd, S_0)$ is stopped and if S^* is its last state then $(S^* y_1, ..., S^* y_q)$ is a solution for the outset $(a_1, ..., a_p)$.

Further, *Fd* is called *correct with respect to the universal* (one-solution) *problem* $\text{Req}_{\text{int}}(x_1, ..., x_p; y_1, ..., y_q)$ if *Fd* is correct with respect to a particular problem $\text{Req}_{\text{int}}(a_1, ..., a_p; y_1, ..., y_q)$ for each outset $(a_1, ..., a_p)$ (with respect to int).

Let us remind explicitly here that the heart of all programming (and of all constructive and computational mathematics), i.e. the *algorithmization* of evaluation problems, is not concerned here at all, where *algorithmization problem* consists in finding an algorithm (or a program or a flow diagram) which is correct with respect to the evaluation problem.

How to prove the correctness of *Fd* with respect to a particular one-solution problem $\text{Req}_{\text{int}}(a_1, ..., a_p; y_1, ..., y_q)$?

By each stopped and consistent (see [36, 11]) *truth branch* $\text{TrBr} = (C_1, t_1, C_2, t_2, ..., C_{s-1}, t_{s-1}, C_s)$ of *Fd* (where $C_i \in \text{Com}$ and $t_i \in \{\text{yes}, \text{no}\}$ is defined by the "edge" (C_i, C_{i+1}) when C_i is relational) the following *situation sequence* $\text{Sit}_{\text{TrBr}} = (\text{Sit}_0, \text{Sit}_1, ..., \text{Sit}_{s-1})$ is determined:

(11.2) (i) $\text{Sit}_0 = \{x_1 = a_1, ..., x_p = a_p\}$;

(ii) $i > 0$ and Sit_{i-1} has been defined already $\Rightarrow$ either (a) $C_i = r^{(n)}(z_1, ..., z_n)$ and then Sit_i arises from Sit_{i-1} by adding of observational formula $r^{(n)}(z_1, ..., z_n)$, $\neg r^{(n)}(z_1, ..., z_n)$ if $t_i = \text{yes}$, $t_i = \text{no}$, respectively; or (b) $C_i = f^{(n)}(z_1, ..., z_n) =: z_0$ and then Sit_i arises from Sit_{i-1} by following changes: 1) from Sit_{i-1} each observational formula must be omitted, in which z_0 occurs as a free variable; 2) the observational formula $f^{(n)}(z_i, ..., z_n) = z_0$ is added, and 3) in all those formulas of Sit_{i-1} which contain the term $f^{(n)}(z_1, ..., z_n)$ each occurrence of this term is replaced by z_0; or (c) $C_i = \text{STOP}$ and then no further situation is defined.

If $\text{Comput}_{\text{int}}(Fd, S_0) = (S_0, S_1, ..., S_{s-1})$ is a stopped computation and $\text{Sit}_{\text{TrBr}_{\text{int}}}(Fd, S_0) = (\text{Sit}_0, \text{Sit}_1, ..., \text{Sit}_{s-1})$ is the corresponding situation sequence, then one can easily see that for each $i = 0, 1, ..., s-1$

(11.3) Sit_i *is satisfied by* S_i, i.e. each observational formula from Sit_i becomes a true assertion (observation) if free variable x is replaced by its value $S_i x$.

Furthermore if $\text{Sit}_{\text{TrBr}} = (\text{Sit}_0, \text{Sit}_1, ..., \text{Sit}_{s-1})$ is a situation sequence of a stopped and consistent truth branch TrBr of Fd such that

(11.4) $\text{Sit}_{s-1} \cup Axi \Rightarrow \text{Req } (x_1, ..., y_q)$ is valid for each interpretation int and each outset $(S_0 x_1, ..., S_0 x_p)$ such that $\text{TrBr}_{\text{int}}(Fd, S_0) = \text{TrBr}$,

then Sit_{TrBr} can be called a *situation proof of correctness for TrBr of Fd with respect to the evaluation problem* Req $(x_1, ..., x_p; y_1, ..., y_q)$ *and with respect to Axi*, because (11.4) follows from

(11.5) $\text{Sit}_{s-1} \cup Axi \vdash \text{Req } (x_1, ..., x_p; y_1, ..., y_q)$

and this can be established by the usual proving methods.

Therefore, if for each stopped and consistent truth branch TrBr of Fd Sit_{TrBr} is a situation proof with respect to the same evaluation problem and to the same set of axioms, then Fd is correct with respect to the universal evaluation problem and to the given axioms (obviously in each interpretation).

If there exists only a finite (and small) number of stopped and consistent truth branches of Fd then the above mentioned *situation proof of correctness* can be applied directly, because only a finite and small number of establishing the logical consequences (11.5) is required. Otherwise suitable dependencies among the infinite number of truth branches must be found in such a way that a finite number of situation proofs of correctness will imply the correctness for all remaining branches. (probably Engeler's normal form [27, 28] seems to be proper for this purpose).

Unfortunately the universal problems are very often unsolvable and from the point of view of applications it is usually sufficient to solve either a particular problem only or some particular problems for those outsets which satisfy certain *assumptions* Assmp $(x_1, ..., x_p)$ only. Then the *problem with assumptions* is the pair $\langle$Assmp$(x_1, ..., x_p)$. Req$(x_1, ..., x_p;$ $y_1, ..., y_q)\rangle$ and is specified as above. Only such outsets $(a_1, ..., a_p)$ are admitted for which Assmp$_{int}(a_i, ..., a_p)$ is true, and in the situation sequence the requirement (11.2) (i) must be replaced by

(11.2) (i*) Sit$_0$ = Assmp $(x_1, ..., x_p)$

(*and therefore no individual constants are needed*).

Situation proofs of correctness were made for simple flow diagrams with respect to the two evaluation problems:

PROBLEM 1 (system of linear equations)
   Req: $\{x_{11}y_1 + x_{12}y_2 = x_{10}, x_{21}y_1 + x_{22}y_2 = x_{20}\}$; Inp$_{Rep}$ = $\{x_{11}, x_{12}, x_{10},$ $x_{21}, x_{22}, x_{20}\}$ and Outp$_{Req}$ = $\{y_1, y_2\}$;
   Assump: $x_{11}x_{22} - x_{12}x_{21} \neq 0$;
   Axi: all the axioms of real number algebra;

PROBLEM 2 (sorting problem)
   Axi: '$\geq$' is antisymmetric, transitive and trichotomic;
   Req: Inp$_{Rep}$ = $\{x_1, x_2, x_3, x_4\}$ = Outp$_{Req}$; for an arbitrary initial state S_0 a result state S^* should be found such that there exists a permutation π of integers $1, 2, 3, 4$ such that $S^*x_{\pi i} = S_0 x_i$ for $i = 1, 2, 3, 4$, and $S^*x_1 \leq S^*x_2 \leq S^*x_3 \leq S^*x_4$;
   Assump: none.

12. DATA BASE AND n-ARY QUANTITIES

If we observe that John's height is 183 units then 'height(John) = 183' is an observation of a new kind, because 'height' is neither a unary relation nor a unary operation, but a unary *quantity*, which is a concept well known in all sciences.

A *data base* [32, 33, 34] is a set of observational formulas with individual constants (thus observations), the great majority of which is concerned with certain quantities (usually the unary quantities only).

Therefore three kinds of things must be distinguished from the point of view of cognitive procedures of human cognition: *objects* from Obj, *truth values* from {true, false} and *numbers* from Num (which are replaced by the corresponding *numerals*). If $n \geq 1$ then each *n-ary relation* $r^{(n)} \in \text{Rel}^{(n)}$, *n-ary operation* $f^{(n)} \in \text{Opr}^{(n)}$, *n-ary quantity* $q^{(n)} \in$ $\text{Qua}^{(n)}$ is intensionally characterized as a cognitive procedure, the extensional characteristics of which is as follows

(12.1) (i) Domain $r^{(n)} \subset \text{Obj}^n$, Range $r^{(n)} \in \{$true, false$\}$;
     (ii) Domain $f^{(n)} \subset \text{Obj}^n$, Range $f^{(n)} \in \text{Obj}$;
     (iii) Domain $q^{(n)} \subset \text{Obj}^n$, Range $q^{(n)} \in \text{Num}$, respectively

Then a structure $\langle \text{Obj}, \{$true, false$\}, \text{Num}, \text{Opr}, \text{Rel}, \text{Qua} \rangle$ represents a *cognitive algebra* which underlies each data base.

An observational formula concerning an *n*-ary quantity symbol $q^{(n)}$ from $\text{SymbQua}^{(n)}$ has the form

(12.2) $(q^{(n)}, x_1, x_2, ..., x_n, x_{n+1})$ where $x_1 \in \text{Loc}$ for $i = 1, 2, ..., n+1$,

and its meaning with respect to an interpretation int of SymbQua into Qua and to a state of storage S is as follows

(12.3) (int $q^{(n)}$) $(Sx_1, ..., Sx_n) = Sx_{n+1}$ where $Sx_i \in \text{Obj}$ for $i = 1, 2, ..., n$, but $Sx_{n+1} \in \text{Num}$.

With respect to two sorts of individuals, objects and numbers, a two-sorted language [35] is required.

ACKNOWLEDGEMENT

'I would like to thank Bourbaki who kindly supported this work.'

BIBLIOGRAPHY

[1]   Tarski, A.: 'Wahrheitsbegriff in den formalisierten Sprachen', *Studia Philosophica* **I** (1935), Lwów, 261–405.
[2]   Carnap, R.: *Introduction to Symbolic Logic and its Applications*, Dover Publ., New York, 1958.
[3]   Russell, B.: *Human Knowledge: Its Scope and Limits*, London–New York, 1948.
[4]   Wittgenstein, L.: *Tractatus logico-philosophicus*, London, 1922.
[5]   Wijngaarden, A. v., Mailoux, B. J., Peck, J. B. L., and Kosters C. H. A.: *ALGOL 68*, Math. Centrum, Amsterdam, 1968.
[6]   Bauer, F. L. and Coos, G.: *Informatik I, II*, Springer, Berlin-Heidelberg-New York, 1971.
[7]   McCarthy, J., Abrahams, P. M., Edwards, D. J., Hart, T. P., and Levin, M. I.,: *LISP 1.5 Programmar's Manual*, MIT Press, Cambridge, 1962, 2nd ed.
[8]   Čulík K.: *Mathematical Theories of Languages* (in print at Mouton from 1969).
[9]   Čulík K.: *Introduction to Mathematical Theory of Languages* (Czech), a mimeographed textbook for a postgraduate course, published by Institute of Technology in Plzoň in June 1975.
[10] Neumann, J. v.: *Collected Works V: Design of Computers, Theory of Automata and Numerical Analysis*, Pergamon Press, Oxford-London-New York-Paris, 1963.
[11] Čulík, K.: 'The Importance of Formal Methods for Definition of Programming Languages and Operating Systems (on Mathematical Theory of Computers)', an invited talk to GAMM-Tagung in Göttingen (in print) 1975.
[12] Čulík, K.: 'On Logical Types of Structured Individuals and their Application in Computer Theory', submitted for IFIP WG 2.2 meeting in Rigi, January 1975 (in print).
[13] Mostowski, A.: *Mathematical Logic* (Polish), University Press, Warszawa-Wróclaw, 1948.
[14] Krotsel, G. and Krivino, Y. L.: *Modelltheorie*, Springer, Berlin-Reidelberg-New York, 1972.
[15] Shoenfield, J. R.: *Degrees of Unsolvability*, McGraw-Hill, New York, 1958.
[16] Petri, C. A.: 'Interpretations of Net Theory', GMD MBH Bonn, Interner Bericht 75-07, 18.07.1975, 34 p., presented at the Symposium on Math. Foundations of Computer Science 1975 in Mariánské Lázně, September 1975.
[17] Čulík, K.: 'Some Notes on Logical Analysis of Programming Languages', *Teorie a metoda* **III/I** (1971), Czech. Academy of Sciences, Prague, 101–111.
[18] McCarthy, J. and Hayes, P. J.: 'Some Philosophical Problems from the Standpoint of Artificial Intelligence, *Machine Intelligence* **4** (ed. by Meltzer, B. and Michie, D.), Edinburgh, 1969, pp. 463–502.
[19] Štěpánková, O. and Havel, I. M.: 'A Logical Theory of Robot Problem Solving, Techn. rep. ÚVT-1/75, Technical University of Prague, Institute of computation technique, January 1975, 68 pp.
[20] Pawlak, Z.: 'Programmed Machines' (Polish), *Algorithmy* **V** (1969), Polish Academy of Sciences, Warsaw, 5–19.
[21] Čulík, K.: 'Syntactical Definitions of Program and Flow Diagram', *Aplikace matematiky* **18** (1973), 280–301.

[22] Čulík, K.: 'Algorithmic Algebras of Computers', *Czech. Math. Jour.* **23** (1973), 670–689, presented at the International Symposium and Summer school on Math. Foundations of Computer Science, Warsaw-Jabłonna, August 1972.

[23] Čulík K.: 'A Note on Complexity of Algorithmic Nets without Cycles', *Aplikace matematiky* **16** (1971), 297–301.

[24] Čulík, K.: 'Structural Similarity of Programs and Some Concepts of Algorithmic Method', 1. Fachtagung der GI über Programiersprachen, *Springer Lecture Notes in Economics and Mathematical Systems* **75** (1975), 240–280.

[25] Manna, Z.: *Program Schemas, in Currents in the Theory of Computing* (ed. by A. V. Aho), Prentice Hall 1973, pp. 90–142.

[26] Čulík, K.: 'Equivalencies of Parallel Courses of Algorithmic Nets and Precedence Flow Diagrams, *Proceedings of symposium and summer school on mathematical foundations of computer science*, High Tatras, September 1973, pp. 27–38.

[27] Engeler, E.: 'Structure and Meaning of Elementary Programs', in Symposium on the Semantics of Algorithmic Languages (ed. by E. Engeler), *Springer Lecture Notes in Math.* **188** (1971).

[28] Engeler, E.: *Introduction to the Theory of Computation*, Academic Press, New York, 1973.

[29] Hladký, M.: *Introduction to Theoretical Cybernetics* (Mathematical theory of languages) (Czech), textbook of the Intitute of Technology of Brno, Prague, SNTL, 1975.

[30] Čulík, K.: 'Extensions of Rooted Trees and their Applications', presented at the Graph Symposium in Prague 1974 (but excluded from the proceedings).

[31] Hajek, P. and Havránek, T.: *A Logic of Automated Discovery*, in preparation.

[32] Engles, R. W.: 'A Tutorial on Data-Base Organisation', IBM TR 00.2004, Syst. Dev. Div., Poughkeepsie, March 20, 1970.

[33] Codd, E. F.: 'Further Normalization of the Data-Base Relational Model', IBM RT RJ 909, August 31, 1971.

[34] Kent, W.: 'A Primer of Normal Forms' (in relational data base), IBM TR 02.600, Syst. Dev. Div., San Jose, December 17, 1973.

[35] Carnap, R.: *Introduction to Symbolic Logic and its Applications*, Dover Publications, New York, 1958.

[36] Luckham, D. S. Park, D. M. P., and Paterson, M. S.: 'On Formalized Computer Programs', *Jour. of Computer and System Sciences* **4** (1970), 220–249.

ERWIN ENGELER

STRUCTURAL RELATIONS BETWEEN
PROGRAMS AND PROBLEMS

ABSTRACT. By a problem in a universal first-order theory Γ we understand a formula $\varphi(y)$ with a free variable y. The problem $\varphi(y)$ is solvable relative to auxiliary problems $\psi_1(v), ..., \psi_s(v)$ if solution algorithms for the ψ_i can be so composed as to yield all solutions of φ in all models of Γ. The complexity of the composite algorithm is the number of times that the auxiliary algorithms have to be called. A lower bound for the complexity of φ is obtained by developing a generalized Galois theory for theories Γ and problems φ satisfying some reasonable restrictions; our lower bound is the logarithm of the order of the 'Galois group' of φ.

1. INTRODUCTION

We begin by placing our investigations into a framework describing various activities by logicians in the field of the theory of computation.

At the basis is the *mathematical structure on which computations are performed*. This may be a relational structure such as the natural number system $\langle N, S, 0, = \rangle$ (for elementary recursion theory), a geometry (for various theories of geometrical constructions), etc. Mathematical structures also arise typically in computing as recursively defined data structures such as trees, lists, records, and the like (where, in contrast to the relational structures mentioned, the body of mathematical knowledge about them is rather meager). We deal here with two aspects of the mathematical structure in question, namely the descriptive aspect, and the computational aspect.

The *descriptive aspect* of relational structures is dealt with by formalisms of classical (or intuitionistic) logic for finite or infinitary languages, for example; variants of λ-calculi have been found to be appropriate for recursively defined data structures. Of course, the choice of a descriptive formalism depends on the uses we foresee for it.

The *computational aspect* is customarily dealt with by various algorithmic (or 'programming') languages, of which the simplest define the classes of flowchart programs operating in relational structures (see Section 3

Butts and Hintikka (eds.), *Logic, Foundations of Mathematics and Computability Theory*, 267–280.

below). Such languages can be enriched by allowing recursive procedures, counters and other features; the investigation of their relative strength is an interesting and important area of research. There is also the need to explain computations on objects of higher type, on infinitary objects and on incompletely determined objects, to deal with notions of non-determinism and (asynchronous) parallel computation.

The goal of *algorithmic logic* is to establish and investigate relations between the formal descriptive and computational aspects of mathematical structures. It is apparent that such relations wait to be established on various levels. One particular example is between flowchart programs and constructive fragments of $L_{\omega_1\omega}$ which we originated and which Rasiowa's group developed into a unified and well understood tool (see e.g. Salwicki's paper for this Congress).

A central concern of algorithmic logic is the investigation of the solvability of *algorithmic problems*. The classical paradigms of an algorithmic problem are geometrical construction problems: Given a figure F (consisting of finitely many points and lines) construct an algorithm π which by repeated use of ruler and compass, produces a figure F' satisfying some prescribed condition $\varphi(F')$. For example, to construct the perpendicular bisector l of two points P, Q we seek an algorithm $\pi(P, Q)$ such that if $l := \pi(P, Q)$ then

$$l \perp (P, Q) \wedge d(P, l) = d(Q, l).$$

It is apparent, that to make the concept of an algorithmic problem precise, we shall have to define precisely what we admit as an 'algorithm' and what language we choose our 'condition φ' from. As it is possible to develop an interesting theory already with the simplest such choices, we shall concentrate here on computing in relational structures by means of flowchart programs and formulating problems by quantifier-free first-order formulas.

Our main goal here is to discuss the *solvability* of algorithmic problems *relative* to some auxiliary capabilities. In geometry we may, for example, envision an algebraic or transcendental curve as given and use it – by intersecting it with constructed lines – to solve problems not solvable without (e.g. trisection of angles, squaring the circle). In algebra, we may admit as additional capabilities the extraction of kth roots. For the classical case, Galois theory is the appropriate tool; our method here is to use a generalization of Galois theory to deal with the more generalized situation as far as possible.

In computation theory we are not only interested in (relative) solvability in principle, but also in the *feasibility of algorithms*. Concretely, we should like to have good upper and lower bounds for the number of operations to be performed for the solution of a given problem. As a rule, upper bounds require intelligent programming, lower bounds generally call for rather non-trivial mathematics. The main result of the present paper is a general theorem giving lower bounds for relative solution algorithms.

2. THE GROUP OF A PROBLEM

We recall briefly some of the definitions and results of [1]. Let L be a first-order language for relational structures of a fixed finite similarity type. L_0 is the set of *basic* (i.e. negated or unnegated atomic) *formulas* of L. If $a_1, ..., a_m$ are new individual constants we let $L(a_1, ..., a_m)$, $L_0(a_1, ..., a_m)$ denote the corresponding augmented languages. Lower case Greek letters $\varphi, \psi, \rho, ...$ denote formulas; sets of formulas are denoted by capital Greek letters. We may indicate the occurrence of additional individual constants in the style $\varphi(a_1, ..., a_m)$, $\Delta(a_1, ..., a_m)$. The *provability* relation is indicated by $\Gamma \vdash \varphi$, the *satisfaction* relation by $\mathfrak{A} \vDash \varphi$, where $\mathfrak{A}$ is a relational structure of the given type and φ a closed formula of L. By a *diagram* in L, respectively $L(a_1, ..., a_m)$, we mean a consistent subset of L_0, respectively $L_0(a_1, ..., a_m)$ which contains, for each atomic formula α, either α or $\neg\alpha$. A *theory* Γ in L is a consistent set of closed formulas of L; it is *universal*, if Γ consists of universally quantified Boolean combinations of atomic formulas exclusively.

A theory Γ is said to have the *amalgamation property* if for any models $\mathfrak{A}$, $\mathfrak{B}_1$, and $\mathfrak{B}_2$ and any injections of $\mathfrak{A}$ into $\mathfrak{B}_1$ and $\mathfrak{B}_2$ there is a model $\mathfrak{C}$ of Γ and injections of $\mathfrak{B}_1$ and $\mathfrak{B}_2$ into $\mathfrak{C}$ such that the diagram

$$\begin{array}{ccc} \mathfrak{A} & \to & \mathfrak{B}_1 \\ \downarrow & & \downarrow \\ \mathfrak{B}_2 & \to & \mathfrak{C} \end{array}$$

commutes. The importance of the amalgamation property for Galois theory has been observed by Jónsson [2], the general concept is due to Fraïssé. The class of theories with the amalgamation property is wide enough to attract a general theory: it includes fields, differential fields, geometries, boolean algebras, lattices, etc. A particularly useful class consists of those theories whose models have the same fixed finite cardinality, where the property is trivially satisfied.

Let Γ be a universal theory. By a *problem*, we understand a quantifier-free formula $\varphi(y_1, ..., y_k)$ in some extension $L(a_1, ..., a_m)$ of L, with exactly the free variables $y_1, ..., y_k$ indicated. For notational convenience we choose $m = k = 1$; the reader can easily supply the modifications in definitions and proofs required below. The *spectrum* of $\varphi(y)$ in Γ is the set of all $n \in \mathbb{N} \cup \{\infty\}$ for which there exists a diagram $\Delta(a)$, consistent with Γ, such that in all models of $\Gamma \cup \Delta(a)$ the problem $\varphi(y)$ has exactly n solutions; n is called the *degree* of $\Delta(a)$ in this case. If $\infty \notin \text{spectrum}(\varphi)$ then the spectrum is finite (by completeness theorem), its maximal element is then called the *degree* of φ. We shall be concerned here exclusively with problems of finite degree. The theory, however, generalizes.

Let $\Delta(a)$ be a diagram in Γ, $\varphi(y)$ a problem and n the degree of $\Delta(a)$ for $\varphi(y)$. Consider new individual constants $b_1, ..., b_n$ and a diagram $\Delta(a, b_1, ..., b_n)$ which extends $\Delta(a)$, and is consistent with Γ. The extended diagram is called a *splitting diagram* (for φ over $\Delta(a)$) if

(i) $\Gamma \cup \Delta(a, b_1, ..., b_n) \vdash \varphi(b_i), \qquad i = 1, ..., n;$

(ii) $b_i \neq b_j \in \Delta(a, b_1, ..., b_n), \qquad i \neq j.$

The *group of the problem* $\varphi(y)$ is defined by

$$G_{\Delta(a)}(\varphi) = \{s \in S_n : \Gamma \cup \Delta(a, b_1, ..., b_n) \vdash \rho \equiv \rho^s$$
$$\text{for all} \quad \rho \in L_0(a, b_1, ..., b_n)\},$$

where S_n is the set of all permutations of $\{1, 2, ..., n\}$ and ρ^s results from ρ by substituting $b_{s(i)}$ for all occurrences of b_i in ρ, $i = 1, 2, ..., n$. If Γ has the amalgamation property then one can show (see [1] for details), that $G_{\Delta(a)}(\varphi)$ is indeed a group and does not depend on the particular choice of a splitting diagram.

The group of a problem can also be characterized, as in classical Galois theory, by a suitable group of automorphisms. Namely, let $\mathfrak{A}(a)$, $\mathfrak{A}(a, b_1, ..., b_n)$ be, respectively, the minimal models of $\Gamma \cup \Delta(a)$ and $\Gamma \cup \Delta(a, b_1, ..., b_n)$. (Such models exist by the universality of Γ.) Let $G(\mathfrak{A}(a, b_1, ..., b_n)/\mathfrak{A}(a))$ be the group of automorphisms of $\mathfrak{A}(a, b_1, ..., b_n)$ which leave $\mathfrak{A}(a)$ pointwise fixed; then (see [1], Corollary 6) we have

$$G_{\Delta(a)}(\varphi) \cong G(\mathfrak{A}(a, b_1, ..., b_n)/\mathfrak{A}(a)).$$

We shall use both aspects of the group of $\varphi(y)$ interchangeably as the case may be.

For applications, it is sometimes convenient, indeed essential, to enlarge the group of $\varphi(y)$ by relaxing the symmetry requirements. Namely, let L^* be a sublanguage of L which still contains the vocabulary of $\varphi(y)$. Then

$$G^*_{\Delta(a)}(\varphi) = \{s \in S_n : \Gamma \cup \Delta(a, b_1, ..., b_n) \vdash \rho \equiv \rho^s$$
$$\text{for all} \quad \rho \in L^*_0(a, b_1, ..., b_n)\}$$

is obviously a (normal) extension of $G_{\Delta(a)}(\varphi)$. It is easy to define the appropriate automorphism group isomorphic to $G^*_{\Delta(a)}(\varphi)$.

Questions on how to actually determine the group of a problem will be treated in Section 4.

3. Relative solvability

Let Γ be a universal theory with the amalgamation property, fixed for the following. Let L be the language of Γ, containing function symbols f^i, predicate symbols R_j and individual constants c_j altogether finitely many. Some, or all, of these symbols will now be associated to *basic capabilities* for performing *constructions* in models $\mathfrak{A}$ of Γ. We express these capabilities, using variables $x_1, x_2, ...$, in the form of basic *instructions*

$$x_i := c_j;$$
$$x_i := f^i(x_{k_1}, ..., x_{k_{nj}});$$
$$\text{if} \quad R_j(x_{k_1}, ..., x_{k_{mj}}) \quad \text{then} \ ... \ \text{else} \ ... \ .$$

Such instructions are composed into some kind of programs, e.g. *flowchart-programs* whose unary nodes are labelled by instructions of the first or second kind and whose binary nodes are labelled by instructions of the third kind. By a *path* through a program we understand a (finite or infinite) sequence of symbols from the set

$$\{i^c j, i^{f^l} k_1, ..., k_{n_i}, j^{R^+} k_1, ..., k_{m_j}, j^{R^-} k_1, ..., k_{m_j}\} \qquad i, j, k_1, ...$$

which describes a legal execution sequence of the program; $i^c j$ denotes execution of $x_i := c_j$, etc. The central requirement of the programming language is the following: For every program and every diagram Δ, consistent with Γ, the set

$$\{w: w \text{ is a finite initial segment of a path through}$$
$$\text{the program taken in some model of } \Gamma \cup \Delta\}$$

is recursive *relative* to Δ. Flowchart programs clearly have this property, so do recursive programs.

We employ programs to construct solutions for problems as follows: Let $\varphi(y)$ be a problem of degree n in L. Let π be a program with *output variables* $y_1, ..., y_n$ whose instructions are chosen from the set of basic instructions. The program π is said to *solve* the problem $\varphi(y)$ if for every diagram Δ consistent with Γ and every model $\mathfrak{A}$ of $\Gamma \cup \Delta$ there is a finite legal path through the program π, such that the execution of π along this path terminates with values $b_1, ..., b_n$ for $y_1, ..., y_n$ such that $\mathfrak{A} \vDash \varphi(b_i)$, $i = 1, ..., n$, and the number of different b_i is equal to the degree of Δ. Clearly, these solutions will be the values of constant terms of L, and hence the group of $\varphi(y)$ is trivial in this case.

The situation becomes more interesting, if we add to the basic capabilities the (hypothetical) solvability of some auxiliary problems with nontrivial groups and consider problems $\varphi(y)$ solvable using these added capabilities. – Concretely, let $\psi(a, v)$ be a problem of degree m in $L(a)$, where a is an individual constant which we shall use as a parameter. We introduce m unary function symbols and add the capabilities

$$\{x_i := g_j(x_k): j = 1, ..., m\}_{i,k}$$

to our programming language. For every diagram $\Delta(a)$ consistent with Γ and every splitting diagram $\Delta(a, b_1, ..., b_m)$ for $\psi(a, v)$ over $\Delta(a)$ and model $\mathfrak{A}(a, b_1, ..., b_n)$ of $\Gamma \cup \Delta(a, b_1, ..., b_n)$ the terms $g_1(a), ..., g_m(a)$ give an enumeration (perhaps with repetition) of the set $\{b_1, ..., b_n\}$. We say that $\varphi(y)$ is *solvable relative* to $\psi(u, v)$ if there exists a program in the extended capabilities which solves $\varphi(y)$.

{To have a concrete example, the reader is invited to think of Γ as the theory of fields of characteristic zero, $\varphi(y)$ a polynomial equation $p(y) = 0$ with rational coefficients and $\psi(u, v)$ the equation $v^m = u$. Relative solvability here would mean solvability of $p(y) = 0$ by mth radicals.}

Let π be a solution program for $\varphi(y)$ using the additional capabilities, and let w be a finite path through π which is consistent with Γ, i.e. such that there is a model of Γ in which execution of π actually progresses along w. Let k be the number of times a problem ψ has to be solved along this path. We construct a sequence of diagrams

$$\Delta \subseteq \Delta(a_{11}, ..., a_{1m_1}) \subseteq \Delta(a_{11}, ..., a_{1m_1}, a_{21}, ..., a_{2m_2})$$
$$\subseteq \cdots \subseteq \Delta(a_{11}, ..., a_{km_k})$$

corresponding to this path as follows: The first time the problem ψ needs to be solved along this path is after some instructions of the original programming language are executed. The problem, therefore, is posed in the form $\psi(\tau_1, v)$, where τ_1 is a constant term of L. Let $\Delta(a_{11}, ..., a_{1m_1})$ be a splitting diagram of $\psi(\tau_1, v)$ over Δ. The later diagrams are found the same way; $\Delta(a_{11}, ..., a_{pm_p}, a_{p+1,1}, ..., a_{p+1,m_{p+1}})$ is a splitting diagram of $\psi(\tau_{p+1}, v)$ over $\Delta(a_{11}, ..., a_{pm_p})$, where τ_{p+1} is a constant term in $L(a_{11}, ..., a_{pm_p})$. The numbers m_p are the respective degrees of the problems $\psi(\tau_p, v)$, $p = 1, ..., k$. Finally, let $y_1, ..., y_n$ be the output variables of π. Since π solves $\varphi(y)$ along w the values of $y_1, ..., y_n$ are determined and can be expressed as constant terms $\sigma_1, ..., \sigma_n$ of $L(a_{11}, ..., a_{km_k})$.

For the thus defined sequence of diagrams we now determine a corresponding sequence of permutation groups. Let P be the set of all permutations t of $\{a_{11}, ..., a_{km_k}\}$ such that $t(a_{ij}) = a_{ip}$ for some p. Then, for each $q = 0, 1, ..., k$, let

$$G_q = \{t \in P: \Delta^t(a_{11}, ..., a_{km_k}) = \Delta(a_{11}, ..., a_{km_k})$$
$$\text{and} \quad t(a_{ij}) = a_{ij} \quad \text{for all} \quad i \leqslant q\}.$$

$\Delta^t(a_{11}, ..., a_{km_k})$ is the result of replacing all occurrences of a_{ij} in formulas of $\Delta(a_{11}, ..., a_{km_k})$ by $t(a_{ij})$; $i = 1, ..., k$, $j = 1, ..., m$. Obviously

$$G_0 \supseteq G_1 \supseteq G_2 \supseteq \cdots \supseteq G_k = \{e\}.$$

This sequence of groups does not depend on the particular choice of the splitting diagrams $\Delta(a_{11}, ..., a_{pm_p})$. This is shown by induction, using the general fact [1, Theorem 1] that any two splitting diagrams are just permutational variants of each other: For every splitting diagram $\Delta^1(a_{11}, ..., a_{1m_1})$ of Δ there exists a permutation s of $\{a_{11}, ..., a_{1m_1}\}$ such that $\Delta^1(a_{11}, ..., a_{1m_1}) = \Delta^s(a_{11}, ..., a_{1m_1})$.

Our central point here is to investigate relations between the groups G_q, $G_\Delta(\varphi)$ and $G_{\Delta(a_{11},...,a_{pm_p})}(\psi(\tau_{p+1}, v))$.

LEMMA 1. $G_\Delta(\varphi) \cong \{s \in S_n : \exists t \in G_0 \text{ such that } \Gamma \cup \Delta(a_{11}, ..., a_{km_k}) \vdash \sigma_{s(i)} = \sigma_i^t, \; i = 1, 2, ..., n\}$.

Proof. Let $\Delta(b_1, ..., b_n)$ be a splitting diagram of $\varphi(y)$ over Δ. Let S be the set of all permutations of $\{b_1, ..., b_n\}$. It is clear from the definition that

$$G = \{s \in S : \Delta(b_1, ..., b_n) = \Delta^s(b_1, ..., b_n)\} \cong G_\Delta(\varphi(y)).$$

Let $\bar{P}$ be the set of all permutations $\bar{t}$ of $\{b_1, ..., b_n, a_{11}, ..., a_{km_k}\}$ such that $\bar{t}$ restricted to $\{a_{11}, ..., a_{km_k}\}$ belongs to P. Let

$$\bar{G} = \{\bar{t} \in \bar{P} : \Delta^{\bar{t}}(b_1, ..., b_n, a_{11}, ..., a_{km_k})$$
$$= \Delta(b_1, ..., b_n, a_{11}, ..., a_{km_k})\}.$$

We define a map $f : \bar{G} \to G$ by

$$f(\bar{t})(b_i) = \bar{t}(b_i).$$

f is a homomorphism of $\bar{G}$ onto G. It is easy to check that f is homomorphic. To prove that f is onto let $s \in G$ be given. Let $\mathfrak{U}(b_1, ..., b_n)$ be the minimal model of $\Gamma \cup \Delta(b_1, ..., b_n)$, $\mathfrak{A}(b_1, ..., b_n, a_{11}, ..., a_{km_k})$ the minimal model of $\Gamma \cup \Delta(b_1, ..., b_n, a_{11}, ..., a_{km_k})$. Consider the following application of the amalgamation property:

$$\mathfrak{A}(b_1, ..., b_n) \rightsquigarrow \mathfrak{A}(b_1, ..., b_n, a_{11}, ..., a_{km_k})$$
$$\downarrow s \qquad\qquad \downarrow \bar{t}_s$$
$$\mathfrak{A}(b_1, ..., b_n) \rightarrow \mathfrak{C}.$$

$\mathfrak{C}$ can be chosen as an extension of $\mathfrak{U}(b_1, ..., b_n, a_{11}, ..., a_{km_k})$. Since $\{b_1, ..., b_n\}$ are all of the possible solutions of $\varphi(y)$, in all extensions, and $\bar{t}_s$ is monomorphic, $\bar{t}_s$ must permute $\{b_1, ..., b_n\}$. By the same reasoning $\bar{t}_s$ permutes each $\{a_{q1}, ..., a_{qm_q}\}$. Thus, $\mathfrak{C}$ can be chosen as $\mathfrak{A}(a_1, ..., b_n, a_{11}, ..., a_{km_k})$ and $\bar{t}_s \in \bar{G}$ is onto.

{We are using the same letters to denote the permutation of the individual constants and the corresponding automorphisms of the models; which meaning is chosen is clear from the context}. Define $g: G_0 \rightarrow G$ by $g(t) = f(\bar{t})$, where $\bar{t}$ is the extension of t to $\{b_1, ..., b_n\}$ defined by $\bar{t}(b_i) = \sigma_i^t$, $i = 1, ..., n$. The map g is obviously a homomorphism onto G. Hence $s \in G$ iff there exists $t \in G_0$ such that $g(t) = s$. Observe $g(t) = s$ iff $g(t)(b_i) = s(b_i)$ for $i = 1, ..., n$, iff $\Gamma \cup \Delta(a_{11}, ..., a_{km_k}) \vdash s(\sigma_i) = \sigma_i^t$ for $i = 1, ..., n$. It follows that

$$G_\Delta(\varphi) \cong G = \{s \in S : \exists t \in G_0 \text{ with}$$
$$\Gamma \cup \Delta(a_{11}, ..., \dot{a}_{km_k}) \vdash s(\sigma_i) = \sigma_i^t\}$$
$$\cong \{s \in S_n : \exists t \in G_0 \text{ with}$$
$$\Gamma \cup \Delta(a_{11}, ..., a_{km_k}) \vdash \sigma_{s(i)} = \sigma_i^t\}.$$

COROLLARY 2. $G_\Delta(\varphi) \lhd G_0$.

LEMMA 3. $G_{q+1} \lhd G_q$, $q = 0, 1, ..., k-1$.

Proof. Let $s \in G_q$, $t \in G_{q+1}$. We show $s^{-1}ts \in G_q$. Clearly $\Delta^{s^{-1}ts}(a_{11}, ..., a_{km_k}) = \Delta(a_{11}, ..., a_{km_k})$. Furthermore, for all q and j there exists p such that

$$s^{-1}ts(a_{q+1j}) = s^{-1}t(a_{q+1p}) = s^{-1}(a_{q+1p}) = a_{q+1j}.$$

LEMMA 4. $G_q/G_{q+1} \cong G_{\Delta(a_{11}, ..., a_{qm_q})}(\psi(\tau_{q+1}, v))$, $q = 0, 1, ..., k-1$.

Proof. Let S be the set of all permutations of

$$\{a_{q+1,1}, ..., a_{q+1,m_{q+1}}\}.$$

Then

$$G_{\Delta(a_{11},...,a_{qm_q})}(\psi(\tau_{q+1}, v))$$

$$= \{s \in S: \Delta^s(a_{11}, ..., a_{q+1m_{q+1}}) = \Delta(a_{11}, ..., a_{q+1m_{q+1}})\}.$$

We construct an isomorphism f of $G_{\Delta(a_{11},...,a_{qm_q})}(\psi(\tau_{q+1}, v))$ onto G_q/G_{q+1} as follows: Let $s \in G_{\Delta(a_{11},...,a_{qm_q})}(\psi(\tau_{q+1}, v))$, and let $\mathfrak{A}(a_{11}, ..., a_{km_k})$ be the minimal model of $\Gamma \cup \Delta(a_{11}, ..., a_{km_k})$. Let t be any extension of s to an automorphism of $\mathfrak{A}(a_{11}, ..., a_{km_k})$ leaving $\mathfrak{A}(a_{11}, ..., a_{qm_q})$ pointwise fixed. Then we define $f(s) = tG_{q+1}$.

(a) The existence of t is proved in the same way as that of $\bar{t}_s$ in the proof of Lemma 1.

(b) f is well-defined, i.e. it does not depend on the choice of t: Let t_1, t_2 be extensions of s. Consider $u = t_2^{-1}t_1$. We need to show $u(x) = x$ for all $x \in \mathfrak{A}(a_{11}, ..., a_{qm_q})$. But $t_2^{-1}t_1(x) = t_2^{-1}s(x) = s^{-1}s(x) = x$, since $s(x) \in \mathfrak{A}(a_{11}, ..., a_{qm_q})$.

(c) f is one-to-one: Assume $f(s_1) = f(s_2)$, i.e. $t_1G_{q+1} = t_2G_{q+1}$. Then $t_1 \in t_2G_{q+1}$ and $t_1 = t_2y$ for some $y \in G_{q+1}$. For an arbitrary element $x \in \mathfrak{A}(a_{11}, ..., a_{q+1m_{q+1}})$ we have then $s_1(x) = t_1(x) = (t_2y)(x) = t_2(y(x)) = t_2(x) = s_2(x)$. Hence $s_1 = s_2$.

(d) f is onto: Let $tG_{q+1} \in G_q/G_{q+1}$, $t \in G_q$ be given. Let t^1 be the restriction of t to $\{a_{11}, ..., a_{q+1m_{q+1}}\}$. Then $\Delta^{t^1}(a_{11}, ..., a_{q+1m_{q+1}}) = \Delta^t(a_{11}, ..., a_{q+1m_{q+1}}) = \Delta(a_{11}, ..., a_{q+1m_{q+1}})$. In addition, $t^1(a_{ij}) = a_{ij}$ for all $i \leqslant q$, since $t \in G_q$. Therefore

$$t^1 \in G(\mathfrak{A}(a_{11}, ..., a_{q+1m_{q+1}})/\mathfrak{A}(a_{11}, ..., a_{qm_q}))$$

$$\cong G_{\Delta(a_{11},...,a_{q+1mq+1})}(\psi(\tau_{q+1}, v)), \quad \text{and} \quad f(t^1) = tG_{q+1}.$$

(e) f is a homomorphism: Because $G_{q+1} \lhd G_q$ we may compute

$$f(s_1) \cdot f(s_2) = t_1G_{q+1}t_2G_{q+1} = t_1(t_2G_{q+1}t_2^{-1})t_2G_{q+1}$$

$$= t_1t_2G_{q+1} = f(s_1 \cdot s_2).$$

The last equality follows from the fact that if t_1, t_2 are extensions of s_1, s_2 then $t_1 \cdot t_2$ is an extension of $s_1 \cdot s_2$. Namely, $t_1t_2(a_{q+1j}) = t_1s_2(a_{q+1j}) = s_1s_2(a_{q+1j})$ because $s_2(a_{q+1j})$ is of the form a_{q+1p}.

4. Complexity

There are two tasks still before us. One, to use the above results to obtain our complexity theorem, the other to discuss effective methods for obtaining the group of a problem. Lemma 4 implies (using the notational conventions of Section 3):

$$|G_q| = |G_{q+1}| \cdot |G_{\Delta(a_{11},...,a_{qmq})}(\psi(\tau_{q+1}, v))|,$$
$$q = 0, 1, ..., k-1,$$

since $|G_k| = |\{e\}| = 1$, we have therefore

$$|G_0| = \prod_{q=0}^{k-1} |G_{\Delta(a_{11},...,a_{qmq})}(\psi(\tau_{q+1}, v))|,$$

and thus, by Corollary 2,

$$|G_\Delta(\varphi)| \leq \prod_{q=0}^{k-1} |G_{\Delta(a_{11},...,a_{qmq})}(\psi(\tau_{q+1}, v))|.$$

We shall use this inequality to obtain a lower bound for the complexity of $\varphi(y)$.

Assume, then, that $\varphi(y)$ can be solved relative to $\psi(u, v)$ in all models of $\Gamma \cup \Delta$ by a program π. The *complexity of π along a path w* is the number of times an auxiliary problem ψ needs to be solved along w until all solutions of $\varphi(y)$ are obtained. {Thus we are looking for worst-case behaviour}. The *complexity of π* is the minimal complexity of π along paths. The *complexity of $\varphi(y)$* is the minimal complexity of solution algorithms π relative to auxiliary problems ψ.

THEOREM.   *The complexity of $\varphi(y)$ has a lower bound* $\log_b |G_\Delta(\varphi)|$, *where* $b = \max \{|G_{\Delta(a)}(\psi(a, v))|: \Gamma \cup \Delta(a) \text{ consistent}, \Delta(a) \supseteq \Delta\} > 1$.

Proof.   Let π be an arbitrary solution algorithm for $\varphi(y)$ relative to $\psi(u, v)$ and w a path through π consistent with $\Gamma \cup \Delta$. Let k be the number of times ψ needs to be solved along w. We are looking for a lower bound to k

independent of π and w. Define $\alpha_q = 1/\log |G_{\Delta(a_{11},...,a_{qmq})}(\psi(\tau_{q+1}, v))|$, $q = 0, 1, ..., k-1$. Then:

$$k = \sum_{q=0}^{k-1} \alpha_q \cdot \log |G_{\Delta(a_{11},...,a_{qmq})}(\psi(\tau_{q+1}, v))|$$

$$\geq \min_{q=0,...,k-1} (\alpha_q) \cdot \sum_{q=0}^{k-1} \log |G_{\Delta(a_{11},...,a_{qmq})}(\psi(\tau_{q+1}, v))|$$

$$= \min_{q=0,...,k-1} (\alpha_q) \cdot \log \prod_{q=0}^{k-1} |G_{\Delta(a_{11},...,a_{qmq})}(\psi(\tau_{q+1}, v))|$$

$$= \min_{q=0,...,k-1} (\alpha_q) \cdot \log |G_0| \geq \min_{q=0,...,k-1} (\alpha_q) \cdot \log |G_\Delta(\varphi)|$$

$$= \log |G_\Delta(\varphi)|/\log \max_{q=0,...,k-1} |G_{\Delta(a_{11},...,a_{qmq})}(\psi(\tau_{q+1}, v))|$$

$$\geq \log |G_\Delta(\varphi)|/\log b = \log_b |G_\Delta(\varphi)|.$$

The theorem obviously generalizes to more than one auxiliary problem and to problems of the form $\varphi(y_1, ..., y_k)$.

How should we determine $G_\Delta(\varphi)$?

It is reasonable to *assume that we are given a solution algorithm π of $\varphi(y)$ relative to $\psi(u, v)$ which terminates in all models of $\Gamma \cup \Delta$. Let us *assume*, in addition, *that the universal theory of $\Gamma \cup \Delta$ is decidable*, i.e. that for every quantifier-free formula $\chi(x_1, ..., x_m)$ of L we can decide whether

$$\Gamma \cup \Delta \vdash \forall x_1, ..., x_m \chi(x_1, ..., x_m).$$

It follows, that for additional constants $a_1, ..., a_m$ we can decide whether

$$\Gamma \cup \Delta \cup \{\chi(a_1, ..., a_m)\}$$

is consistent. This fact can be used to effectively find a consistent finite path through π.

The initial part of the path, up to the first time a problem ψ has to be solved is obviously determined by Δ alone. Let $\psi(\tau_1, v)$ be the first problem, τ_1 a constant term of L. We need to know the degree of Δ for

$\psi(\tau_1, v)$. This is the maximal m_1 such that

$$\Gamma \cup \Delta \cup \{\psi(\tau_1, a_{1j})\}_{j=1}^{m_1} \cup \{a_{1j} \neq a_{1k}\}_{j \neq k}$$

is consistent. By assumption, this number m_1 can be effectively found. Progressing further along the path, we need to decide relations $R(\mu_1, ..., \mu_k)$, where the μ_j are constant terms of $L(a_{11}, ..., a_{1m_1})$. Using the assumption again, we are able to augment the previously extended set consistently in such a fashion that the resulting set of formulas determines the path up to the next time ψ needs to be solved, say for $\psi(\tau_2, v)$, τ_2 a constant term of $L(a_{11}, ..., a_{1m_1})$. Again, we determine m_2, etc. This process is repeated, finitely often by assumption on the termination of π, until we reach the end of the path. At that point we have collected, by conjunction, a finite quantifier-free formula $\delta(a_{11}, ..., a_{km_k})$ and have determined constant terms $\sigma_1, ..., \sigma_n$ of $L(a_{11}, ..., a_{km_k})$ such that

$\Gamma \cup \Delta \cup \{\delta(a_{11}, ..., a_{km_k})\}$ is consistent;

$\Gamma \cup \Delta \cup \{\delta(a_{11}, ..., a_{km_k})\} \vdash \varphi(\sigma_i)$, $i = 1, ..., n$;

$\Gamma \cup \Delta \cup \{\delta(a_{11}, ..., a_{km_k})\} \vdash \sigma_i \neq \sigma_j$, $i \neq j$.

We use Lemma 1 to compute $G_\Delta(\varphi)$. For this we consider

$$G_0 = \{t \in P: \Delta^t(a_{11}, ..., a_{km_k}) = \Delta(a_{11}, ..., a_{km_k})\}.$$

Of $\Delta(a_{11}, ..., a_{km_k})$ which enters the definition of G_0 we only know the finite subset whose conjunction is $\delta(a_{11}, ..., a_{km_k})$. This formula, however determines $\Delta(a_{11}, ..., a_{km_k})$ completely (up to a permutation of the a_{ij}) because it forces the $\Delta(a_{11}, ..., a_{qm_q})$ to be splitting diagrams. Thus G_0 is completely determined by

$$\Gamma \cup \Delta \cup \{\delta(a_{11}, ..., a_{km_k})\}.$$

Assume that G_0 is known. To obtain $G_\Delta(\varphi)$ we extend the last set of formulas consistently until, for all i, $j \leq n$ and all $t \in G_0$ we have decided the equation

$$\sigma_i = \sigma_j^t.$$

Altogether, we have shown, that *under the stated assumptions the group* $G_\Delta(\varphi)$ *can be effectively determined.* We have not claimed that our algorithm is always the best one, or even feasible. The assumptions, however, seem reasonable: That a relative solution algorithm be known, that the theory have the basic decidability property stated and that the groups of the auxiliary problems $\psi(\tau, v)$ be known for 'algebraic' elements τ.

5. REMARKS

(1) The theorem gives an exact formulation to the following observation: A problem grows harder to solve if less properties can be found which distinguish its individual solutions. This suggests, that in the search for large lower complexity bounds it is reasonable to formulate those special cases where the set of solutions admits as many symmetries as possible.

(2) We are possibly losing some of the complexity information by computing $\log_b |G_\Delta(\varphi)|$ instead of $\log_b |G_0|$, since $G_\Delta(\varphi) \lhd G_0$. The index of $G_\Delta(\varphi)$ in G_0 is a measure for the *directness* of the relative solution algorithm. If this index is one, then each admissible permutation of the a_{ij} is determined by an admissible permutation of the b_i. This is true, like in the classical case of Galois theory, if the a_{ij} can be expressed as terms in the b_i.

(3) It is easy to produce examples which show that the full 'fundamental theorem of Galois theory' does not generalize without additional assumptions: Not to every (normal) subgroup of the group of a problem does there exist a corresponding submodel.

8092 Zürich, Switzerland

BIBLIOGRAPHY

[1] Engeler, E.: 1975, 'On the Solvability of Algorithmic Problems', in H. E. Rose and J. C. Shepherdson (eds.), *Logic Colloquium '73,* North-Holland Publ. Co., 1975, pp. 231–251.
[2] Jónsson, B.: 1962, 'Algebraic Extensions of Relational Systems', *Mathematica Scandinavica* **11,** 179–205.

A. SALWICKI

ALGORITHMIC LOGIC, A TOOL FOR
INVESTIGATIONS OF PROGRAMS

The paper shows a method of connecting formulas (of new type) with the properties of a program. Metamathematical studies of the algorithmic logic obtained in this way lead to a uniform approach to all the methodological problems connected with programming.

1. INTRODUCTION

The main aim: for every program and given algorithmic property:

(1) to find a formula that expresses the property of the program,
(2) to study the satisfiability (validity, stability, etc.) of this formula in order to find out about the properties of the program.

The very first approach in this direction was made by Thiele (1966); then E. Engeler (1967) proposed to use a sublanguage of $\mathscr{L}_{\omega_1 \omega}$. As can easily be shown, the classical first-order predicate calculus is not appropriate for the aims mentioned above.

Let us look at the following example:

$$while\ a \neq b\ do$$
$$begin$$
$$while\ a > b\ do\ a := a - b;$$
$$while\ b > a\ do\ b := b - a;$$
$$end$$

and ask what kind of computation this program implies? No one is able to give an answer before he is informed about the nature of the objects associated with the variables and the meaning of the relational and functional signs. Variables may denote real numbers, polynomials, segments on Euclidean plane or other objects.

Butts and Hintikka (eds.), Logic, Foundations of Mathematics and Computability Theory, 281–295.
Copyright © 1977 by D. Reidel Publishing Company, Dordrecht-Holland. All Rights Reserved.

From that the observation follows; programs are expressions without meaning. More precisely, an interpretation of language induces the interpretation of a given program.

REMARK.   On the other hand, if it is known that the objects of the universe in question satisfy axioms of Archimedean ring then it can be proved that the program computes the greatest common divisor if it exists or 'loops' otherwise, regardless of the nature of the objects.

Providing interpretations for programs is then the first aim in the theory of programs. Let us note that the tools used in programming can be stratified into the following levels containing:

(1^1) assignments as atomic programs. Programs form an algebra closed under three operations: composition $\circ$, branching $\vee$ and iteration $*$.

(2^1) procedures i.e. implicit definitions. The set of generators of the algebra of programs is enriched by atomic programs calling procedures: procedure statements.

(3^1) coroutines i.e. cooperating in quasi parallel way programs.

(4^1) processes i.e. programs concurrently performed.

Every one of the three first levels is contained in the next one.

On the first level, programs form a free algebra and hence the interpretation can be described in a usual way (cf. Rasiowa and Sikorski, 1963) (of interpreting terms and formulas) adapted to the new set of expressions. The details can be found in Rasiowa (1974a), Mirkowska (1971), Kreczmar (1974), and Banachowski (1975c).

The remaining levels contain more complicated objects. Programs with procedures do not form a free algebra. Their interpretation must be described by other means.

This is why the notion of computation is to be studied. At present we can give a number of computation definitions with interesting properties exploited in Salwicki (1976). Most of them concern the level 2. Studying computations for the levels 3 and 4 should be the first step in the process of learning of properties of more complicated programs with coroutines and/or parallel processes.

For every level: if the notion of computation is 'properly' introduced then the mapping associating the initial valuation of variables occurring

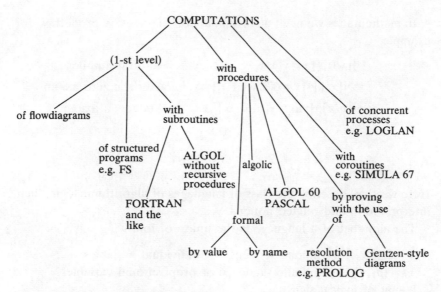

in a program K with the valuation obtained as a result of computation will be called the realization K_R – interpretation – of the program K. Denotations: $v' = $ val $K, (v)$ or $v' = K_R(v)$.

For every pair (K, α) composed of a program K and a formula α the expression: $K\alpha$ will be called a formula. Its logical value is defined as follows

$$\text{val}\,(K\alpha, v) = \begin{cases} \text{val}\,(\alpha, \text{val}\,(K, v)) & \text{if } \text{val}\,(K, v) \text{ is defined} \\[2mm] \mathbf{0} & \text{otherwise} \end{cases}$$

Formulas built from these, of the form $K\alpha$, and which are atomic describe the properties of programs and algorithmic properties of structures:

$K1$ obviously describes the halting property

$\alpha \Rightarrow K\beta$ describes the correctness property. We shall say that the program K is correct with regard to the formulas α and β (Read v as data and $v' = $ val (K, v) as results of the program K then $\alpha \Rightarrow K\beta$ is read as: if the data satisfy α then the results are defined and satisfy β).

In mathematics we meet algorithmic, non-elementary properties, for example

$\circ[[x/0]*[x \neq y[x/x+1]]]\mathbf{1}$: y is a natural number

$\neg\circ[[x/1]*[x \neq 0[x/x+1]]]\mathbf{1}$:   characteristic zero axiom

$\circ[[u/x]*[\neg u > y[u/u+x]]]\mathbf{1}$:   Archimedean axiom

2. DEFINITIONS

Here we shall describe the class of languages of algorithmic logic, their interpretation and related notions.

The alphabet of a language is the union of following sets:

$V = \{x_0, x_1, ...\}$ – infinite sequence of individual variables
$V_0 = \{p_1, p_2, ...\}$ – infinite sequence of propositional variables,
L – set of logical signs

Φ – family of sets of n-argument functors $\Phi = \bigcup\limits_{n=0}^{\infty} \Phi_n$

P – family of sets of m-argument predicates $P = \bigcup\limits_{m=1}^{\infty} P_m$

$\Pi = \{\circ, \underline{\vee}, *\}$ – set of program connectives
$\{\bigcup, \bigcap\}$ – iteration quantifiers
U – set of auxiliary symbols, contains /, [,] and others

The set of well-formed expressions of a language contains

T – set of terms,
F – set of quantifier-free formulas.

These sets are well known from any textbook on mathematical logic whereas the following ones are an original part of algorithmic logic.

S – set of parallel substitutions,
FS – set of programs,
FSF – set of formulas.

Every expression of the form

$$[z_1/w_1 ... z_n/w_n], \qquad n = 0, 1, ...$$

where $z_1 \ldots z_n$ are distinct variables and w_i is a term iff z_i is an individual variable – otherwise z_i is a propositional variable and w_i is a quantifier-free formula. Substitutions are denoted by the letter s with indices and can be read as

$\quad$ *begin* $z_1 := w_1$ *and* (simultaneously) $\ldots z_n := w_n$ *end*

The set of programs FS is the least set containing S and closed under program connectives. Hence if K, $M \in$ FS and $\alpha \in$ F then the expressions $\circ[KM]$, $\underline{\vee}[\alpha KM]$, $*[\alpha K]$ are programs. This may be written in an ALGOLic way as follows

$\quad\quad$ *begin* K; M *end*
$\quad\quad$ *if* α *then* K *else* M
$\quad\quad$ *while* α *do* K $\quad\quad$ respectively.

FSF is the set of formulas. It is defined as the least set containing F, closed under propositional connectives and the following rules of forming new formulas:

$\quad$ if α is a formula and $K \in$ FS then the expressions $K\alpha$, $\bigcup K\alpha$, $\bigcap K\alpha$ also belong to FSF,
$\quad$ if α is a formula, x is a free individual variable in α then $(\exists x)\alpha$ and $(\forall x)\alpha$ are formulas.

Note, that the notions of free variable and input (output) variables may be defined, however they need too much space to be cited here (cf. Salwicki, 1976).

We assume the reader is familiar with the notion of realization of language. Our notation follows Rasiowa and Sikorski (1963). A realization of functors and predicates will be denoted by R. A valuation of variables will be denoted by v. Values of terms and formulas will be defined in the usual way. We use $\tau_R(v)$, $\alpha_R(v)$ to denote them (also val (τ, v) or $\underset{R}{\vDash} \alpha[v]$ may be used).

$s_R(v)$ denotes the valuation v' which differs from v only in points $z_1 \ldots z_n$, $v'(z_i) = w_R(v)$ $i = 1, \ldots, n$.

The interpretation K_R of a program K may be defined in an obvious way. Let us mention another possibility based on the fact that FS is a free algebra generated by substitutions (cf. Mirkowska, 1972; Salwicki, 1970a).

Let W denote the set of all valuations. For every program K the mapping K_R is a transformation (may be partial) of the set W into itself.

The realization of formulas is as usual. Formulas of the form $K\alpha$ obtain values according to the scheme

$$(K\alpha)_R(v) = \begin{cases} \alpha_R(K_R(v)) & \text{when} \quad v' = K_R(v) \text{ is defined} \\ \mathbf{0} & \text{otherwise} \end{cases}$$

$$(\bigcup K\alpha)_R(v) = \underset{i\in\omega}{\text{l.u.b.}} \ (K^i\alpha)_R(v)$$

$$(\bigcap K\alpha)_R(v) = \underset{i\in\omega}{\text{g.l.b.}} \ (K^i\alpha)_R(v)$$

The notions of model and tautology are introduced as usual.

With these definitions in mind the reader can check the sense of algorithmic properties listed at the end of the preceding section.

3. ALGOLIC COMPUTATIONS

Here, we introduce another notion of the computation of a program. Let K be a program. By an algolic computation we shall understand a sequence of pairs $(v, M_1 \ldots M_n)$

> where v is a valuation, $M_1 \cdots M_n$ a finite sequence of programs

which satisfies:

(c0) (v, K) is the first pair of the sequence,

(c1)   for every two consecutive pairs in the sequence (to be denoted by arrow $\mapsto$) one of the following conditions holds

     a. $(v, sM_1 \ldots M_n) \mapsto (s_R(v), M_1 \ldots M_n)$

     b. $(v, \circ[KM]M_1 \ldots M_n] \mapsto (v, KMM_1 \ldots M_n)$

     c. $(v, \underline{\vee}[\alpha KM]M_1 \ldots M_n) \mapsto (v, KM_1 \ldots M_n)$ if $\alpha_R(v) = \mathbf{1}$

              $(v, \vee[\alpha KM]M_1 \ldots M_n) \mapsto (v, MM_1 \ldots M_n)$ if $\alpha_R(v) = \mathbf{0}$

     d. $(v, *[\alpha K]M_1 \ldots M_n) \mapsto (v, \underline{\vee}[\alpha \circ [K*[\alpha K]][\ ]]M_1 \ldots M_n)$

(c2)   the pair $(v, \varnothing)$ does not possess any successor.

The sequence $M_1 \ldots M_n$ of the programs can be treated as a queue of tasks to be performed in order to finish the calculations.

An algolic computation is finite if it contains $(v, \varnothing)$ and is infinite otherwise.

It can be shown that

$$v' = K_R(v) \quad \text{iff} \quad (v, K) \overset{*}{\mapsto} (v', \varnothing)$$

i.e. iff a computation starting from (v, K) ends with the result v'.

4. Procedures

One type of procedure (the most general case) is the non-functional procedure. To define computation of a program which make use of procedures we must introduce a few definitions.

Let L be an algorithmic language. Consider expressions of the form

$$K_i(x_1 \ldots x_{n_i}; z_1 \ldots z_{m_i}), \qquad i = 1, \ldots, p$$

where K_i is a symbol of a program (not a program itself) and $x_1 \ldots x_{n_i}$, $z_1 \ldots z_{m_i}$ are variables.

We modify the definition of FS-expressions as follows: pro-FS′ – the set of pro-FS′ expressions – is the least set containing substitutions, expressions of the form

$$K_i(\tau_1 \ldots \tau_{n_i}; u_1 \ldots u_{m_i})$$

(where $\tau_1 \cdots \tau_{n_i}$ are terms and $u_1 \cdots u_{m_i}$ are variables) and closed under the rules of formation of new pro-FS′ expressions by the use of $\circ$, $\vee$, $*$ program functors.

A system of non-functional procedures is a set of pairs of pro-FS′ expressions

$$K_i(x_1 \ldots x_{n_i}; z_1 \ldots z_{m_i}) \sim M_i, \qquad i = 1, \ldots, p$$

where: M_i is a pro-FS′ expression, all input variables of M_i form exactly the set $\{x_1 \ldots x_{n_i}\}$, variables $z_1 \ldots z_{m_i}$ belong to the output variables of

M_i and finally output sets for M_i $(i = 1, ..., p)$ are pairwise disjoint and no expressions considered in the sequel contain any variable common with the output set for any M_i $(i = 1, ..., p)$.

By an FS' expression we shall mean a pair composed of a pro-FS' expression and a system of procedures which defines all program symbols used in this pair.

For the purposes of this section we can conceive of the set of variables as partitioned into a denumerable family of denumerable classes. Every class can be represented as an infinite sequence

$$x = \{x^{(1)}, x^{(2)}, ..., x^{(n)}, ...\}.$$

All variables of a class are of consistent type, individual or propositional. Element $x^{(i)}$ will be called the ith copy of the variable x. Let ω be an expression in the language. We shall assume the expressions are built primarily from the first copies of variables. By $\omega^{(i)}$ we shall denote the result of simultaneous replacement of all variables by their ith copies. The expression $\omega^{(i)}$ will be called the ith copy of ω.

The definition of an algolic computation for an FS'-expression program with procedures is a modification of that from the previous section.

The point (c1) is enlarged by

e.   let $K(x_1 ... x_n ; z_1 ... z_m) \sim M$ be a procedure then

$$(v, K(\tau_1 ... \tau_n ; y_1 ... y_m)M_1 ... M_n)$$
$$\mapsto (v, [x_1^{(i)}/\tau_1 ... x_n^{(i)}/\tau_n]M^{(i)}[y_1/z_1^{(i)} ... y_m/z_m^{(i)}]M_1 ... M_n)$$

where $i - 1$ is the greatest integer number of a copy existing in the sequence $M_1 ... M_n$.

5. BASIC LAWS

Consider the set FSF of formulas and their interpretation induced by algolic computations [cf. Introduction]. To be more precise we construct formulas from atomic ones and also from the formulas of the form $K\alpha$ (where the program K *may contain procedures*).

The question what are the tautologies of this calculus, is important since the answer gives us the tools for analyzing programs with procedures that can be written in ALGOL 60, PASCAL and similar languages.

The case when programs do not contain procedures is completely solved in Mirkowska (1972), Kreczmar (1974), and Banachowski (1975c). Hence the only thing that is to be done is an axiomatization of the procedures.

We admit the following axioms:

1st group – the axioms of the propositional calculus,

2nd group – let s be a substitution $[z_1/w_1 \ldots z_n/w_n]$ and δ a quantifier-free formula then

$$s\delta \Leftrightarrow \mathrm{Sub}^{z_1 \ldots z_n}_{w_1 \ldots w_n} \delta$$

$$K(\alpha \vee \beta) \Leftrightarrow K\alpha \vee K\beta$$

$$K(\alpha \wedge \beta) \Leftrightarrow K\alpha \wedge K\beta$$

$$K\neg\alpha \Rightarrow \neg K\alpha$$

$$K1 \Rightarrow (\neg K\alpha \Rightarrow K\neg\alpha)$$

$$K(\alpha \Rightarrow \beta) \Rightarrow (K\alpha \Rightarrow K\beta)$$

$$\circ[KM]\alpha \Leftrightarrow KM\alpha$$

$$\underline{\vee}[\delta KM]\alpha \Leftrightarrow \delta \wedge K\alpha \vee \neg\delta \wedge M\alpha$$

$$*[\delta K]\alpha \Leftrightarrow \bigcup \underline{\vee}[\delta K[\ ]](\alpha \wedge \neg\delta)$$

$$M \cup K\alpha \Leftrightarrow M\alpha \vee M \cup K(K\alpha)$$

$$M \cap K\alpha \Leftrightarrow M\alpha \wedge M \cap K(K\alpha)$$

3rd group – this consists of all formulas of the form

$$M\alpha \Leftrightarrow M'\alpha,$$

where (1) program M contains a procedure of the form

$$K(x_1 \ldots x_n; z_1 \ldots z_m) \sim N$$

(2) program M' results from M by replacing an occurrence of

subexpression of the form $K(\tau_1 \ldots \tau_n; y_1 \ldots y_m)$ contained in M by

$$\circ[[x_1^{(i)}/\tau_1 \ldots x_n^{(i)}/\tau_n]N^{(i)}[y_1/z_1^{(i)} \ldots y_m/z_m^{(i)}]]$$

here i is the smallest integral number of a copy of N not existing in M, and the following inference rules

$$\frac{\alpha, \alpha \Rightarrow \beta}{\beta} \qquad \frac{\alpha \Rightarrow \beta}{K\alpha \Rightarrow K\beta}$$

$$\frac{\{MK^i\alpha \Rightarrow \beta\}_{i \in \omega}}{M \cup K\alpha \Rightarrow \beta} \qquad \frac{\{\alpha \Rightarrow MK^i\beta\}_{i \in \omega}}{\alpha \Rightarrow M \cap K\beta}$$

Since the inference rules and axioms of the first two groups were checked previously (Mirkowska, 1972) it remains to prove the adequacy of the remaining axioms. This is easy and follows from the definition of computation. Proof of completeness is difficult. In fact the only way I know is by Gentzen-style axiomatization which is so detailed that it can not be given here (cf. Mirkowska, 1971, 1972).

The question whether the infinitistic rules of inference are necessary was studied in Kreczmar (1971) and the answer is: yes.

Having a completeness theorem, one can replace the verification of semantic properties – e.g. halting, correctness properties etc. – by examination of whether the formulas describing these properties are provable or not.

6. Short review of works in algorithmic logic

As was related earlier the origins of algorithmic logic are to be connected with the works by H. Thiele and E. Engeler.

Work at Warsaw University was started in 1968: and the early results are contained in Salwicki (1970a, b, c). One of the first theorems asserts that go to instructions are unnecessary – which implies that flow-diagram computations can be replaced by computations for FS-programs. Studies in programmability were continued in Salwicki

(1976) and it was observed that Post's theorem should be stated as follows: a relation r is strongly programmable iff both r and its complement are programmable relations.

This provokes the question: what are the connections between programmability and recursiveness? It can be argued that programmability is a generalization of recursiveness. A class of structures is distinguished for which both notions coincide. A structure – relational system – of this class is called constructive.

Formalized axiomatic systems of algorithmic logic were studied in Mirkowska (1972). It included the theorem on the normal form of program once again showing that programmability is recursive computability generalized to the case of an arbitrary relational system. Further results: the compactness property does not hold for algorithmic logic, completeness theorem – proved by algebraical methods initiated by Rasiowa and Sikorski – and also Gentzen-style axiomatization of algorithmic logic. The last result may be helpful in mechanical proving which can be used both for analyzing a program and for supplying it with an interpretation (cf. Salwicki, 1976). An analogue of Herbrand's theorem has been proved. This fact is used in the estimation of degrees of unsolvability for important classes of algorithmic properties. For example an immediate corollary of the theorem is: the set of programs which always terminate their computations is recursively enumerable.

In Mirkowska (1972) the language of algorithmic logic does not contain classical quantifiers. An extension of the axiom system to the case with classical quantifiers can be found in Banachowski (1975c, d) and Kreczmar (1973, 1974).

Effectivity problems in algorithmic logic are studied in Kreczmar (1973). This paper contains: a theorem stating that the set of all tautologies of algorithmic logic is recursively isomorphic to the set of formulas valid in the standard model of arithmetic; a theorem that the set of all consequences of a set A is hyperarithmetical with respect to A; exact estimations of the degrees of unsolvability for the following properties of programs: stop property, correctness, equivalence of two programs made for distinct classes of realizations; a set of natural or real numbers, class of all realizations, class of all finite realizations etc. Results were obtained in a simple way through metamathematical and algebraical observations with the complete elimination of Goedel's arithmetization and Turing machines.

Problems concerning the correctness of programs and their modular properties are investigated in Banachowski (1975d). Here by using four quantifiers the author is able to give the one complete theory of modular properties of programs – which contain the theories of Floyd, Manna, Hoare and others. The results are used in experiments with proving the properties of programs interpreted in the set of integers (Banachowski *et al.*, 1974).

Programs with only propositional variables were studied in Grabowski (1972). It was shown that unexpectedly this case is decidable, and equivalent in a sense to classical propositional calculus. It may have some influence on bitwise computations and their programs; on the other hand it is strongly related to the theory of Yanov (1958).

Programs with tables are studied in Dańko (1975). Apart from axioms, the paper contains a comparative study of the different tools of programming – among them a simple example of a non-programmable function which is defined by a procedure (Dańko, 1974). The structure in the example is obviously not a constructive one; all kinds of programs describe the same set of functions provided that the realizations are in constructive structures.

The work on axioms which completely characterize computations in fields of complex and real numbers was initiated by Engeler. An elegant algebraic proof that all algorithmic properties of programs in the field C of complex numbers are provable from the axioms of fields of characteristic zero was found by A. Kreczmar. This enabled him to prove that certain simple functions are not programmable in C. More difficult is the result found by A. Kreczmar in cooperation with T. Mostowski that the algorithmic properties of programs in the field of real numbers are provable from the axioms of formally real fields. Properties valid in the field of real numbers with ordering are consequences of the Archimedean axiom.

The formalized ω^+-valued algorithmic logic was studied by Rasiowa (1972, 1973a, b, 1974a, b, c, d). The motivation can be justified by the case statements

$$case\langle\text{arithm. expr.}\rangle \ of \ begin \ K_1; \ldots; K_n \ end, \qquad n = 2, 3, \ldots$$

which occur in programming languages. The ω^+-valued algorithmic logic takes a generalized Post algebra P as its semantic basis. This approach allows us to investigate procedures, coroutines and push-down algorithms (cf. Rasiowa, 1974d) in an original way.

Procedures can be treated also as implicit definitions in algorithmic logic which is based on two-element Boolean algebra. The formal computations introduced in Salwicki (1976) are an effective tool for the investigation of procedures. Formal computations and also diagrams in Gentzen-style axiomatization may be used in order to estabilish the semantics of procedures. Formal computations can be used in order to get the least model of a given set of procedures – it is another formulation of McCarthy's principle of recursion induction – when diagrams give a method of extending this semantics.

A yet still more detailed survey of results in algorithmic logic will be published in the Proceedings of Computer Science Semester held at Banach Centre in 1974 (PWN Publishers, Moscow).

BIBLIOGRAPHY

Banachowski, L.: 1974a, 'Formalization of the notion of data structures', *Reports of Warsaw University Comp. Centre* **46**, in Polish.

Banachowski, L.: 1974b, 'Modular Approach to the Logical Theory of Programs', *Proc. Symp. Math. Found. Comp. Sci. Lecture Notes in Comp. Sci.* **28**, Springer Verlag, Berlin.

Banachowski, L., Szczepańska, D., Wasersztrum, W., and Zurek, J.: 1974, *Automatic Verification of Program Correctness*, Warsaw University, Dept. Math. in Polish.

Banachowski, L.: 1975a, 'An Axiomatic Approach to the Theory of Data Structures', *Bull. Acad. Pol. Sci., Ser. Math. Astr. Phys.* **23**, 315–323.

Banachowski, L.: 1975b, 'Modular Properties of Programs', *Bull. Acad. Pol. Sci., Ser. Math. Astr. Phys.* **23**, 331–337.

Banachowski, L.: 1975c, 'Extended Algorithmic Logic and Properties of Programs', *Bull. Acad. Pol. Sci., Ser. Math. Astr. Phys.* **23**, 325–330.

Banachowski, L.: 1975d, 'Investigations of Properties of Programs by Means of the Extended Algorithmic Logic', Ph.D. thesis Warsaw University Dept. Math.

Dańko, W.: 1974, 'Not Programmable Function Defined by a Procedure', *Bull. Acad. Pol. Sci., Ser. Math. Astr. Phys.* **22**, 587–594.

Dańko, W.: 1975, 'Programs with Tables and Procedures', Ph.D. thesis, Warsaw University, Dept. Math., in Polish.

Engeler, E.: 1967, 'Algorithmic Properties of Structures, *Math. System Theory* **1**, 183–195.

Floyd, R. W.: 1967, 'Assigning Meanings to Programs', *Proc. Symp. Appl. Math.*, AMS **19**, 47–53.

Góraj, A., Mirkowska, G., and Paliszkiewicz, A.: 1970, 'On the Notion of Description of Program', *Bull. Acad. Pol. Sci., Ser. Math. Astr. Phys.* **18,** 499–505.

Grabowski, M.: 1972, 'The Set of Tautologies of Zero-order Algorithmic Logic is Decidable', *Bull. Acad. Pol. Sci., Ser. Math. Astr. Phys.* **20,** 575–582.

Hoare, C. A. R.: 1969, 'An Axiomatic Basis for Computer Programming', *CACM* **12,** 576–583.

Kreczmar, A.: 1971, 'The Set of All Tautologies of Algorithmic Logic is Hyperarithmetical', *Bull. Acad. Pol. Sci., Ser. Math. Astr. Phys.* **19,** 781–783.

Kreczmar, A.: 1972, 'Degree of Recursive Unsolvability of Algorithmic Logic', *Bull. Acad. Pol. Sci., Ser. Math. Astr. Phys.* **20,** 615–617.

Kreczmar, A.: 1973, 'Effectivity Problems of Algorithmic Logic', Ph.D. thesis, Warsaw University Dept. Math.

Kreczmar, A.: 1974, 'Effectivity Problems of Algorithmic Logic, *Automata Languages, and Programming, Lecture Notes Comp. Sci.* **14,** 584–600, Springer Verlag, Berlin.

Manna, Z.: 1969, 'The Correctness of Programs', *JCSS* **3,** 119–127.

Mazurkiewicz, A.: 1972, 'Recursive Algorithms and Formal Languages', *Bull. Acad. Pol. Sci., Ser. Math. Astr. Phys.* **20,** 793–799.

Mirkowska, G.: 1971, 'On Formalized Systems of Algorithmic Logic', *Bull. Acad. Pol. Sci., Ser. Math. Astr. Phys.* **19,** 421–428.

Mirkowska, G.: 1972, 'Algorithmic Logic and Its Applications in Program Theory', Ph.D. thesis, Warsaw University Dept. Math. in Polish.

Mirkowska, G.: 1974, 'Herbrand Theorem in Algorithmic Logic', *Bull. Acad. Pol. Sci., Ser. Math. Astr. Phys.* **22,** 539–543.

Perkowska, E.: 1972, 'On Algorithmic m-valued Logics', *Bull. Acad. Pol. Sci., Ser. Math. Astr. Phys.* **20,** 717–719.

Perkowska, E.: 1974, 'Theorem on Normal Form of a Program', *Bull. Acad. Pol. Sci., Ser. Math. Astr. Phys.* **22,** 439–441.

Rasiowa, H.: 1972, 'On Logical Structure of Programs', *Bull. Acad. Pol. Sci., Ser. Math. Astr. Phys.* **20,** 319–324.

Rasiowa, H.: 1973a, 'On Logical Structure of Mix-valued Programs and ω^+-valued Algorithmic Logic', *Bull. Acad. Pol. Sci., Ser. Math. Astr. Phys.* **21,** 451–458.

Rasiowa, H.: 1973b, 'Formalized ω^+-valued Algorithmic Logic', *Bull. Acad. Pol. Sci., Ser. Math. Astr. Phys.* **21,** 559–565.

Rasiowa, H.: 1974a, 'A Simplified Formalization of ω^+-valued Algorithmic Logic', *Bull Acad. Pol. Sci., Ser. Math. Astr. Phys,* **22,** 595–603.

Rasiowa, H.: 1974b, 'Extended ω^+-valued Algorithmic Logic', *Bull. Acad. Pol. Sci., Ser. Math. Astr. Phys.* **22,** 605–610.

Rasiowa, H.: 1974c, 'On ω^+-valued Algorithmic Logic and Related Problems', CC PAS reports, Warsaw.

Rasiowa, H.: 1974d, 'ω^+-valued Algorithmic Logic as a Tool to Investigate Procedures', *Proc. Symp. Math. Found. Comp. Sci., Lecture Notes in Comp. Sci.* **28,** Springer Verlag, Berlin.

Rasiowa, H. and Sikorski, R.: 1963, *Mathematics of metamathematics*, PWN, Warsaw.

Salwicki, A.: 1970a, 'Formalized Algorithmic Languages, *Bull. Acad. Pol. Sci., Ser. Math. Astr. Phys.* **18,** 227–232.

Salwicki, A.: 1970b, 'On the Equivalence of FS-Expressions and Programs', *Bull. Acad. Pol. Sci., Ser. Math. Astr. Phys.* **22,** 275–278.

Salwicki, A.: 1970c, 'On Predicate Calculi with Iteration Quantifiers', *Bull. Acad. Pol. Sci., Ser. Math. Astr. Phys.* **22**, 275–285.

Salwicki, A.: 1974, 'Procedures, Formal Computations and Models', *Proc. Symp. Math. Found. Comp. Sci., Lecture Notes in Comp. Sci.* **28**, 464–484, Springer Verlag, Berlin.

Salwicki, A.: 1976, 'Programmability and Recursiveness' (an application of algorithmic logic to procedures), Diss. Mathematicae, to appear.

Thiele, H.: 1966, *Wissensschaftsteoretische Untersuchungen in Algorithmische Sprachen*, Berlin.

Yanov, J.: 1958, 'On Logical Schemes of Algorithms', *Probl. Cyb.* **1**, 75–127.

V

PHILOSOPHY OF LOGIC AND MATHEMATICS

A. A. MARKOV

ON A SEMANTICAL LANGUAGE HIERARCHY IN A CONSTRUCTIVE MATHEMATICAL LOGIC

The subject of this paper is a constructive mathematical logic, i.e. a logic suitable for constructive mathematics. Constructive mathematics differs radically from classical mathematics in the understanding of existence and disjunction. We say in constructive mathematics that there exists an object satisfying some condition iff we possess a construction of such an object. We say that the disjunction 'A or B' takes place (A and B are propositions) iff we possess a construction of a true proposition which coincides either with A or with B. Whatever might be the classical understanding of existence and disjunction, it differs from the constructive one inasmuch as such constructions are not needed according to the classicist's point of view.

Thus constructive mathematics cannot be controlled by classical logic and requires a special logic – the constructive mathematical logic.

It is natural to attack the problem of building constructive mathematical logic by proposing adequate semantics for the traditional logical signs: &, $\vee$, $\exists$, $\forall$, $\supset$ and $\neg$. The first three present no difficulties. '&' can be understood as the conjunction 'and' and we already have explained our understanding of '$\vee$' and '$\exists$'.

Let us consider the quantifier '$\forall$' equivalent to 'whatever is'. It is used in universal propositions which state that every object of a given type satisfies some condition. What constructive meaning can be assigned to a proposition of this sort? It asserts of course that we possess a general method of proving every particular example of the universal proposition i.e. a method, which for every object of the type in question would prove that this object satisfies the condition. Let us ascribe to universal propositions exactly this meaning.

It remains now to ascribe a constructive meaning to the signs '$\supset$' and '$\neg$'. For this purpose we construct a 'tower' of languages

$$(1) \qquad Я_0, Я_1, Я_2, ..., Я_N, ..., Я_\omega, Я_{\omega+1}.$$

Butts and Hintikka (eds.), Logic, Foundations of Mathematics and Computability Theory, 299–306.
Copyright © 1977 by D. Reidel Publishing Company, Dordrecht-Holland. All Rights Reserved.

Each of them has an exact syntax i.e. a well defined notion of a formula. Each of them has a definite semantics dealing with constructive objects called *literoids* and *verboids*. A literoid is a string of the form $(a \cdots a)$; a verboid is composed of literoids. The literoids form a sort of infinite alphabet; verboids are 'words' in this alphabet.

The language $Я_0$ – the basement level of the tower – is a very simple language suitable for talking about verboids, their equality and diversity, their prefixes and suffixes, about occurrences of verboids in other verboids (Markov, 1974a).

The alphabet of $Я_0$ contains the signs: &, $\vee$, $\forall$ and $\exists$. The semantics of the sign & is ordinary; the semantics of the disjunction sign has been already explained. The signs $\forall$ and $\exists$ are used only as restricted quantifiers i.e. for formalizing such propositions as 'for every verboidal prefix X of the verboid Y...', 'there exists such a verboidal suffix X of the verboid Y that...'.

The alphabet of $Я_0$ contains also the equality and diversity signs $=$ and $\neq$, which can be written between two terms denoting verboids.

The signs $\supset$ and $\neg$ are not used in $Я_0$. However negation can be introduced as an operation $\mathfrak{N}$ which transforms formulae into formulae. For every closed formula A of $Я_0$ we have the true formula

(2) $\vee A \mathfrak{N}(A)$

and the false formula

$\& A \mathfrak{N}(A)$

(Here and in the sequel we use the Polish notation.) The law of the excluded middle (2) is constructive, i.e. there exists an algorithm S_0 which finds out for every closed formula A, whether A is true. In this sense the language $Я_0$ is decidable.

Implication can also be introduced in $Я_0$ as material implication:

$\mathrm{Imp}\,(A, B) \rightleftharpoons \vee \mathfrak{N}(A)B.$

The next language $Я_1$ is obtained from $Я_0$ by introducing the unrestricted existential quantifier (Markov, 1974b).

The applicability of a normal algorithm to a string can be expressed by a formula of $Я_1$. As a consequence no normal algorithm is possible which finds out for every closed formula of $Я_1$, whether this formula is true.

At the same time a formal calculus S_1 can be so constructed that any closed formula A of $Я_1$ is true if and only if it is deducible in S_1. In other words the calculus S_1 is semantically valid and semantically complete for $Я_1$.

There is no negation, no implication and no universal quantification in $Я_1$. However the calculus S_1 gives rise to a sort of implication which can join two closed formulae of $Я_1$. Namely the proposition

if A then B,

where A and B are closed formulae of $Я_1$ can be understood as

(3)   whatever be the string Z, if Z is a proof of A in S_1 then B.

Here the implication under the universal quantifier can be understood as material, since the predicate

Z is a proof of A in S_1

is decidable. The meaning of the universal quantifier in (3) has been already explained.

The language $Я_2$ (Markov, 1974c) is obtained from $Я_1$ by adjoining implication and the universal quantifier. The implication sign can join two formulae of $Я_1$ only. The universal quantifier $\forall$ can be applied to every formula. The conjunction sign – to every two formulae. As to the signs $\vee$ and $\exists$ their occurrences in formulae of $Я_2$ can be caused only by their occurrences in formulae of $Я_1$.

The meaning of & is ordinary; the meaning of the universal quantifier was explained earlier – a closed formula which begins with $\forall$ means the presence of a general method of proving every example of the formula in question. The implication $\supset AB$ where A and B are closed formulae of $Я_1$ was explained above.

It is impossible to have a formal calculus which would be valid and complete for $Я_2$. We propose however a semiformal system S_2 which determines deducibility of a closed formula of $Я_2$ from another closed formula of $Я_2$. This system consists of 13 rules and possesses the following property of semantical validity: every formula deducible in S_2 from a true formula is itself true.

Given any closed formula A of $Я_2$ every rule of S_2 permits us to obtain under some circumstances a closed formula deducible from A. For example rule 1 states that every true closed formula of $Я_2$ can be considered as deducible from A. Rule 2 states that A is deducible from A. Rule 3 states that if B and C are closed formulae of $Я_1$ such that both formulae B and $\supset BC$ are deducible from A, then the formula C can be considered as deducible from A.

Rules 2 and 3 are finitary. There are many other finitary rules in S_2. Rule 1 is not finitary, nor is Rule 11 which states: whenever we have a general method enabling us to establish that $\mathfrak{F}\lfloor XHQ \rfloor$ is deducible from A whatever the verboid Q, then the closed formula $\forall XH$ can be considered as deducible from A. Here H is a formula of $Я_2$ containing no free occurrences of variables other than X and $\mathfrak{F}\lfloor XHQ \rfloor$ denotes the result of substitution of the verboid Q for X in H.

Thanks to the inductive definition of the deducibility relation the method of induction can be applied for proving theorems about formulae deducible from a given closed formula A of $Я_2$. Let $\mathfrak{P}$ be some property which can be possessed by closed formulae of $Я_2$. We define the A-*inductivity* of $\mathfrak{P}$ by 13 A-inductivity conditions which are in a natural one-to-one correspondence with the 13 deducibility rules. For example to deducibility Rule 1 corresponds A-inductivity Condition 1: every true closed formulae possess the property $\mathfrak{P}$; to Rule 2 corresponds Condition 2: A possesses the property $\mathfrak{P}$; to Rule 3 corresponds Condition 3: whenever B and C are closed formulae of $Я_2$ such that B and $\supset BC$ possess $\mathfrak{P}$, C possesses $\mathfrak{P}$. We have the following induction theorem.

If the property $\mathfrak{P}$ is A-inductive, then every formula deducible from A possesses this property.

By means of this theorem we prove in particular that S_2 is semantically valid.

Now we can pass to the language $Я_3$ (Markov, 1974d). Its alphabet is obtained from the alphabet of $Я_2$ by adjoining the stroke $|$. With the aid of the stroke the implication sign of order one $\supset|$ is formed. Formulae of $Я_3$ of the form $\supset|AB$ where A and B are formulae of $Я_2$, are called *implications of order one*. If A and B are closed formulae, then $\supset|AB$ means the deducibility of B from A in the sense explained above.

Thus there are in $Я_3$ two distinct implications: the implication of order zero whose meaning was explained earlier and the just now introduced implication of order one. The former can join two formulae of $Я_1$ only; the later can join two arbitrary formulae of $Я_2$. If A and B are two closed formulae of $Я_1$ then one can form both implications $\supset AB$ and $\supset|AB$. Fortunately they are equivalent: whenever one of them is true the other is true also. The proof of this fact is not quite simple. It can be accomplished by the method of induction.

We propose a semiformal system S_3 determining deducibility of a closed formula of $Я_3$ from another closed formula of $Я_3$. This system possesses the property of semantical validity: every formula deducible from a true formula of $Я_3$ is itself true. With the aid of S_3 we build a language $Я_4$ in which implication of order two appears. $\supset\|AB$ where A and B are closed formulae of $Я_3$ means that B is deducible from A in S_3.

The construction of the hierarchy can be continued further (Markov, 1974e) in the same manner.

There are in the language $Я_N$ many implications, but they are all in accordance with each other: as soon as $\supset KAB$ is true and $\supset LAB$ meaningful, $\supset LAB$ is true (K and L are here natural numbers i.e. strings of strokes). This permits us to unite all the languages $Я_N$ (N natural number) into one language $Я_\omega$ (Markov, 1974f). Formulae of $Я_\omega$ arise from formulae of the languages $Я_N$ by simply dropping the strokes. For every formula A of $Я_\omega$ a natural number N and a formula B of $Я_N$ can be so determined that A arises from B by the procedure of dropping strokes. If A is closed then B is also closed and the meaning of A in $Я_\omega$ can be defined as identical with the meaning of B in $Я_N$.

The language $Я_\omega$ is closed with respect to the operations symbolized by $\supset$ and &. Negation can be naturally introduced by the definition

$$\daleth A \rightleftharpoons \supset A\,(0 \neq 0),$$

where 0 means the empty verboid.

There exists an important relation between the language $Я_\omega$ and the classical predicate calculus (CPC). Let A be a closed formula of CPC without the signs $\vee$ and $\exists$. Let us say that A is *valid in* $Я_\omega$ provided that every formula of $Я_\omega$ which arises from A by substituting predicates of $Я_\omega$ for predicate symbols and verboids for constants, is true in $Я_\omega$. We have the following theorem.

A closed formula of CPC is valid in $Я_\omega$ *if and only if it is deducible in* CPC (Markov, 1974h).

This theorem characterizes the place of CPC in constructive mathematics. As a corollary the law of double negation

$$(4) \qquad \supset \daleth\daleth A A$$

holds in $Я_\omega$. Here A is an arbitrary closed formula of $Я_\omega$. In particular A can be the formula of $Я_1$ which expresses the applicability of a normal algorithm $\mathfrak{A}$ to a string P. $\daleth A$ would then express the non-applicability of $\mathfrak{A}$ to P i.e. would say that the process of application of $\mathfrak{A}$ to P never ends. $\daleth\daleth A$ would express the refutability of $\daleth A$ in our hierarchy i.e. it would say that this process can not be prolonged arbitrarily far. The implication (4) assures that in this case the process finishes. This is an exact form of the principle of constructive selection (of the so called Markov principle). In this form the principle is proved.

The language $Я_\omega$ is not closed relative to the logical operations of disjunction and existence. The next and last language of our hierarchy $Я_{\omega+1}$ is obtained by adjoining to $Я_\omega$ these operations (Markov, 1974g). The semantics of $Я_{\omega+1}$ can be introduced by means of an algorithm which transforms every formula of $Я_{\omega+1}$ into a formula of the form

$$\exists X A,$$

where A is a formula of $Я_\omega$. This algorithm is analogous to Shanin's algorithm for the constructive deciphering of formulas (Shanin, 1958).

The language $Я_{\omega+1}$ is closed relative to all traditional logical operations. It is suitable for developing some portion of constructive analysis. It seems on the other hand that $Я_{\omega+1}$ is not convenient enough for further hierarchy building.

The hierarchy (1) has been the object of an interesting investigation of Kanovich (1974). Let us say with Kanovich that a language L is *reducible* to a language M iff there exists a normal algorithm which transforms every closed formula A of L into a closed formula of M equivalent to A. It is evident that $Я_N$ is reducible to $Я_{N+1}$ and to $Я_\omega$ whatever be the natural number N. Kanovich proved that on the other hand neither $Я_{N+1}$ nor $Я_\omega$ is reducible to $Я_N$. The proof is based on a normal form for formulae of $Я_N$ established by Kanovich.

Kanovich introduced an extension operation for languages and studied its application to languages of the system (1). Let L be some language, P – the record of a normal algorithm $\mathfrak{A}$ which transforms every natural number into a closed formula of L. We introduce then a formula of new type $\forall P$ which says that $\mathfrak{A}(K)$ is true whatever be the natural number K. Thus we construct an extension $[L]$ of the language L. There are in $[L]$ two types of formulae: formulae of L and formulae of the new type.

Kanovich proved that $[Я_\omega]$ is not reducible to $Я_\omega$ and not even to $Я_{\omega+1}$. On the other hand $[[Я_\omega]]$ is equivalent to $[Я_\omega]$ and $[Я_N]$ is equivalent to $Я_N$ for every natural number N greater than 1.

Starting with

$$Я_1^1 \rightleftharpoons [Я_\omega]$$

we can build a hierarchy

$$Я_1^1, Я_2^1, ..., Я_N^1, ..., Я_\omega^1,$$

where every $Я_{N+1}^1$ is obtained from $Я_N$ by a procedure analoguous to construction of $Я_4$ on the base of $Я_3$. This gives the hierarchy

$$Я_0, Я_1, ..., Я_N, ..., Я_\omega, Я_1^1, Я_2^1, ..., Я_N^1, ..., Я_\omega^1.$$

This process can be continued into constructive transfinite.

ACKNOWLEDGEMENT

It is my pleasant debt to thank Mrs Julia Robinson for her valuable assistance in the preparation of this paper for print.

University of Moscow

BIBLIOGRAPHY

Kanovich, M. I.: 1974, *On an Extension of the Stairwise Semantical System of A. A. Markov. Algorithm Theory and Mathematical Logic*, Comp. Center Akad. Nauk SSSR (in Russ.), pp. 62–70.

Markov, A. A.: 1974a, 'On the Language Я_0', *Dokl. Akad. Nauk SSSR*, **214,** No. 1, English translation *Soviet Math. Dokl.* **15,** 1974, No. 1.

Markov, A. A.: 1974b, 'On the Language Я_1', *Dokl. Akad. Nauk SSSR*, **214,** No. 2, English translation *Soviet Math. Dokl.* **15,** 1974, No. 1.

Markov, A. A.: 1974c, 'On the Language Я_2', *Dokl. Akad. Nauk SSSR*, **214,** No. 3, English translation *Soviet Math. Dokl.* **15,** 1974, No. 1.

Markov, A. A.: 1974d, 'On the Language Я_3', *Dokl. Akad. Nauk SSSR*, **214,** No. 4, English translation *Soviet Math. Dokl.* **15,** 1974, No. 1.

Markov, A. A.: 1974e, 'On the Languages Я_4, Я_5, ...', *Dokl. Akad. Nauk SSSR*, **214,** No. 5, English translation *Soviet Math. Dokl.* **15,** 1974, No. 1.

Markov, A. A.: 1974f, 'On the Language Я_ω', *Dokl. Akad. Nauk SSSR*, **214,** No. 6, English translation *Soviet Math. Dokl.* **15,** 1974, No. 1.

Markov, A. A.: 1974g, 'On the Language $\text{Я}_{\omega+1}$', *Dokl. Akad. Nauk SSSR*, **215,** No. 1, English translation *Soviet Math. Dokl.* **15,** 1974, No. 2.

Markov, A. A.: 1974h, 'On the Completeness of the Classical Predicate Calculus in Constructive Mathematical Logic', *Dokl. Akad. Nauk SSSR* **215,** No. 2, English translation *Soviet Math. Dokl.* **15,** 1974, No. 2.

Shanin, N. A.: 1958, 'On Constructive Understanding of Mathematical Judgements', *Trudy Math. Inst. V. A. Steklov* **52,** 226–311. (In Russian.)

VI

ON THE CONCEPT OF A SET

LARGE SETS

This paper deals with the search for new axioms in set theory from the objectivistic point of view. A description of objectivism is followed by a general discussion of this search for new axioms. The second half of the paper concentrates on an examination of the reflection principle and attempts to use it to introduce new axioms sufficient to justify the existence of measurable cardinals.

Owing to the time limit and many unforeseen ramifications of the issues involved, more general philosophical considerations envisaged originally have been left out. In particular, I had intended to consider the place of mathematics in human knowledge and the limitations of what the mathematical mode can contribute to general philosophy. For example, I believe that the relation of a set to its elements as a uniform abstract limit of the fertile interplay between the one and the many lends special philosophical significance to the reduction of mathematics to set theory, and gives mathematics its natural place in the realm of human knowledge by the transition from organic wholes and concrete concepts to sets and formal concepts.[1]

1. OBJECTIVISM OF SETS

According to Gödel, a set is a unity of which the elements are constituents. Objects are unities and sets are objects. He considers the iterative concept of set the correct concept and describes it in the following manner.

This concept of set, however, according to which a set is something obtainable from the integers (or some other well-defined objects) by iterated application of the operation 'set of', not something obtained by dividing the totality of all existing things into two categories, has never led to any antinomy whatsoever; that is, the perfectly 'naive' and uncritical working with this concept of set has so far proved completely consistent. It

Butts and Hintikka (eds.), Logic, Foundations of Mathematics and Computability Theory, 309–333.
Copyright © 1977 by D. Reidel Publishing Company, Dordrecht-Holland. All Rights Reserved.

follows at once from this explanation of the term 'set' that a set of all sets or other sets of a similar extension cannot exist since every set obtained in this way immediately gives rise to further applications of the operation 'set of' and, therefore, to the existence of larger sets. (Gödel, 1964b, pp. 262–263.)

Gödel states explicitly that he "considers mathematical objects to exist independently of our constructions and of our having an intuition of them individually" and "requires only that the general mathematical concepts must be sufficiently clear for us to be able to recognize their soundness and the truth of the axioms concerning them" (Gödel, 1964b, p. 262).

In accordance with this position, a set presupposes for its existence (the existence of) all its elements but not for its knowability (the knowability of) all its elements individually (a distinction introduced in Gödel, 1964a, p. 219). For example, the power set of the set of natural numbers is known to exist even though we do not know all its elements (viz. all sets of natural numbers) individually. This matter of presupposition, so far as existence is concerned, is not a question of temporal priority. Objects have to exist in order for their unity to exist. A multitude of objects having the property that the unity of the objects in the multitude exists is a set.

The multitude V of all sets does not have this property. The concept of set contains the component that sets are wholes each with its elements as constituents and therefore rules out the possibility of a set belonging to itself because it would then be its own constituent.[2] This has the consequence that the multitude V of all sets, being on a higher level than every set, cannot be a unity. It presupposes for its existence (the existence of) all sets and therefore cannot itself be a set because no such whole can presuppose (the existence of) itself (as one of its constituents) for its existence. Of course it follows that no multitude can be a set if its being a set would by justifiable axioms of set theory compel V to be a set also. Each multitude like V is what Cantor calls an inconsistent multitude: "For a multitude can be such that the assumption of a 'being-together' of *all* its elements leads to a contradiction, so that it is impossible to conceive of the multitude as a unity, as 'one finished thing'." (Cantor, 1932, p. 443). One might compare this situation with the non-existence of a largest natural number or a largest countable ordinal. Since, relying on familiar definitions in set theory, ω is the set of all natural numbers, it cannot be one of them since no set can belong to itself; similarly with ω_1 and countable ordinals. They differ from V in that they are sets. The universe of all sets is the extension of the concept of set and is a multitude (many) and not a unity (one). By appealing to generally accepted principles of set theory, we can also give auxiliary arguments to show that V cannot be a set. For example, the multitude of all subsets of a set is again a set and a larger set. If V were a set, the multitude of all its subsets would be a larger set, contradicting the fact that no multitude of sets could be larger than V.

It is only through our knowledge obtained in studying mathematics (and in particular set theory) that the view of sets as existing independently of our knowledge is reached. We are led to recognize the existence of a lot of sets. But we are a long way from knowing the full extension of the universe of all sets. Moreover, once we recognize the interaction of knowledge and existence, it becomes an almost evident truth that the totality of all sets is open-ended relative to our possible knowledge. This is not much different in the aspect of complete knowledge from the situation with the physical world: we know more and more about it but we shall never achieve complete knowledge which would imply a form of death of our consciousness and of the world. Or we may wish to compare the situation with Hegel's Absolute construed properly in its incompletability. In other words, our existing knowledge leads by extrapolation to the above objective view of the totality of sets, and once we look at the knowledge we have in its generality (i.e. in its projection to all the knowledge we can possibly have), we get a check on the possibilities of multitudes forming unities. Even though we do not possess a sharp boundary of what unities can exist, we do see that V and extension like V cannot exist as unities.

If we try to visualize the universe of all sets, we have a choice of taking in the realm of non-sets including the physical world (and therewith those sets such as the present inhabitants of Beijing) or leaving it out. If we leave out the objects which are not sets, we have a peculiar tree (or rather a mess of many trees) with the null set[3] as the root so that each set is a node and two sets are joined by a branch if one belongs to the other. To get some order out of this chaos, one uses the power set operation. Clearly the null set must be at the bottom. If

we consider the power set of the null set, we get all possible subsets of
the null set, and so on. In the original mess of trees representing V,
every node except the root has some branch going downwards. Hence,
every node must eventually lead back to the null set. But if we use a
power set at each successor stage and take union at every limit stage,
we should be able to exhaust all possible sets on the way up so that
each node will be included in this one-dimensional hierarchy.

It seems surprising that arbitrary collections of objects into wholes
should receive such a neat order. Yet it is not easy to think of any
non-artificial situation that would defeat this order.

Thus far we have tried to keep the knowing subject out as far as
possible and to reach the iterative concept of set without reference to
any temporal or subjective priority. Once we have arrived at this
framework of what is commonly known as the rank hierarchy, we can
conveniently speak of collecting a multitude of given sets into a new
set. This now deals with our knowledge about sets and does not
concern the ontological status of sets. Also, the suggestive term 'genetic concept of set' belongs to the same level (viz. of knowledge rather
than of existence).

Of course, for our knowledge, there are many difficulties about this
iterative concept or rank hierarchy. We are led to the objective
viewpoint by extrapolation from our limited knowledge, and once we
get the objective picture which is necessarily blurred in places, we are
obliged to be more explicit in order to convince ourselves that axioms
of set theory are true for this framework. For example, it is left vague
as to how far we can continue the iteration process, assuming implicitly
that we continue 'as far as possible'. This brings us to the question of
ordinal numbers as the indices of the stages of iterations. Of course, we
do not assume all ordinals given in advance since, for our knowledge,
they do not come from heaven. Rather the formation of sets and stages
can interact so that, for example, the axiom of replacement can yield
sets which lead to new stages.

Generally it should be clear that there are diverse ways of introducing and justifying axioms. The justifications can only to a greater or
lesser degree serve to indicate that the axioms under consideration are
true for the iterative concept of set. Elsewhere Gödel has listed a
heterogeneous and not mutually exclusive group of five ways of

justifying axioms of set theory (Wang, 1974, pp. 189–190). Of these five ways, the first one (viz. that of overviewing a multitude as an intuitive range of variability) occupies a special place and is the only one acceptable to the subjectivists. It is sufficient to yield enough of set theory as a foundation for classical mathematics and has in fact been applied (Wang, 1974, pp. 181–190) to justify all the axioms of ZF, which go beyond what is needed for classical mathematics.[4] In contrast with it, the other four ways (including, in particular, the reflection principle) are more objective and may be spoken of as being abstract rather than intuitive. For example, as will be sketched below, justifying Ackermann's axiom by the reflection principle goes beyond the use of intuitive ranges.

The word 'only' should be deleted from the statement "we can form a set from a multitude only in case the range of variability of this multitude is in some sense intuitive" (Wang, 1974, p. 182). Otherwise it would be too strong and imply that the other four ways of justifying axioms of set theory all are reducible to the way of discovering that the ranges of variability are intuitive. In applying the idea of having an overview of a range, we are helped by a contrast between the power set of ω and the totality V of all sets obtainable by the iterative concept. Admittedly the intuition of the former calls for a strong idealization as is seen from the fact that we have no constructive consistency proof of the assumption of its existence. But clearly we do not have even a similarly weak intuition of the range of V.

2. THE SEARCH FOR NEW AXIOMS

It is remarkable that natural questions such as the continuum hypothesis and Souslin's hypothesis which arise at an early stage of the development of set theory are demonstrably undecidable by the commonly accepted axioms (viz. those of ZF and beyond). This exhibits a transparent incompleteness of these axioms that calls for extensions beyond them.

If we think of the axioms of set theory as characterizing the concept of set and contrast them with (say) the way elements and species are characterized in chemistry and biology, we notice a surprising feature in that little is said about the properties which sets are to possess. The only common properties of all sets are that they are extensional and well-founded. All the other axioms tell us how to generate new sets (related in appropriate ways to certain given sets). Much of the strength of set theory comes from the richness of the universe of sets and the stipulated interrelationship among these sets. It is, therefore, not unnatural that the search for new axioms concentrates on finding new principles of generating sets.

In the abstract, when the crucial issue is the determination of the universe of discourse, there are two opposite ways of rendering axioms more complete: by adding more to the universe of discourse or by excluding the possibility of further extensions of it. In each concrete situation, there are reasons for considering which way is the appropriate one. In the study of natural numbers, it is relatively easy to generate all the numbers we want, and the appropriate completion is to say that the numbers generated are all. The principle of mathematical induction is meant to accomplish just this purpose by trying not entirely successfully to stipulate that the desired set is the intersection of all sets which contain 0 and are closed with respect to the successor operation. Already, when we come to the real numbers, we experience the difficulty of generating enough of them. For example, if we generate only the algebraic numbers and stipulate that they are all the real numbers, we make the mistake of excluding the familiar real numbers e and π.

For set theory, axiom systems such as those of ZF and its diverse extensions generally possess 'natural' models each of which is just a particular rank R_α in the rank hierarchy. Two questions naturally arise when one attempts to approximate the intended grand (natural) model by explicit axioms. First, for each α, how rich (wide) a rank R_α is required by the axioms? Second, how many ranks or how long a hierarchy is required by the axioms?

If we begin with the axioms of ZF, it is well known that we could consistently add two axioms to restrict respectively the length and the width of the rank model. The length can be restricted by asserting that there are no inaccessible cardinal numbers. The width of each rank can be restricted by saying that all sets are constructible, i.e. $V = L$. The resulting system is of course more categorical and more complete than

ZF. Just the second proposition is sufficient to prove the generalized continuum hypothesis and refute Souslin's hypothesis.

Nonetheless, it is generally thought that neither proposition is true according to the intended vague concept of set. The task at present is to examine principles for generating more sets to increase the length of the whole hierarchy and the width of the ranks. Already there appears to be an open-endedness with regard to the countable ordinals (Cantor's second number class); there is a feeling that there are many countable ordinals not generated by known axioms. Similarly, there is the problem of finding principles of generating directly more real numbers (or sets of natural numbers). So far, there do not seem to be promising candidates of new axioms aimed directly at generating more real numbers or more countable ordinals. (By the way, strong axioms of infinity are of course not the only possible road towards the solution of the continuum problem; there is also an attempt to find directly suitable structural properties of the continuum.)

The candidates currently under consideration are directed towards introducing large sets to extend the length of the rank hierarchy, some of which incidentally as a byproduct also broaden certain ranks. Hence, the central concern today may be said to be that of looking for large sets.

Epistemologically the general phenomenon of new axioms is significant but elusive. It finds in the current state of set theory a particularly illuminating special case which is less difficult to study with more concentrated attention. For example, this special case illustrates the distinction between discovering and justifying new axioms, between description of principles currently employed for justifying new axioms and justification of these principles.

New axioms emerge in diverse ways and are recognized then or later as falling under a few broad categories: by analogy (especially with ω, e.g. Ramsey and measurable and strongly compact cardinals), by closure of iterating extensions of known operations (e.g. inaccessible numbers and Mahlo numbers[5]), by appeal to regimentations of formal logic (e.g. indescribable numbers), etc. Sometimes these possibly idealized origins are taken as their justification. For example, Gödel believes that the principle of uniformity or clear analogy "makes the existence of strongly compact cardinals very plausible, due to the fact that there should exist generalization of Stone's representation theorem for ordinary Boolean algebras to Boolean algebras with infinite sums and products" (Wang, 1974, p. 190).

Apart from the appeal to principles, there are two types of less theoretical justification at least with regard to plausibility: viz. familiarity and application. Familiarity can eliminate certain inconsistent 'axioms' and increase our confidence in the others. Application strengthens our belief in an 'axiom' through its plausible consequences. Measurable cardinals are often thought to be plausible more on account of their plausible consequences than on the ground of arguments from theoretical principles. Some logicians go as far as saying that there must be measurable cardinals because so many interesting (rather than or in addition to being plausible) theorems have been proved about them. We shall not enter further into these considerations.

We shall make no attempt to pursue systematically the large cardinals defined so far in the literature but confine our attention to discussing a small number of them in relation to some reflection principles and their motivations.[6]

The search for stronger axioms of infinity is a lively issue in the philosophy of mathematics today, especially from the viewpoint of so-called 'substantial factualism' suggested in Wang (1974). It has a 'material' base in current mathematical practice and bears on several traditional problems including the one on the nature of axioms in mathematics.

Relative to general philosophical issues, the concept of (pure) set ordinarily employed in set theory seems to bring out the problem of universals (one and many) in its purest and simplest form. The contradictions in the concept of a universal set (that it is complete and yet not even completable, that it is a set and cannot be a set, that it is larger than every set and yet always ends up being some set) seems to offer a rich yet abstract context in which contradictions serve as the driving force in the development of the concept as concentrated in the search for large cardinals. Moreover, set theory also illustrates in an area least 'tainted' by and remote from studies directly bearing on the physical world how unfinished (despite the exceptional coherence in this case) human knowledge is. It is clear that many propositions which are labelled axioms in set theory are by no means self-evident and, for all we know, may be false and lead to contradictions.

In the search for new axioms in set theory there is an interplay of theory and social practice in a very peculiar corner of the totality of human activities. The social practice consists largely of introducing definitions and proving theorems about what is defined by them and postulating as axioms that there exist sets satisfying such and such definitions. Theory tries to supply, by an appeal to our imprecise concept of set, some plausible story as a reason or rationalization for believing in such and such 'axioms'. Or to speak in a more academic manner: Candidates for new axioms turn up here and there in the cooperative efforts of mathematical investigation; attempts at theoretical justification gradually emerge, suggesting additional candidates for new axioms; these candidates are examined with respect to their consequences and generally to their relation to the contemporary framework of set theory. And the process continues. The problem in placing this special activity in the larger context of central social practice is peculiarly involved.

Set theory presupposes a good deal of idealization and extrapolation. Extrapolation clearly implies an element of arbitrariness or at least choice from different alternatives. Idealization also contains a sense of proceeding by analogy. To eliminate alternatives involves a complex historical process of successions of experimentations, consensi, and rationalizations. We learn to consider certain consensi self-evident and we may be wrong even as to the truth of what is taken to be self-evident. But in the normal course of events, we choose to concentrate on a fixed alternative that has usually been a consensus perhaps for an extended period of time and shun speculating idly on other alternatives. There are also different levels of idealization from the simple infinity to the inaccessible, etc. to the 'absolutely infinite'. Each time we move a level higher, there is room for doubt of additional assumptions.

If we compare set theory with chess, we see that the rules in set theory are less arbitrary and they do not remain fixed once and for all. If we take first order logic for granted and consider the axioms of set theory as merely hypotheses from which we try to draw conclusions, we are unable to explain why this set of hypotheses is so much more interesting than a million other sets, nor how these hypotheses can grow and yield more hypotheses. Set theory is not like group theory where we study many instances of an abstract structure in parallel; it is rather the interrelation (or the hierarchy) of the different sets that is the central concern in set theory. Doing set theory has been compared with climbing mountains. True, we also create new paths in climbing mountains. But the difficult thing is just to explain that there is indeed a mountain there; there is indeed a rock bottom. Here we cannot help recognizing that there is a basic difference between physics and set theory as far as we are interested in a ready or at least stable answer to the question: What objects does it study?[7]

3. REFLECTION PRINCIPLES

It was noted before that no set can belong to itself because it presupposes its elements for existence. Hence, in particular, the universe V of all sets cannot be a set. Another property of V is that it is undefinable (by any 'structural property'). This is the basic reflection principle. It has many ramifications which can be seen from considering its meaning and justification. It is not known whether there is any set which is not definable (by any property or concept) in the same sense. It is possible that every set is definable (or at least the extension of some property).

Suppose a multitude A of given sets exists. If it is a set, then its submultitudes are sets, independently of whether we have adequate linguistic means to represent a property as the criterion for selecting the elements from A in each case. Consider all properties P which satisfy the following condition: if only sets can have the property P, then the multitude of all x having the property P is also a set. Call these properties 'structural properties'. Generally for each property P, there are three possibilities: (1) it is true of certain sets as well as certain non-sets; (2) it is only true of sets and determines a set; (3) it is only true of sets but does not determine a set. A structural property is any property for which either (1) or (2) is true.

Since V is not a set, by this definition, no structural property can characterize V uniquely. This conclusion of course says very little because the burden of applying it depends on knowing that properties in certain appropriate collections are indeed structural properties. Obviously, the property of being a set cannot be a structural property. Or, if the universe of discourse is confined to sets, then, for instance, the property of not belonging to itself cannot be a structural property.

The power and justification of the reflection principle depend on the following consideration. For our knowledge, the 'iterative' process means that we can use all conceivable means to continue the process, that the process is 'unlimited'. But the range of what is conceivable is inexhaustible. We cannot see exactly how we can possibly reach V beginning with the null set according to the rank hierarchy; there are a lot of gaps inbetween. Therefore, we can never capture the full content of V and any property which characterizes V uniquely can only do this 'by default' and does not tell us much about the elements of V. For example, self-identity would characterize V if we agree beforehand that the universe of discourse is V. Generalizing, one reaches the belief that a property is likely to be a structural property unless we have fairly direct reasons for realizing that it is not a structural property. This permits the use of a lot of structural properties and lends much power to the reflection principle. An illustration of applying this consideration is the justification of Ackermann's axiom to be discussed below.

In the special case when the universe of discourse is restricted to sets only, we have of course a simpler situation since in this case any property can only be true of sets and possibility (1) is ruled out. The reflection principle then takes on the special form that no structural property can characterize V uniquely in the sense of separating it from its segments. Given a proposed structural characterization of V, there is a set already satisfying it. In fact, owing to the special nature of the rank hierarchy, there are many sets satisfying it.[8]

In order to apply the concept of structural property to obtain many sets and develop set theory, the common practice is to use axiom systems formulated in the framework of first order logic. A basic choice is between confining the universe of discourse to sets or using a larger universe of discourse which contains, beyond sets, also higher entities such as 'supersets', concepts, or properties. One familiar use of structural properties is to pull given sets together. For example, given a set of acceptable axioms for generating sets, we need a means for pulling together the sets thus generated in order to formulate a new axiom that would carry out the continuation of the iterative process. More generally, we can also introduce axioms (such as the existence of measurable cardinals) which require that sets possessing certain properties exist. We do not know that they lead to contradictions and believe that they do not lead to contradictions. We believe that the properties involved are structural. But these axioms generate large sets which go far beyond the multitude of given sets (say in ZF) and can be seen to be true for the iterative concept of set only after a good deal of study (if at all), because we do not see directly the stages involved in going from the null set up to (say) the smallest measurable cardinal.

We are so far speaking of a property P as characterizing the collection of objects having the property P. An alternative way of viewing the relation of a property to the entity it characterizes is to say that a property P characterizes A if A is the only thing (or smallest thing in some suitable sense of 'small') having the property P. It is easily seen that if A is characterized uniquely by a property in one sense, it is characterized uniquely by a simply related property in the other sense.

Consider first the case when the universe of discourse in the object language is confined to sets, using as usual the language of ZF. The simplest reflection principle commonly considered is the following set of theorems in ZF:[9]

R1.   If a sentence $F(x, y)$ of ZF is true in V, then there is some rank R_α in which $F(x, y)$ is true.

A natural idea of strengthening this is by uniformatization: instead of one rank for each sentence, we require a single rank for all sentences. But if we add simply such an axiom schema to ZF, we obtain an outright contradiction because we would get: there is some α (the index of the required rank) such that, for all x, x is an ordinal if and only if $x \in \alpha$. But it is provable in ZF that the above statement is false. To avoid this complication, Levy (1959) expands the language of ZF by adding a new constant, say M. The required principle is then formulated as follows (F^M be ing the usual relativization of F to M).

R1U.   If $F(x, y)$ does not contain M and $x, y \in M$, then $F(x, y) \leftrightarrow$
        $F^M(x, y)$.

This interesting principle says explicitly that all sentences true of the universe are already true of a set M; we suppose also included a companion axiom saying that M is a rank, viz. $\exists \alpha(M = R_\alpha)$. When it is added to ZF, the theorems not containing M remain the same. This shows that we can consistently say of ZF that there is a set M such that everything true of the universe is true of M. In other words, relative to the language of ZF, M is like the universe. It is easy to view R1U as an application of the reflection principle that if V has a structural property P, then there is already a set M having P. For A to have the property P in question means simply that any sentence of ZF is true if and only if it is true with quantifiers ranging over A. Since the universe of discourse for ZF is intended to be V, of course V has this property.

The system obtained from ZF by adding R1U also has natural models one of which is, say R_β. Within such a model, M is represented by a rank R_α, $\alpha < \beta$. Here we do have the situation where two sets (two ranks) which are both like the universe are like one another. In fact, it is an easy theorem that if A and B are both elementary submodels of C and A is included in B, then A is an elementary submodel of B. Clearly as long as we are thinking of natural (or rank) models, we can in this way get many models which are alike in the specified sense.

This leads to the model-theoretic form of R1U:

R1M.   There are α, β such that $R_\alpha < R_\beta$ (R_α is an elementary
        submodel of R_β).

In other words, the set $M(= R_\alpha)$ is so self-sufficient that for any sentence (with parameters confined to sets in M) to be true in the larger universe R_β of sets, we need only the sets in M to satisfy all the quantifiers (or their Skolem functions).

Once R1U is accepted, there is a natural desire to further strengthen it. Given the familiar hierarchy of the simple theory of types and the need and the wish to extensionalize the move from definable to arbitrary functions over M, we are led to the second-order reflection principle proposed by Bernays (1961):

R2. $F(X) \to \exists \alpha R_{\alpha+1} \vDash F(X \cap R_\alpha)$.

This principle deals with given sentences and given parameters. It cannot be uniformized by a single R_α because otherwise if the universe has a model $R_{\beta+1}$, $\beta > \alpha$, we could take $\exists y(y \in X)$ as $F(X)$ and $R_\beta - R_\alpha$ as X. Hence, no single α would make R2 true for all F and X. In fact, to think of finding a suitable $R_{\alpha+1}$ for each $F(X)$ we should consider X and F simultaneously with regard to the closure of Skolem functions and make adjustments with respect not only to the quantifiers in F but also to the given X. To answer this question of uniformizing R2 requires a different viewpoint. In fact, initially W. Reinhardt and J. Silver were led to introduce 1-extendible cardinals to get an appropriate uniformization of R2. Roughly speaking, we have to use elementary embeddings rather than elementary submodels (see below). On the other hand, there are obvious generalizations of R2 in its local form to higher orders and these are closely related to the indescribable numbers.

4. Measurable cardinals

The principle R1U was introduced originally in a study of Ackermann's system (Ackermann, 1956) which can also be viewed as using a language obtained from that of ZF by adding a constant M. The central axiom is a schema which says that every property expressible without using M determines a member of M if it applies only to members of M.[10] The axiom was introduced as a codification of Cantor's famous definition: By a 'set' we shall understand any collection into a whole A of definite, well-distinguished objects a (which we will call the 'elements' of A) of our (sense) intuition or our thought (1895, see Cantor, 1932, p. 282). Construing M as the universe V of all sets, Ackermann specifies that the collection (in the definition) is to be done by arbitrary properties expressible in the language of ZF.

Let us look at Ackermann's idea more closely. A natural approach to Cantor's definition is to say that from among the given sets at each stage, those with a common property can be collected into a (new) set. This differs from Frege's notion in the qualification 'given'. Thus, we do not get a direct contradiction by considering the property of not belonging to itself. Rather we get just a new set collecting together all the 'given' sets. The concept of given sets contains an ambiguity coming from its inductive character: as new sets get generated, they become given sets. Given a fixed category K of property, we would like to think of the closure M of the process of getting sets by all these properties (beginning say with the null set as the initially given set). We can express the situation by a generalization of Ackermann's axiom: if P is in K and $P(x)$ implies $x \in M$, then $\{x \mid P(x)\} \in M$. But this can only work if M itself cannot be got this way by a property P in K. Otherwise we would get $M \in M$ which is contrary to our intuition and generally yields a contradiction quickly. This does not exclude the possibility of capturing M (relative to K) by some property not in K and establishing M as a set.

We are left with the problem of finding an adequate supply of properties. This is similar to the problem of explicating Zermelo's idea of definite property which is commonly thought to be solved (at least in part) by Skolem's proposal (of using all first-order open ε-sentences) now adopted in ZF; the difference is that Zermelo asks for separation from each given set, while we are now looking for separation from the totality of all given sets. The natural idea suggested by Ackermann is again that we use properties expressed by open sentences in the first order language built up with the membership relation (plus $=$ or with $=$ defined).

However, once Ackermann's axiom is formulated explicitly, we encounter a serious problem of interpreting his system, because M is thought of as V and at the same time taken as one thing in the universe of discourse. If the variables range over sets only, then M cannot be V because M belongs to the range. If M is to be taken as V, then the variables must range over a larger domain containing things which are not sets.

One possibility is to interpret Ackermann's axiom as we have done with R1U. The range of variables includes indeed only sets but M is

also a set which shares a property P with V. The property P is simply
the conditions on M put down in Ackermann's system. We are led to
these conditions because they are satisfied by V. But by the reflection
principle, there is already a set M possessing this property. We shall
soon call attention to what seems to be an insuperable difficulty in this
interpretation.

Alternatively, we can view M as V and think of Ackermann's axiom
as following directly from the reflection principle that V is not definable (by any structural property) in the larger universe of discourse. In
that case, habit would have us think of M as a (proper) class. But then
we also have to envisage certain classes of classes, etc. in Ackermann's
system.

In either interpretation, it seems desirable to look more closely at
those objects which do not belong to M. Ackermann does not do this
in any explicit manner, except for including a principle saying that
arbitrary properties in his system determine subclasses (or subsets) of
M. However we wish to interpret the situation, we have the contrast
between M and the more inclusive universe of discourse U, as well as
the understanding that M to a certain extent corresponds to V, the
universe of all sets. There is a choice of viewing M as V or viewing U
as V. In either case, there is something special about M since it is a
special property to be the universe of sets or even to be a good
approximation to it.

The difficulty in interpreting Ackermann's axiom by taking U as V
seems to be this. The property P (viz. Ackermann's conditions on M),
when taken as a property of V, has to talk about things outside V ($=$ U
in this interpretation). Thus, in Ackermann's axiom an open sentence,
e.g., $x = x$, can be satisfied by things not in M. But if U is taken as V
and P is thought of a property of V, we must go beyond U to get an
interpretation of P. Hence, in order to understand Ackermann's
axiom, we have to envisage a larger universe than V in any case. This,
I take it, is the reason why a proper interpretation of Ackermann's
axiom calls for taking U as more inclusive than V and an appeal to the
reflection principle that V ($= M$) is undefinable by any structurable
property (in particular, any property expressible in the language with
variables ranging over U but not containing the constant M itself). For
brevity, we shall speak of members of U as classes.

Two equivalent systems have been proposed by Powell (1972)[11] and Reinhardt (1974a)[12] which extend Ackermann's system and in which a lot of measurable cardinals can be proven to exist. As Powell and Reinhardt have observed,[13] their systems can be reformulated by substituting in each case Ackermann's axiom for one of their two crucial axioms. Gödel thinks that this is the reasonable formulation. When thus reformulated, their systems go essentially beyond Ackermann's by a new axiom (S3.3) (Reinhardt, 1974a, p. 15) which as Gödel remarks, says that all subclasses of $M(=V)$ can be defined without reference to M, i.e. M can be eliminated from such definitions. Gödel thinks that it is this formulation which gives a certain degree of plausibility to this system. Generally Gödel believes that in the last analysis every axiom of infinity should be derived from the (extremely plausible) principle that V is undefinable, where definability is to be taken in more and more generalized and idealized sense.

If we think of properties as defined over the whole range U by open sentences, the Powell-Reinhardt system contains a strong form of extensionality with regard to properties. Two properties P and Q are identical, provided they are true of the same sets (i.e. members of M). Hence, each property is determined uniquely by its behavior on M, i.e. for every property P, there is a subclass X of M such that $X = P \cap M = iP$, and that $P = Q$ if $iP = iQ$. Moreover, since every subclass X of M is definable by a property, we have, for every X, a projection jX so that $jX = jY$ if $X = Y$.

The following second order reflection principle is provable in the Powell-Reinhardt system (see Reinhardt, 1974b, p. 196):

R2U. $(\forall x \in M)(\forall X \subseteq M)(F^{M^+}(x, X) \leftrightarrow F(x, jX))$, where M^+ is the power class of M.

We have here finally arrived at the uniformization of R2. Here the old problem about arbitrary parameters is avoided. If R_β is a model of U and R_α is the model of M, $\alpha < \beta$, then the subset $R_\beta - R_\alpha$ cannot be a jX, so that we would not have the situation that jX is not empty but $X = jX \cap R_\alpha$ is. This means that we are no longer requiring that $R_{\alpha+1}$ be an elementary submodel of $R_{\beta+1}$ but rather just that we have an elementary embedding j of $R_{\alpha+1}$ into $R_{\beta+1}$. It is however also clear that for all classes X which are elements of M, $jX = X$, since otherwise X would contain a member that is not an element (of M). But this says exactly that α is a 1-extendible cardinal. In fact, the model-theoretical version of R2U contains the definition of 1-extendibility as applied to α.

R2M.   There are ranks R_α and R_β, $\alpha < \beta$ and an elementary embedding $j: R_{\alpha+1} \to R_{\beta+1}$ such that for all X in R_α, $j(X) = X$.

It follows immediately that $j(\alpha) = \beta$ and α is the first ordinal moved. For the embedding j, the value of α is important and is sometimes called the critical point.

It is familiar that 1-extendible cardinals are measurable. Given the intended model of the set theory described above, it is not surprising that it can be proved in the system that there are measurable cardinals. In fact, the proof is quite straightforward. The class α of all ordinals in M is a measurable cardinal with the required ultrafilter consisting of all subclasses X of α such that $\exists P(\alpha \varepsilon P \wedge X = P \cap M)$. (See Reinhardt, 1974a, p. 22).

For several years the only justification of measurable cardinals from reflection principles was by way of 1-extendible cardinals or, to speak in terms of more basic concepts, elementary embeddings of the special kind just mentioned. It was felt that the reflection principles did not give any plausible justification of such tools and that the stringent conditions on R_α in saying that $R_{\alpha+1}$ is like $R_{\beta+1}$ is too unmotivated and beyond reasonable expectations. The set theory described above has now given a more plausible story about justifying measurable cardinals by reflection principles, i.e. by appealing to the idea that since R_α stands for V, it is reasonable to expect it to have unusual properties. Even though this is done not be avoiding but by using those elementary embeddings in its very conception, one is nonetheless inclined to feel that a novel way is introduced by the new theory for looking at the complex notions involved.[14]

NOTES

[1] In preparing this paper, Gödel, Martin and Powell have generously given me their time to help me straighten out a number of points unclear to me. I have read an early draft of

Parson's contribution to this symposium and reformulated certain passages in this revised version to meet his objections to several points in Wang, 1974.

[2] On p. 221 of Gödel (1964a), Gödel declares such a state of affairs to be possible for concepts (in opposition to objects). He now thinks that this is unnecessary for the justification of intensional impredicative definitions. But he still believes that it may occur in certain other instances.

[3] One familiar objection to objectivism of sets is that the null set and unit sets are not adequately accounted for. The following observation by Gödel seems to meet the objection. "Russell adduces two reasons against the extensional view of classes, namely the existence of (1) the null class, which cannot very well be a collection, and (2) the unit classes, which would have to be identical with their single elements. But it seems to me that these arguments could, if anything, at most prove that the null class and the unit classes (as distinct from their only element) are fictions (introduced to simplify the calculus like the points at infinity in geometry), not that all classes are fictions". (Gödel, 1964a, p. 223.)

[4] The three main assumptions of ZF are: (1) there is some infinite set; (2) it is permissible to move from a set to its power set; (3) it is permissible to form any set b of given sets provided b is no larger than a given set c. Each assumption appears plausible enough. They correspond respectively to the axioms of infinity, power set, and replacement.

For example, it is reasonable to think of all natural numbers as forming a set; this is the most basic step of idealization in all infinite mathematics accepted by everybody who pursues set theory. Since 0 is a set and the successor of a set is a set, we obtain a multitude of given sets including all natural numbers. Since we have a definite base (viz. 0) and a sharp operation of generating a new set (the successor) from a given set, it is relatively easy to move to a multitude containing them and thereby get an infinite set. Given a set b, it seems possible to think of all possible ways of deleting certain elements from b; certainly each result of deletion remains a set. The assumption of the formation of the power set of b then says that all these results taken together again make up a set. There is certainly room for closer examination of this assumption. But the uneasiness toward the power set (say of ω) comes usually from having in mind definitions rather than objects. Finally, given a lot of sets including a particular set c, replacing elements of c by other given sets does not seem to lead to a multitude which fails to possess a unity. Hence, any replacement yields from c a new set b that can again be overviewed. Once these three assumptions are accepted, it is easy to see that the universe V of all sets must contain as members an infinite set (say ω), the power set of each member of it, and each result obtained by replacement from a member of it.

An objection to the above justification of full replacement concerns the way in which 'a lot of sets' are given. In the general case, quantifiers over the universe of all sets are employed so that in order to get the set b from a given set c, we may have to face first a collection of sets of which we do not have an overview to begin with. This objection does not affect the contention that the idea of intuitive range justifies a large enough segment of ZF sufficient for doing ordinary mathematics. In fact, it is well known that we can obtain by replacement with quantifiers restricted to previously given sets all the ranks with ω, ω_ω, ω_{ω_ω}, etc. as indices. In ordinary mathematics, we use only much lower ranks. For example, the necessary use of the ω_1th rank in a recent proof of Borel determinacy (Friedman, 1971; Martin, 1975) is viewed as a striking instance where an unusually high rank is needed for proving a theorem more or less in the realm of ordinary mathematics.

The distinction between intuition of objects and intuition of propositions does not seem important. There is. no clear separation because in either case something general comes in anyhow. In particular, to see that the range of a multitude is intuitive involves an intuition of both the objects of the multitude and a proposition, viz. that the objects in the multitude form a unity.

[5] In terms of our previous discussion of ZF, it is easy to see that we have a natural model R_γ of ZF if γ is the order type of the collection On of all the ordinals obtainable in ZF. Moreover, On is determined by three conditions analogous to those for sets: ω belongs to On, 2^α belongs to On if α does, β belongs to On if α belongs to it and β is the limit of no more than α ordinals. This says that when On itself is viewed as an ordinal number, it can be thought of as being not greater than an inaccessible number. If we yield to the temptation of stopping at this apparently neat stage, we encounter a sort of contradiction in our conception. On the one hand, we would like to have V contain all conceivable sets. On the other hand, since we can conceive of the whole of V and On thus specified as a finished totality, we can conceive of adding more ranks on top of V just as we can continue extending any R_α for α an ordinary ordinal number. This contradiction is commonly resolved by choosing to say that there are sets which are inaccessible numbers (a first example of strong axioms of infinity).

It is natural to look for ways of stating that there are many inaccessible cardinals. One obvious idea would be the statement that for every ordinal α, there is the αth inaccessible cardinal. This turns out to be equivalent to saying that there are arbitrarily large inaccessibles or that they form a proper or unbounded class. More than sixty years ago, Mahlo introduced a general principle which generates many more cardinals. Call a class of ordinals closed if it contains all limits of ordinals in it. Mahlo's principle says that every closed unbounded class of ordinals contains a regular cardinal. If we take the class of all limit ordinals α such that if $\beta < \alpha$, $2^\beta < \alpha$, then the principle yields an inaccessible cardinal. Moreover, if we delete any inaccessible cardinals and all smaller ordinals, we get another inaccessible cardinal. Hence, there is a proper class of inaccessibles. If we take all inaccessibles and add their limits, the principle gives an inaccessible α which is the limit of α inaccessibles. The process can be repeated. If we call a cardinal Mahlo when every closed unbounded subset of it contains a regular cardinal, we know that it is larger than every cardinal obtainable by Mahlo's principle. But then we can apply the principle to the class of Mahlo cardinals. And so on.

[6] For useful expositions of large cardinals, the reader is referred to Drake (1974) for general information, and Kanamori (1975) for cardinals larger than measurable cardinals centered around elementary embeddings (to be discussed below).

We mention here the principal categories of large cardinals roughly in order of increasing strength. (1) Inaccessible numbers, Mahlo numbers, etc. which we have briefly considered in the preceding footnote. (2) Indescribable cardinals which we shall not consider except to note that the second order case is essentially Bernays' reflection principle R2 given below (Tharp, 1967). (3) Cardinals based on partition properties (including Ramsey cardinals and an alternative definition of weakly compact cardinals) which we shall not consider. (4) Measurable cardinals. (5) 1-extendible cardinals. (6) Strongly compact cardinals. (7) α-extendible cardinals (for $\alpha > 1$). (8) Supercompact cardinal numbers. (9) Extendible cardinals (α-extendible for all α). (10) Cardinals α such that R_α has a (non-identical) embedding in itself.

[7] Objectification is a universal mode of human thinking and communication. It is in itself neutral. Objectification in religion and in commodity fetishness has led to alienation not

because it is objectification in these cases but because it is false objectification, it is myth-making. One must not conclude hastily by analogy from these demonstrably false objectifications (demonstrable in the strongest sense appropriate to such questions) that a constructivistic view of mathematics is the true one. It is necessary to examine the case on its own merits. Perhaps one of the most significant developments of abstract thinking was the move from the use value of a commodity to its exchange value. One was led to think of the use value as the particular and the phenomenal and the exchange value as the universal and the essential. Just because this produced the false consciousness under which humanity has been suffering so much does not negate the value of abstract thinking in science.

If the progress of science is to resolve contradictions discovered by the scientist in his consciousness of the external world and a work of art is to resolve contradictions discovered by the artist in his consciousness of 'the internal world', there is a temptation to consider pure mathematics a branch of the arts (as is done officially in the University of Cambridge). There is of course a grain of truth in this classification. But this has some of the flavor of sayings like 'life is a drama, or a dream, or a game', etc.

It is important to remember the autonomy of set theory, while not denying that all mathematics arose initially from the needs of men. Once a subject becomes independent, the internal criterion dominates the practice of it to a large extent so that, for instance, no mathematician waits for the physicist to verify a theorem before considering it proved. It is the empiricists who find abstract ideas so threatening and feel it necessary to assign to mathematics special places such as taking it as an instrument for other sciences or an ideal language, or attempting to found it on the non-mathematical, etc. It is they who find such an unbridgeable gap between the concrete and the abstract. It is subjectivists and empiricists who tend to introduce ideological distortions to meddle with the sound scientific practice in mathematics.

There is quite a different question which is political in nature. At a particular historical stage in a particular society whether to encourage the pursuit of subjects like set theory would depend on an examination of the total historical situation at the time and an evaluation of the possible social consequences of such encouragement. On the other hand, ideological misuses of set theory (such as using it to argue for the existence of God or for racial discrimination) can usually be demolished once we have a more correct overall view on the more important questions (viz. those having to do with human welfare). In this connection, it is also relevant to find out what societies and what stations in society and what traditions tend to generate people with strong interests in set theory.

[8] This is a more complex assertion and does not follow immediately from the reflection principle as just stated. When a proposed characterization of V is given, it is of course natural to try to capture as much of the properties of V as we can so that to the extent we succeed in this direction, the set V_1 (say) thus obtained is like V in many respects. But for the result V_2 of another attempt to be like V_1, they must be like V in the same aspects. And this can only happen when there are sufficiently strong interconnections between the two attempts. In practice, on account of a common framework (the rank hierarchy and the fixed meaning of likeness in terms of sentences in the same languages) for the different attempts, we can argue plausibly for the presence of such likeness between different (necessarily unsuccessful) candidates for V.

[9] More explicitly:

R1.   If $V \vDash F(x, y)$, then there is an ordinal α, such that for all x, y in R_α, $R_\alpha \vDash F(x, y)$.

In other words, if a sentence is true in the universe of discourse, it is already true in some rank. The proof that all cases of R1 are theorems of ZF is based on a familiar fact which is very plausible from considerations involving the use of Skolem functions. Given any sentence $F(x)$ in the language of set theory, we can find some β such that, for $x \in R_\beta$, the truth or falsity of $F(x)$ is determined already if we relativize $F(x)$ to R_β, i.e. take all the quantifiers as ranging over the members of R_β (instead of ranging over the universe of all sets). Roughly speaking, every set satisfying a given formula belongs to some fixed rank. Hence, in certain sufficiently high ranks R_β, all the sets which make a difference to the truth of $F(x)$ are already present. The theorem R1 of ZF can also be extended to say that for every sentence $F(x)$ of the language of ZF, there are arbitrarily high ranks β such that, for $x \in R_\beta$, $F(x)$ if and only if $F_\beta(x)$, i.e. $F(x)$ relativized to R_β. (See, e.g. Drake, 1974, p. 99.)

This reflection principle which is provable in ZF is known to be equivalent to the axioms of infinity plus replacement. Extensions of this principle were first studied by Levy and Bernays with a view of getting stronger axioms of infinity. In order to understand these extensions, we look at the inaccessible numbers more closely.

One important feature of inaccessible numbers is that they are regular: α is regular if α is the limit of no less than α ordinals; every regular ordinal is a cardinal. Hence, as argued in note 5, there are arbitrarily large inaccessible cardinals. But then it is also reasonable to extend the principle R1 to:

R1L.   For every formula F of ZF, there are arbitrarily large inaccessible α such that F if and only if F_α (Levy, 1960).

Since R_α gives a natural model of ZF when α is any given inaccessible cardinal, we can see from the outside of ZF that, for each such α, $F \leftrightarrow F_\alpha$ is true for all F. But once we extend ZF to include also (an axiom of) inaccessibles, this is no longer true and is indeed not asserted by R1L. What is asserted by R1L is merely that for every F and every β, there is some larger inaccessible cardinal α with F true in R_α. Since such ranks possess strong closure properties, the principle R1L is plausible. It turns out that Mahlo's principle in its original form is equivalent to R1L which can also be extended by substituting 'Mahlo' for 'inaccessible', etc.
[10] More explicitly, Ackermann's axiom is:

(A)   Let $F(x, y, z)$ be an open sentence not containing M. For all y and z in M, if $\forall x(F(x, y, z) \rightarrow x \in M)$, then $\exists u[u \in M \land \forall x(F(x, y, z) \leftrightarrow x \in u)]$.

In other words, for any parameters y, z in M, if $F(x, y, z)$ is satisfied only by x in M, then we can collect all x satisfying $F(x, y, z)$ into a member of M relative to y and z.

The other axioms of Ackermann's system are extensionality, comprehension for forming subclasses of M, completeness of M (i.e. $y \in M$ if $y \in x \in M$ or $y \subseteq x \in M$), and regularity (left out originally by Ackermann). For brevity, we shall speak of everything in the universe of discourse as classes and members of M as sets.

Let $F(x)$ be a sentence not containing M. If $F(x) \leftrightarrow x \in M$, we would get from (A): $\exists u(u \in M \land u = M)$, and so $M \in M$. Hence, we have:

(B)   If $F(x)$ does not contain M, then $\rightarrow (F(x) \leftrightarrow x \in M)$.

This has the effect of generating new classes beyond M. Roughly speaking, this says that M (or its extension) is not definable by any predicate (or open sentence) not containing M. Every such predicate determines either a smaller extension than M and thereby defining a set, or a larger extension than M and thereby generating classes not in M. For example, we can find such a predicate $F(x)$ which says that x belongs to the first rank R without a power class (i.e. all elements of R and elements of elements of R etc. have power classes). By (A), if $F(x) \to x \in M$, then there is a set which has no power set contradicting the axiom about the completeness of M. But $F(x)$ cannot define M. Hence, M must have a power class. Generally, we can choose $F(x)$ to obtain the power class of the power class of M, etc.

According to Ackermann's own interpretation of his system, M is a proper class and, by the comprehension principle, any open sentence determines a subclass of M. This is contrary to the familiar practice which conditions us to think of proper classes as at the top of the iteration so that there are no further iterations. On the other hand, there is also a natural inclination to go beyond these classes. There is something amiss about the use of classes as distinct from sets, even though this is sometimes convenient. We naturally think of them as sets again; or alternatively, we are unhappy because we cannot have a clear idea of proper classes if they are taken seriously as larger than all sets since we do not have any good grasp of the totality of all sets. Somehow we feel that we can at best only think of these large classes as extensions or ranges of properties (or intensional characterizations).

Later in the text, we shall try to justify Ackermann's system by the reflection principle that V is not definable. In this connection, it is of interest to mention exact mathematical results proving the undefinability as a technical result about models of Ackermann's system: Rudolf Grewe, 'Natural models of Ackermann's set theory', J. *Symbolic Logic* **34** (1969), 481–488 (e.g. Theorem 3.2 on p. 484).

[11] In addition to classes (i.e. proper classes and sets) Powell envisages also properties or predicates (let us call them domains) defined over all classes. A distinction is made between membership (symbolized by ε) and predication or falling under (symbolized by η). Only sets can belong to a domain; both sets and proper classes can fall under a domain. However, two domains are identical if the same sets belong to them because it is assumed that the extension of a predicate is determined by the sets belonging to it. This is a strong principle with surprising consequences such as: if any proper class falls under a domain, then unbounded many do. Relative to the universe of discourse, it is assumed that, on account of the richness of M and the poverty of the collection of available predicates, no two predicates which agree on M can differ elsewhere. Of course, this requires a precarious balance between M and the collection of available predicates.

Given such a principle of 'extensionality', we can derive strong principles of reflection which enable us to move back and forth between sets and classes. For example, if all sets fall under a domain, then all classes do; if there is a class falling under a domain, there is a set too. (Powell, 1972, p. 21.) More generally, we have the strong reflection principle which says essentially that M is an elementary submodel of the universe of discourse (p. 20 and p. 22):

RP. $\forall x \in M(F^M(P, x) \leftrightarrow F(P, x))$.

The predicates are given essentially by the first-order open sentences not containing M, with sets as parameters. As a first attempt at formulating this theory, we shall use P,

Q, etc. to stand schematically for $\hat{x}F(x, a)$, etc. with a ranging over M, obtained from such sentences by abstraction. If we treat predicates in this way, we can limit ourselves to the following axioms of Powell: choice and regularity for M, transitivity of M (viz. $x \in M$ if $x \in y \in M$), identification (if $x \in M$, $x \in P \leftrightarrow x\eta P$), and the crucial axiom:

(P1)   Axiom of extensionality. If $\forall x \in M(x\eta Q \leftrightarrow x\eta P)$, then $Q = P$.

In this context, if P and Q stand for $\hat{x}F(x)$ and $\hat{x}G(x)$, we would have to construe $P = Q$ as $F(x) \leftrightarrow G(x)$, so that the two sentences can be substituted for one another in any given sentence containing one or the other as parts.

This way of treating properties is quite feasible in dealing with proper classes in the Bernays-Gödel system which is a conservative extension of ZF. But in Powell's system, the predicates play a much more essential role so that we have to extensionalize them by bringing them into the range of variables in order to codify adequately the original intentions. Hence, Powell introduces domain variables P, Q, etc. by a contextual definition:

$$\forall P\, F(P) \quad \text{for} \quad \forall x[(\forall y(y \in x) \to y \in M) \to F(x)].$$

This again emphasizes that these domains are determined in their extensions by their intersections with M or, as far as the ε-relation is concerned, they are subclasses of M. But these domains are not pure extensions, they are like properties relative to proper classes.

With P, Q so construed, Powell's system is strengthened by a domain existence axiom:

(P2) $\forall x \in M\, \exists Q\, \forall y(y\eta Q \leftrightarrow F(P, x, y))$, where neither M nor any bounded domain variable occur in F and the free domain variables P, etc. only occur in contexts like $u\eta P$.

It may be noted that conceptually $x \in P$ is just another way of saying that $x\eta P \wedge x \in M$. A natural completion of these axioms accepted by both Powell and Reinhardt is the unrestricted separation axiom:

(P3) $\forall x\, \exists y\, \forall z(z \in y \leftrightarrow (z \in x \wedge F(z)))$, for any open sentence $F(z)$.

[12] Reinhardt's formulation differs from Powell's in eliminating the extra-predicate η so that domain variables range over extensions on all classes rather than on sets only. So the same principle (P1) gets a somewhat different interpretation (with η replaced by ε). It is then necessary to add an additional principle, i.e. (S3.3) on p. 15 of Reinhardt (1974a), to say that all subclasses of M are defined (or are included among the domains):

(P2*) $x \subseteq M \to \exists Q\, \forall y \in M(y \in Q \leftrightarrow y \in x)$.

This is perhaps not as natural as Powell's formulation but has the advantage of enabling us to state briefly a uniformization of the second-order reflection principle R2.

[13] Let T be the system suggested by Reinhardt on p. 23 of Reinhardt (1974a) as an alternative formulation of his system S^*. Then any natural model of T yields one for S^* and T is equiconsistent with S^*. Powell points out that by adding Ackermann's axiom, his strong axiom of extensionality can be replaced by the usual one, viz. $P = Q$ if $\forall x(x\eta P \leftrightarrow x\eta Q)$.

[14] From our earlier remarks on proper classes, it is natural to try to extend R2U and think of the domains in the above theory again as sets. In fact, Reinhardt (1974b) has suggested such extensions at least with a view of getting enough extendibility to yield supercompact cardinals. But Powell has pointed out two difficulties. In the first place, there seems to be an equivocation between E and E_0 in Definition 6.2 (p. 199): the justification seems to be directed at E_0 while the derivation of supercompact cardinals seems to need E. In the second place, the transitions from 1-extendible to extendible, and then to the embedding of the universe of sets into itself (which is known to lead to contradictions) (Kunen, 1971) seem so gradual that one might even be led to doubt the weaker end and be sceptical about 1-extendible cardinals and the set theory described above.

BIBLIOGRAPHY

Ackermann, W.: 1956, 'Zur Axiomatik der Mengenlehre', *Math. Ann.* **131,** 336–345.

Bernays, P.: 1961, 'Zur Frage der Unendlichkeitsschemata in der axiomatischen Mengenlehre', *Essays on the Foundations of Mathematics*, Magnes Press, Hebrew Univ., Jerusalem, pp. 3–45.

Cantor, G.: 1932, *Gesammelte Abhandlungen*, Berlin.

Drake, F. R.: 1974, *Set Theory – An Introduction to Large Cardinals*.

Friedman, H.: 1971, 'Higher Set Theory and Mathematical Practice', *Annals of Math. Logic* **2,** 326–357.

Gödel, K.: 1964a, 'Russell's Mathematical Logic', *Philosophy of Bertrand Russell*, 1944, pp. 125–153. Reprinted in *Philosophy of Mathematics* (ed. by P. Benacerraf and H. Putnam), pp. 211–232.

Gödel, K.: 1964b, 'What is Cantor's Continuum Problem?', *Am. Math. Monthly* **54,** 515–525. An extended version is in the collection just cited, pp. 258–273.

Kanamori, A.: 1975, *Large Large Cardinals*, Lecture notes at The University of Cambridge, 77pp.

Kunen, K.: 1971, 'Elementary Embeddings and Infinitary Combinatorics', *J. Symbolic Logic* **36,** 407–413.

Levy, A.: 1959, 'On Ackermann's Set Theory', *J. Symbolic Logic* **24,** 154–166.

Levy, A.: 1960, 'Axiom Schemata of Strong Infinity in Axiomatic Set Theory', *Pacific J. Math.* **10,** 223–238.

Martin, D. A.: 1975, 'Borel Determinacy', *Annals of Mathematics* **102,** 363–371.

Powell, W. C.: 1972, *Set Theory with Predication*, Ph.D. dissertation at S.U.N.Y. Buffalo, 111pp.

Reinhardt, W.: 1974a, 'Set Existence Principles of Shoenfield, Ackermann and Powell', *Fund. Math.* **84,** 12–41.

Reinhardt, W.: 1974b, 'Remarks on Reflection Principles, Large Cardinals, and Elementary Embeddings', *Axiomatic Set Theory, Part II, AMS Proceedings*, pp. 189–206.

Tharp, L. H.: 1967, 'On a Set Theory of Bernays', *J. Symbolic Logic* **32,** 319–321.

Wang, Hao: 1974, *From Mathematics to Philosophy*, Routledge and Kegan Paul, London.

WHAT IS THE ITERATIVE
CONCEPTION OF SET?

I intend to raise here some questions about what is nowadays called the 'iterative conception of set'. Examination of the literature will show that it is not so clear as it should be what this conception *is*.

Some expositions of the iterative conception rest on a 'genetic' or 'constructive' conception of the existence of sets. An example is the subtle and interesting treatment of Professor Wang.[1] This conception is more metaphysical, and in particular more idealistic, than I would expect most set theorists to be comfortable with. In my discussion I shall raise some difficulties for it.

In the last part of the paper I introduce an alternative based on some hints of Cantor and on the Russellian idea of typical ambiguity. This is not less metaphysical though it is intended to be less idealistic. I see no way to obtain philosophical understanding of set theory while avoiding metaphysics; the only alternative I can see is a positivistic conception of set theory. Perhaps the latter would attract some who agree with the critical part of my argument.

However, the positive part of the paper will concentrate on the notion of proper class and the meaning of unrestricted quantifiers in set theory. That these issues are closely related is evident since in Zermelo-type set theories the universe is a proper class.

The concept of set is also intimately related to that of *ordinal*. Although this relation will be remarked on in several places, a more complete account of it, and thus of the more properly *iterative* aspect of the iterative conception, will have to be postponed until another occasion.

I

One can state in approximately neutral fashion what is essential to the 'iterative' conception: sets form a well-founded hierarchy in which the

Butts and Hintikka (eds.), Logic, Foundations of Mathematics and Computability Theory, 335–367.

elements of a set precede the set itself. In axiomatic set theory, this idea is most directly expressed by the axiom of foundation, which says that any non-empty set has an '∈-minimal' element.[2] But what makes it possible to use such an assumption in *motivating* the axioms of set theory is that other evident or persuasive principles of set existence are compatible with it and even suggest it, as is indicated by the von Neumann relative consistency proof for the axiom of foundation.

On the 'genetic' conception that I will discuss shortly, the hierarchy arises because sets are taken to be 'formed' or 'constituted' from previously given objects, sets or individuals.[2] But one can speak more abstractly and generally of the elements of a set as being *prior* to the set. In axiomatic set theory with foundation, this receives a mathematically explicit formulation, in which the relation of priority is assumed to be well-founded.

For motivation and justification of set theory, it is important to ask in what this 'priority' consists. However, for the practice of set theory from there on, only the abstract structure of the relation matters. Here we should recall that the hierarchy of sets can be 'linearized' in that each set can be assigned an ordinal as its *rank*. Individuals, and for smoothness of theory the empty set,[3] obtain rank 0. In general, the rank of a set is the least ordinal greater than the ranks of all its elements.[4]

It should be observed that the notion of well-foundedness is prima facie second-order and thus is not totally captured by the first-order axiom of foundation.[5] ZF with foundation has models in which the relation representing membership is not well-founded. However, it can be seen that in such a model there is a (not first-order definable) binary relation on the universe for which *replacement* fails. The axioms of separation and replacement are also prima facie second-order, and the fault for such failure of well-foundedness lies in the fact that their full content is not captured by the first-order schemata. But then the problem of stating clearly the iterative conception of set is bound up with the problem of the relation of set theory to second-order conceptions. This problem was already present at the historical beginning of axiomatic set theory with Zermelo's use of the notion of 'definite property'.

The idea that the elements of a set are prior to the set is highly persuasive as an approach to the paradoxes. If we suppose that the elements of a set must be 'given' before the set, then no set can be an element of itself, and there can be no universal set. The reasoning leading to the Russell and Cantor paradoxes is cut off.

However, one does not deal so directly with the Burali-Forti paradox. Why should it not be that all ordinals are individuals and therefore 'prior' to all sets, so that there is no obstacle of this kind to the existence of a set of all ordinals? To be sure, once we look at things in this way it becomes persuasive to view the Burali-Forti argument as just a proof that there is no set of all ordinals. Moreover, the conception of ordinals as order types of well-ordered sets would suggest that for any ordinal there is at least one set of that order type to which it is not prior, so that the existence of a set of all ordinals would imply that later in the priority ordering no new order types could arise. But if there were a set of all ordinals, W, then $W \cup \{W\}$ would have just such an order type.

That ordinals need to fit into a priority ordering with sets and indeed be 'cofinal' with them seems to have been neglected in discussion of iterative set theory, perhaps because in the formal theory ordinals are construed as sets, so that this happens automatically.

One would like to maintain that the requirement of priority is the *only* principle limiting the existence of sets, so that at a given position 'arbitrary multitudes' of objects which are at earlier positions form sets. Although it is difficult to make sense of this, it at least should imply a comprehension principle: given a predicate 'F' which is definitely true or false of each object prior to the position in question, there is *at* the position a set whose elements are just the prior objects satisfying 'F'. In particular, the axiom of separation follows: since the elements of x are prior to x, $\{z : z \in x \wedge Fz\}$ exists and is not posterior to x. But to apply this idea more generally, some way of marking positions is needed. The genetic approach in effect assumes such to be given. Conversely, if set theory is assumed, the ordinals offer such a marking.

II

We have now gone about as far as we can without explaining what I have called the genetic approach. Put most generally, it supposes that sets are 'formed', 'constructed', or 'collected' from their elements in a succession of stages. The first part of this idea has some plausibility as an interpretation of some of Cantor's preliminary remarks about what a set is. Thus Cantor's famous 'definition' of 1895:

By a 'set' we understand any collection M into a whole of definite, well-distinguished objects of our intuition or our thought (which will be called the 'elements' of M).[6]

If we were to take 'collection into a whole' quite literally as an operation, then the priority of the elements of a set to the set would simply be priority in order of construction. Cantor's language suggests rather that 'collection' (*Zusammenfassung*) is an operation of the *mind*; in this case the requirement would be that the objects be represented to the *mind* before the operation of collection is performed. However, it will be clear that as it stands this temporal reading is too crude.

It may seem that these notions belong only to the early history of set theory, and in particular that they would have disappeared with the discrediting of logical psychologism at the end of the last century. But the fact is that they are to be found in the contemporary literature. Thus Shoenfield writes (1967, p. 238):

A closer examination of the paradox [Russell's] shows that it does not really contradict the intuitive notion of a set. According to this notion, a set A is formed by gathering together certain objects to form a single object, which is the set A. Thus before the set A is formed, we must have available all of the objects which are to be members of A.

Although Schoenfield says that we form sets 'in successive stages', he does not offer an interpretation, temporal or otherwise, of the stages, although he does use 'earlier' to express their order.

Wang writes, "The set is a single object formed by collecting the members together" (1974, p. 181). He recognizes that the concept of collecting is highly problematic and makes an interesting attempt to explain it; he interprets it as an operation of the mind. We shall discuss his views shortly.

We now have the familiar conception of sets as formed in a well-ordered sequence of stages, where a set can be formed at a given stage only from sets formed at earlier stages and from whatever objects were available at the outset.

The language of Cantor, Shoenfield, and Wang invites regarding the intuitive concept of set as analogous to the concepts of constructive mathematics, where one also uses the idea of mathematical objects as constructed in successive stages, and where there is no stage at which all constructions are complete. An immediately obvious limitation of the analogy is that in the typical constructive case (e.g. orthodox intuitionism) the succession of stages is simply succession in *time*, and incompletability arises from the fact that the theory is a theory of an idealized finite mind which is located at some point in time and has available only what it has constructed in the past and its intentional attitudes toward the future. The same interpretation of iterative set theory would require that the stages be thought of as a kind of 'super-time' of a structure richer than can be represented in time on any intelligible account of construction in time. It is hard to see what the conception of an idealized mind is that would fit here; it would differ not only from finite minds but also from the divine mind as conceived in philosophical theology, for the latter is thought of either as in time, and therefore as doing things in an order with the same structure as that in which finite beings operate, or its eternity is interpreted as complete liberation from succession.

It may seem that there is a much more obvious conflict between iterative set theory and a constructive interpretation of it: set theory is the very paradigm of a *platonistic* theory. As is customary in discussing the foundations of mathematics, platonism means here not just accepting abstract entities or universals but epistemological or metaphysical realism with respect to them. Thus a platonistic interpretation of a theory of mathematical objects takes the truth or falsity of statements of the theory, in particular statements of existence, to be objectively determined independently of the possibilities of our knowing this truth or falsity. Contrast, for example, the traditional intuitionist conception of a mathematical statement as an indication of a 'mental construction' that constitutes a proof of the statement.

Perhaps it would be rash to rule out an interpretation of set theory that would not be platonistic.[7] But in any case it seems that a platonistic interpretation is flatly incompatible with viewing the 'formation' of sets as an operation of the mind. However, that there is not a direct contradiction should be evident when we observe that we are concerned in set theory with what formations of sets are *possible*. In contrast to the situation with intuitionism, we do not require that a statement to the effect that it is possible to 'construct' a set satisfying a certain condition should be itself an indication of a construction. Even if we construe the formation of sets as a mental operation, what is possible with respect to such formation can be viewed independently of our knowledge. Thus there is a *prima facie* resolution of the difficulty posed by platonism.

However, we have not reckoned yet with the actual content of the set-theoretic principles that seem to require a platonistic interpretation, such as the combination of classical logic with the postulation of a set of all sets of integers. In an iterative account, the individual steps of iteration are in Wang's word "maximum" (1974, p. 183). Namely we regard as available at any given stage any set that *could* have been formed earlier. We could represent this assumption as that if a set *can* be formed at a given stage, then it *is* formed (or at least that it *exists* at that stage). This of course has effects on what can be formed later, since every *possibility* of set formation at stage α is such that its result is available at later stages and can therefore enter into further constructions.

We can illustrate this by the manner in which these ideas are used to justify the power set axiom. At a given stage α, any 'multitude' of available objects can be formed into a set. Let x be a set formed at stage α, and let y be a subset of x. Since the elements of y are all elements of x, they must have been available at stage α. Hence y *could have been formed* at stage α. x is available from stage $\alpha + 1$, on; from our assumption it follows that y is available as well. Thus at stage $\alpha + 1$, *every* subset x is available and $\mathfrak{P}(x)$ can be formed.

We should distinguish two principles that are playing a role here, and which can be confused with one another. One is the 'arbitrary' nature of sets, which, following Wang, we have expressed (provisionally) by saying that any 'multitude' of available objects can be formed into a set. The other is the principle that allows the transition from possibility of formation at stage α to availability at stage $\alpha + 1$, perhaps by way of existence at stage α. Both principles may be taken to arise from the idea we expressed above that the priority requirement, here interpreted to mean that sets are formed from *available* objects, is the *only* constraint on the existence of sets. But the second principle begins to undercut the idea of sets as *formed* from available objects, since the successive stages of formation are required *only* because a set must be formed from available objects, and not because of any successiveness in the process of formation itself. The question arises whether the interpretation of the priority of the elements of a set to the set in terms of order of construction does not reduce to viewing this priority as a matter of *constitution*: the elements are prior because they *constitute* the set (to use a more abstract phrase than they are its *parts*, which would invite inferences inappropriate to set theory). This view is close to what I advocate below, but it is quite different from the conception of set formation as an operation of the mind.

Wang makes an interesting attempt to develop the latter idea. He says that a multitude can be formed into a set only if its "range of variability" is "in some sense intuitive" (1974, p. 182). I shall for the time being accept the notion of 'multitude'; the problems concerning it are related to the question of the notion of *class* in set theory. Wang indicates that to form a set is to "look through or run through or collect together" all the objects in the multitude.[8] Thus a condition for a multitude to form a set is that it should be possible thus to 'overview' it. This overviewing is a kind of intuition, presumably analogous to perception. Of course infinite multitudes can be 'overviewed'.

Clearly Wang does not maintain that human beings have the capacity to "run through" infinite collections. He speaks of overviewing "in an idealized sense" (1974, p. 182). In other words, he has a highly abstract conception of the possibilities of intuition. In constructive conceptions of the arbitrary finite, we already disregard the actual bounds of human capacities, in the sense that if a sequence of steps has been performed we always *can* perform a futher step, and any operation can be iterated. Wang's idealized overviewing carries such abstraction further in that finitude and even the limitations posed by the continuous structure of space-time (as the setting of the objects of perception and even of the mind itself) are disregarded.

The question arises what force it still has, on Wang's level of abstraction, to treat the possibilities involved in this kind of motivation of the axioms of set theory as possibilities of *intuition*. The analogy with sense-perception which is central to the constructive conception of intuition in Brouwer or to Hilbert's distinction between intuitive and formal mathematics seems to be almost totally lost. Consider Wang's remarks on the axioms of separation and power set. The former is stated thus: If a multitude A is included in a set x, then A is a set.

Since x is a given set, we can run through all members of x, and, therefore, we can do so with arbitrary omissions. In particular, we can in an idealized sense check against A and delete only those members of x which are not in A. In this way, we obtain an overview of all the objects in A and recognize A as a set (1974, p. 184).

The idealization seems to include something like omniscience: A may be given in some way that does not independently of the axiom assure us that it is a set, and yet we can use it in order to 'choose' the members of x that are in A. A may of course be given to us by a predicate containing quantifiers that do not range over a set; in deciding whether an element y of x is to be deleted, we cannot 'run through' the values of the bound variables as part of the process of checking y against A. It is not clear what more structured account of 'idealized checking' would yield the result Wang needs. An alternative would be to view subsets as run through not by verification but by arbitrary selection. But if a predicate is given, how are we to 'select' just those elements to x that satisfy it unless we can decide which ones do?[9]

More strain on the concept of intuition appears in Wang's treatment of the power set axiom (1974, p. 184):

We have $\cdots$ an intuitive idea of running through with omissions. This general notion $\cdots$ provides us with an overview of all cases of AS [separation] as applied to x.

By saying that we have an "intuitive idea" of running through with omissions, he does not only mean that a *case* of such running through is intuitable, for that would not yield the result. Rather, for a given set x the concept of such runnings through is intuitive in the sense that 'we can' run through *all* cases of it. Something of the content of the idea of intuitive running through seems to be lost here. Clearly in the case of small finite sets of manageable objects, we really do see all the elements 'as a unity' in a way that preserves the articulation of the individual elements. Somewhat larger sets can be seen by a completable succession of steps of bringing one (or a few) objects under one's purview. If we consider arbitrary iterations of such steps, there is no longer any limit to how large a (finite) set can be thus intuited. We also have a simple and clear generative rule for sets such as the natural numbers, though the process of sensibly intuiting is in this case incompletable, so that the *givenness* to us of such a set depends in a more essential way on conception. But regarding the natural numbers as intuitable as a whole amounts just to abstracting from the above incompletability. There is another qualitative leap in dealing with all sets of integers, as has often been remarked on in the literature (as indeed Wang himself emphasizes when he stresses the importance of impredicativity).[10] Here, however we understand the notion of an 'arbitrary set' of integers, say by some picture of arbitrary selection, we do not have the conceptual grasp of what the totality contains that would be given by some method of generating them. The divorce from *sensible* intuition involved in treating this totality as 'intuitable' seems complete, unless perception is used only as a source of quite remote analogies. Two mathematical symptoms of this situation are the absence of a definable well-ordering of the continuum and our inability to solve the continuum problem.

I ought to make clear that I understand by 'intuition' a quasi-perceptual manner in which an object is presented to the mind. In this I follow Kant. The word 'intuition' is also used in the philosophy of mathematics and otherwise for any manner by which propositions can be known where this knowledge is not largely accounted for by deductive or inductive reasoning. There is a tendency to confuse these two senses. As for the appropriateness of my sense of intuition to Wang, I should point out that the other concept of intuition does not distinguish sets from 'multitudes' or other primitive notions that might enter into evident set-theoretic axioms. Moreover, intuition in the latter sense is purely *de dicto*, intuition *that* certain propositions are true, while Wang clearly requires intuition *de re*, intuition *of* sets.[11]

I no longer understand Wang's talk of 'intuitively running through' where it is applied to the set of all sets of integers. In the above I have perhaps connected intuition more closely with the senses (more abstractly Kant's "sensibility") than Wang would find acceptable. But even quite abstract marks of sensibility, such as the structure of time, are lost in this case.

However, there might be an interpretation on which Wang's hypothesis that a 'multitude' is a set if and only if it is an object of intuition would be defensible. The concept of intuition would be logical and ontological rather than perceptual and epistemic. To be an object of intuition would be simply to be an object rather than a

Fregean "concept" or perhaps a property. In other words, Kant's contrast of intuitions as "singular representations" with concepts as "general representations" (*Logic*, § 1) would give virtually the only essential mark of intuition.[12] Although this interpretation would bring Wang's hypothesis into accord with the views I express below, I shall not pursue it further in this paper.

I want now to turn to the axiom of replacement, about which Wang has most interesting things to say. Wang writes:

Once we adopt the viewpoint that we can in an idealized sense run through all members of a given set, the justification of SAR[13] is immediate. That is, if, for each element of the set, we put some other given object there, we are able to run through the resulting multitude as well. In this manner, we are justified in forming new sets by arbitrary replacements. If, however, one does not have this idea of running through all members of a given set, the justification of the replacement axiom is more complex (1974, p. 186).

The picture here is marvelously persuasive; for me, it expresses very well why the axiom of replacement seems obvious. But something like the omniscience assumption of his discussion of separation is present in the remark that "*we put* some other given object there" and "are able to run through the resulting multitude". What is much more revealing is that the objects seem to have no relevant internal structure: Our ability to 'run through' a multitude is preserved if we replace its elements by *any* other objects, for example by much larger sets. It is as if the objects were given only as wholes, or at least that any internal structure would not affect the possibility of running through the totality. A model for this (conceptual rather than intuitive) is the case where the objects are given only by names. Wang seems to be making an hypothesis here, although I do not feel the same qualms as in the case of the power set about taking it as an hypothesis about what is intuitable. It is of course the *combination* of the replacement with the power set axiom that yields sets of very high ranks.

Wang expresses by his picture the idea, present in the earliest intimation of the axiom of replacement (Cantor, 1932, p. 444; cf. Wang, 1974, p. 211) that whether a multitude forms a set depends only on its cardinality and not on the 'internal constitution' or relations of its elements. Put in this way, the axiom is not a principle of *iteration* of set-formation, in line with the conclusion of Boolos (1971, p. 228–9) that it does not follow from the iterative conception of set.[14] The most direct justification of replacement by appeal to ideas about stages seems to me somewhat circular.[15] That of Gödel cited by Wang (1974, p. 186 and n. 5, p. 221) I find less immediate and persuasive.

Although I admit that Wang's picture (apart from the question of omniscience) offers a plausible hypothesis about what is intuitable,[16] it seems to me to be equally plausible as an hypothesis about what can be thought or about what can *be*, and the latter interpretations fit better the case of power set. I want now to pursue the genetic conception of sets in this direction.

III

In the preceding section we saw a number of difficulties with the idea that sets are 'formed' from their elements, in particular by an activity of running through in intuition. I want now to suggest a more 'ontological' view of the hierarchy of sets.

The earliest attempt that we know to explain the paradoxes of set theory and to develop set theory in a way that avoids them is in Cantor's famous letter to Dedeking of July 28, 1899 (1932, pp. 443–7). Cantor there presupposes his earlier 'many into one' characterizations of the notion of set, such as that of 1895 cited above. He begins (p. 443) with "the concept of a definite multiplicity (*Vielheit*)". What he calls an *inconsistent* multiplicity is one such that "the assumption of a 'being together' (*Zusammensein*) of *all* its elements leads to a contradiction". A consistent multiplicity or set is one whose "being collected together to 'one thing' is possible". It is noteworthy that Cantor here identifies the possibility of all the elements of a multiplicity *being together* with the possibility of their being collected together into one thing. This intimates the more recent conception that a 'multiplicity' that does not constitute a set is *merely potential*, according to which one can distinguish potential from actual being in some way so that it is impossible that *all* the elements of an inconsistent multiplicity should be actual.

I am here interpreting Cantor to mean that where there is an essential obstacle to a multiplicity's being collected into a unity, this is due to the fact that in a certain sense the multiplicity does not exist. It does not exist as a totality of its elements; if it did, they would form a unity or could at least be *collected* into a unity. But in the case of inconsistent multiplicities, this is impossible. The sense of this non-existence needs some further elucidation which Cantor does not supply. The language of potentiality and actuality is not in the text, though Cantor may have been suggesting it in calling an inconsistent multiplicity *absolutely infinite* (p. 443).

What seems to me of interest in the present connection in these hints of Cantor is that he seems to be trying to distinguish sets from "inconsistent multiplicities" without real use of any metaphor of *process* according to which sets are those multiplicities whose 'formation' can be 'completed'.[17] Such a metaphor makes the idea of an inconsistent multiplicity as a merely potential totality rather easy. I suggest interpreting Cantor by means of a modal language with quantifiers, where within a modal operator a quantifier always ranges over a set (not, however, one that is explicitly given or even that exists in the 'possible world' it might be taken to range over). Then it is not possible that all elements of, say, Russell's class exist, although for any element, it is possible that *it* exists. As it stands this conception requires it to be meaningful to talk of *any* set (or any object), even though the range of *this* quantifier does not constitute a unity; the elements of its range cannot all 'exist together'. However, at least some such talk can be replaced by ordinary quantification behind necessity.

What one would like to obtain from this conception is some interpretation of the stages of the iterative conception that also does not depend on the metaphor of process. However, I intend first to look at Cantor's conception of a multiplicity. Wang seems to use "multitude" in the same sense, although he does not use it to translate Cantor's *Vielheit* when he discusses Cantor's 1899 correspondence with Dedekind (1974; p. 211). These notions are among a number which occur in the literature on logic and set theory and which purport to be more comprehensive than the notion of set. The most respectable of these notions is that of (proper) class. We should also mention Frege's "concept", Zermelo's "definite property", and Shoenfield's "collection". Gödel's "property of sets" (1964, p. 264 n. 18) presumably also belongs on this list.

Of all these notions, perhaps the most developed from the philosophical side is Frege's notion of a concept; I shall use it for purposes of comparison. What is then striking is that neither Cantor's nor Wang's notion seems to be derived from predication as Frege's is. Since Cantor's notion is one of the prototypes of the notion of proper class, this fact seems to clash with the actual use of the notion of class in set theory (perhaps with exceptions; see below) according to which classes are derived from predication; Zermelo's "definite property" is a more immediate prototype than Cantor's *Vielheit*. I myself have suggested (1974c) that sets are not derived from predication while classes are.

Cantor in 1899 apparently thought of sets as a species of the genus multiplicity, and then perhaps the non-predicative (if not impredicative!) character of multiplicities in general was needed in order to preserve the 'arbitrariness' of sets against its being restricted by what we might express in language. Frege seems to have obtained the same freedom by his realism about concepts. However, for Frege the nature of a concept could apparently only be explained by appeal to predication (more generally, to 'unsaturated' expressions). The sharp distinction between concepts and objects is a shadow of the syntactical difference between expressions with and without argument places. This difference is then 'inherited' by concepts that are not denoted by expressions of any language we use or understand.

I want to suggest that predication plays a constitutive role in the explanation of Cantor's notion of multiplicity as well and that at least an "inconsistent multiplicity" must resemble a Fregean concept in not being straightforwardly an object. In the Cantorian context, predication seems to be essential in explaining how a multiplicity can be given to us not as a unity, that is as a set. Much the clearest case of this is understanding a predicate. Understanding 'x is an ordinal' is a kind of consciousness or knowledge of ordinals that does not so far 'take them as one' in such a way that they constitute an object. We might abstract from language and speak with Kant of knowledge through concepts, but whatever we make of this the predicational structure is still present.

The philosophy of Kant might suggest another way in which a multiplicity might be given not as a unity, namely as an 'unsynthesized manifold'. It seems clear that in the cases Kant actually envisaged, the objects involved would have the definiteness necessary to constitute a

Cantorian 'multiplicity' only if they are a set. Even if we generalize the notion in some way, I do not see how such a "manifold' can be taken up into explicit consciousness except perceptually (intuitively) or conceptually.

The idea that to be an *object* and to be a *unity* are the same thing is very tempting and has deep roots in the history of philosophy. An object is something whose identity with itself (represented in different ways) and difference from other objects can be meaningfully talked about; it is then subject to at least rudimentary application of *number*. This line of reasoning inclines us to identify Cantor's "multiplicity" with Frege's concept at least in that a multiplicity which is not a set is not an object. Some such assumption seems necessary to cut off the question why there are not multiplicities whose elements are not sets or individuals: multiplicities are multiplicities of *objects*, and under that condition there are no restrictions on the existence of multiplicities (although possibly on the use of quantifiers over them), but if a multiplicity is an object, then it is a set.

However, we have to deal with the fact that in Frege the gulf between concepts and objects comes from the structure of predication itself, so that a concept is irremediably not an object, even if only one object falls under it. Cantor evidently holds that some multiplicities just *are* sets, in particular those that are not too large. This may seem not a very essential difference: if a concept F is such that there is a set y of all x such that Fx, then the distinction between F and y is just the distinction between $(\ ) \in y$ and y.[18] For an inconsistent multiplicity there is no such 'reducibility'. In view of Russell's paradox, the idea of the predicative nature of the concept will motivate the idea that there should *be* inconsistent multiplicities, but it does not seem to motivate Cantor's particular principles as to what multiplicities are 'consistent'. For reasons which will become clear later, I do not think we have yet captured the sense in which an inconsistent multiplicity is not an object.

Let us look for a moment at the well-known difficulties of Frege's theory of concepts. The conception has the great attraction that it enables us to generalize predicate places without introducing nominalized predicates that purport to denote objects (classes or attributes) – something that has to be restricted on pain of Russell's paradox. But the temptation to nominalize is irresistible, as Frege himself discovered on two fronts. His construction of mathematics required an 'official' nominalization in postulating extensions. 'Unofficial' nominalizations cropped up repeatedly in his own informal talk about concepts and gave rise to the paradox that the concept *horse* is an object, not a concept.[19] At the end of his life Frege decided the temptation was to be resisted and that neither the expression "the concept F" nor the expression "the extension of the concept F" really denotes anything.[20] However, perhaps yielding to temptation can in one way or another be legitimized, even where the extensions postulated are not sets (cf. my, 1974c).

Does the Cantorian concept of "multiplicity" have to be understood realistically? To the extent that *sets* are understood realistically, of course "consistent multiplicities" are mind-independent in the corresponding sense. However, obviously it does not follow from the fact that we allow classical logic and impredicative reasoning about sets that we have to allow either about classes or other more general entities. The suggestion made below that such entities are at bottom *intensions* would imply, if we think of an intension in the traditional way as a meaning entertained by, and in some sense constructed by, the mind, that realism about them is inappropriate. However, in view of the interest of impredicative conceptions of classes for large cardinals, both predicative and impredicative conceptions should be pursued.[21]

Let us now return to Cantor's suggestion that the elements of an "inconsistent multiplicity" cannot all exist together. I do not conclude that an inconsistent multiplicity does not exist in any sense; even the hypothesis that it is not an object will have to be qualified. However, one implication is clear: it is not a totality of its elements; it is not 'constituted' in a definite way by its elements. Its existence cannot require the prior existence of all its elements, because there is no such prior existence.

I wish to explicate the difference between sets and classes by means of some intensional principles about them. From the idea that a set is constituted by its elements, it is reasonable to conclude that it is *essential* to a set to have just the elements that it has and that the *existence* of a set requires that of each of its elements. Exactly how one states these principles depends on how one treats existence in modal languages. I shall assume that the truth of $x \in y$ requires that y exists (Ey). Then we have:

(1) $x \in y \rightarrow Ex \wedge Ey$

(2) $x \in y \rightarrow \Box(Ey \rightarrow x \in y)$

(3) $x \notin y \wedge Ey \rightarrow \Box(x \notin y).$[22]

My proposal is that these principles should fail in some way if y is an "inconsistent multiplicity" or proper class. Indeed Reinhardt has suggested that proper classes differ from sets in that under counterfactual conditions they might have different elements (1974a, p. 196). I am endorsing this suggestion as an explication of the intuitions about "inconsistent multiplicities" that I have been discussing.

As before, for a given 'possible world' we should think of the bound variables as ranging over a set, perhaps an R_α (see note 4); but the sets that exist in that world are elements of the domain, while classes are arbitrary subsets of the domain.

Reinhardt does not use an intensional language; in his formulation the actual world V is part of a counterfactual 'projected universe' which is the domain of the quantifiers. He assumes a mapping j on $\mathfrak{P}(V)$ such that if $x \in V$, $jx = x$. We can think of jx as a 'counterpart' of x in the projected universe, in the sense of Lewis (1968). Thus sets are their own counterparts and can be strictly reidentified in alternative possible worlds. A set y can have no new elements in the projected universe and its only new non-elements are all x such that $x \notin V$. This accords with (1)–(3).

For a class P, jP can have additional elements in the projected universe, so that it violates (3), although jP agrees with P for 'actual' objects (elements of V).[23] Reinhardt's extensional language must distinguish jP from P; hence my thinking of jP as a 'counterpart'. A complication is that P itself occurs in the projected universe, though now as a set. Reinhardt himself suggests an alternative reading, by which a class x is an *intension*, so that P in the actual world and jP in the projected universe are the 'values' in two different possible worlds of the same intension. A formal language in which this reading might be formulated is the second-order modal language of Montague

(1970), where the first-order variables range over objects and are interpreted (in the manner usual for modal logic) rigidly across possible worlds, and the second-order variables range over intensions, which in the semantics are functions from possible worlds to extensions of the appropriate type; but see page 353.[24]

Now doubt what is most interesting about Reinhardt's idea is the impredicative use of proper classes that he combines with it, with the result that the ordinals in V are, in the projected universe, a measurable cardinal (1974b, p. 22; or Wang, 1977, p. 327). However, in discussing the idea of proper classes as intensions I want to keep the predicative interpretation in mind as well.

IV

Cantor's conception suggests a more radical view than we have drawn from it so far, namely that one can in a sense not meaningfully quantify over absolutely all sets. In (1974a) and (1974c) I sketched all too briefly a 'relativistic' conception of quantifiers in set theory. The idea was that an interpretation that assigns to a sentence of set theory a definite sense would take its quantifiers to range over a set (presumably R_α for some large α), but that normally such a sentence would be so used as to be 'systematically ambiguous' as to *what* set the quantifiers ranged over.[25] A merit of this view was (1974c, pp. 8, 10–11) that it yielded a kind of reduction of classes to sets. Since I have here followed Cantor and Reinhardt in viewing classes as quite different from sets, here I can defend this relativistic position only in a modified form. I shall now present my present understanding of the matter.

What *does* follow from the thesis that the elements of an inconsistent multiplicity cannot all exist together is that quantification over all sets does not obey the classical correspondence theory of truth. The totality of sets is not 'there' to constitute any 'fact' by virtue of which a sentence involving quantifiers over all sets would be true. The usual model-theoretic conception of logical validity thus leaves out the 'absolute' reading of quantifiers in set theory.

If the only constraint on an interpretation of a discourse in the language of set theory is that it should make statements proved in first-order logic from axioms accepted by the interpreter *true*, then the interpreter needs only a minimally stronger theory than that applied in the discourse to interpret it so that the quantifiers range over some R_α.[26] However, it seems clear that this condition is too weak, and it remains so even if the interpreter seeks to capture not just one particular discourse but what the set theorist he is interpreting might be taken to be *disposed* to assent to. This is the case envisaged in (1974c, p. 10), where we suppose that the interpreter takes the quantifiers to range over R_α for an α with an inaccessibility property undreamed of by the speaker.

Let us suppose this inaccessibility property to be P and that the interpreter chooses the least such α. One weakness of the reading is that it takes $(\exists \alpha)P\alpha$ to be false; although the supposition that the speaker is not disposed to assent to it is reasonable, the result is arbitrary in that we have no reason to suppose the speaker disposed to dissent from it.

A more decisive objection is that if the interpreter gets the speaker to understand P and convinces him that there is a cardinal satisfying it, then he must attribute to the speaker a meaning change brought about by this persuasion: previously his 'concept of set' excluded P-cardinals; now it admits them.

The speaker, however, can (going outside the language of set theory and taking of himself and his intentions) question this and say that P-cardinals are cardinals in just the sense in which he previously talked of cardinals; he will presumably reinforce this by assenting to a numer of statements that follow directly from the existence of a P-cardinal by axioms or theorems of set theory he accepted previously.

The idea that quantifiers in set theory are systematically ambiguous was meant to meet this kind of objection by saying that the interpretation of the speaker's quantifiers as ranging over a single R_α cannot be an exactly correct interpretation, since it fixes the sense of statements whose sense is not fixed by their use to this degree. However, it still seems to imply that the speaker who is convinced of the existence of a P-cardinal undergoes a meaning change in a weaker sense, in that the ambiguity of his quantifiers is reduced by 'raising the ante' as to what degree of inaccessiblity an α has to have so that R_α will 'do'.

We should observe that assertions in pure mathematics are made with a presumption of *necessity*; if we attribute this to our speaker we can see how *P*-cardinals are immediately captured by his previous set theory, since necessary generalizations are not limited in their force to what 'actually' exists. We can see the 'meaning change' in accepting *P*-cardinals as analogous to the speaker's considering a different possible world or range of possible worlds.

The force of this analogy is limited, as we can see by a little further reflection on the conception of 'inconsistent multiplicities' as intensions. It seems that we cannot consider a proper class as given even by an *intension* that is definite in the sense of, say, possible-world semantics, as a function from possible worlds to extensions. To begin with, it is only by an interpretation external to a discourse that one can speak in full generality of the range of its quantifiers and the extensions of its predicates. The systematic ambiguity of the language of set theory arises from the fact that such an interpretation can itself be mapped back into the language of set theory when stronger assumptions are made. Thus we should think of predicates whose 'extensions' are proper classes as really not having *fixed* extensions.[27]

This situation does not change if we enlarge the language of set theory to an intensional language. Here we are able to express the 'potentiality of the totality of sets' in that it is necessarily true that the domain of a bound variable *possibly* exists as a set. But however such an intensional language is formulated, it will still be possible to read it in a set-theoretic possible world semantics, and even if on the most straightforward reading the union of the domains for all possible worlds is all sets, the assumption that there is a set that realizes the properties of this union presumably has the same plausibility that other such reflection principles have. In such a model, of course a second-order intension will be represented by a set.[28]

We should not be surprised at this; it is really a consequence of the general nature of true 'systematic ambiguity', where there is no general concept of 'possible interpretation' which is not either inadequate or infected with the same difficulties as the language it interprets. Otherwise one could resolve the ambiguity as generality (meaning by *A*, '*A* on every possible interpretation') or indexically, by some contextual device or convention indicating which interpretation is meant. Russell's "typical ambiguity" was essential in that according to his theory of meaning there was no way of expressing by a single generalization *all* instances of a formula where the variables were understood as typically ambiguous. In (1974b) I handled semantical paradoxes by observing that paradoxical sentences could be taken to have no truth-value or not to express propositions on the interpretations presupposed by the semantic concepts occurring in them, while obtaining definite truth-values or coming to express propositions on interpretations 'from outside'. But at some point there must on this account be systematic ambiguity, or else one could generate 'super' paradoxes such as 'this sentence is not true on any interpretation'.

Thus although it is true to say that a proper class is given to us only 'in intension', this statement does not have quite its ordinary meaning. Obviously what is lacking is not just its being given to *the mind* 'in extension'; that is lacking for most sets as well. What is lacking has to do, one might say, with being, and moreover if the underlying intension had the fixed, completed existence a proper class lacks then the class would have it as well. However, some general ideas about intensional concepts do have application to this case. If we think of classes as given only by our understanding of the (perhaps indefinitely extendible) language of set theory, then the assumption that impredicative reasoning about classes is valid is rather arbitrary. This way of looking at classes corresponds to thinking of intensions as meanings and of meanings as constructions of the mind. This is the conception that is appropriate to applying intensional logic to propositional attitudes. Alternatively (and here we have a clearer theory) intensions are thought of as individuated by modal conditions, as 'functions from possible worlds to extensions'. This is the conception appropriate to modal logic. It seems neutral with respect to the question of impredicativity.

Let me make a final comment on the *predicative* conception of classes. If we understand ·a second-order language containing set theory in this way then set existence does for us the work of the axiom of reducibility in Russell's theory of types. For predicates which are high in a ramified hierarchy or which more generally are expressible only by 'logically complex' means, the existence of a set $\{x: Fx\}$ provides a simple equivalent $x \in a$ for a a *name* of $\{x: Fx\}$. Clearly it serves as an equivalent only *extensionally*. In the intensional situations envisaged above the equivalence of $x \in a$ and Fx will not be necessary even if the name a has been introduced by stipulation ('*a priori*' in Kripke's sense; cf. Tharp (1975)), and therefore the two predicates will behave differently in intensional contexts. Thus the license for impredicativity given by assuming the existence of sets does not nullify the predicative conception of intensions even for intensions that have sets as extensions. Of course this 'reducibility' does not obtain for predicates that do not have sets as extensions.[29]

In conclusion, I would claim that the above discussion had added something to the explication of the idea that an 'inconsistent multiplicity' is not really an object, since even as an intension it is systematically ambiguous. The task remains to explain whether the ideas of the last two sections are helpful in understanding the 'stages' of the genetic conception and the underlying priority of the elements of a set to the set.

V.

In the last two sections we sought to avoid using either epistemic concepts or the metaphor of process in trying to understand the conditions for the existence of sets. However, we concentrated largely on the distinction between sets and classes or 'multiplicities' and on discourse about absolutely all sets.

The idea that any available objects can be formed into a set is, I believe, correct, provided that it is expressed abstractly enough, so that 'availability' has neither the force of existence at a particular *time* nor of giveness to the human mind, and formation is not thought of as an action or Husserlian *Akt*. What we need to do is to replace the language of time and activity by the more bloodless language of potentiality and actuality.

Objects that exist together *can* constitute a set. However, we do have to distinguish between 'existing together' and 'constituting a set'. A multiplicity of objects that exist together *can* constitute a set, but it is not necessary that they *do*. Given the elements of a set, it is not necessary that the set exists together with them. If it is possible that there should be objects satisfying some condition, then the realization of this possibility is not as such the realization *also* of the possibility that there be a set of such objects. However, the converse does hold and is expressed by the principle that the existence of a set implies that of all its elements.

The same idea would be expressed in semantic terms by the supposition that we can use quantifiers and predicates in such a way that the range of the quantifiers and the objects satisfying any one of the predicates can constitute single objects, but these objects are not already captured by our discourse. However, this way of putting the matter might be taken to rule out too categorically an 'absolute' use of quantifiers and predicates. Without returning to an ontological characterization such as the Cantorian language of 'existing together', we can say that this is the condition under which quantifiers and predicates obtain definiteness of sense.

Above we suggested that the axiom of power set rests on a sort of principle of plentitude, according to which all the possible subsets of a given set are capable of existing 'at once'. Against what we have just said one might object that there is no intrinsic reason why the 'potentiality' of a set relative to its elements should not be nullified in our theory by a similar principle of plentitude.

The short answer to this objection is that such treatment would lead to contradictions, Russell's paradox in particular. We could apparently consistently assume (as in New Foundations) that the domain of discourse is a set in the domain, but then of course there will be other 'multiplicities' of elements of the domain that are not in it.[30]

A further point is that there seems to be an intrinsic ordering of 'relative possibility' in the element-set relation that is lacking for the arbitrary subsets of a given set. A set is an *immediate* possibility given its elements, the sets of which it is an element are at least at another remove. We do of course have conceptions of the 'simultaneous' realization even of infinite hierarchies in this ordering, but such a conception gives the possibility of sets that are still higher.

This observation should remind us that more is involved in the 'iterative conception' of set than the priority of element to set, since in Gödel's words we think of arbitrary sets as obtained by *iteration* of the "operation 'set of'" starting with individuals, and we have not yet dealt with the concept of iteration. To do so adequately would be beyond the scope of this paper. I shall make a few remarks.

First, our strategy has been to use modal concepts in order to save the idea that *any* multiplicity of objects can constitute a set; one makes only the proviso that they 'can exist together', and this proviso I take to be already given by the meaning of the quantifiers unless they are used in a 'systematically ambiguous' way. One saves thereby the universal comprehension axiom as well, though in a form that hardly seems 'naive' any more: In the second-order modal language it would have to be expressed by the statement that for every attribute P there is an attribute Q that is the rigidification of P (note 24 above) and such that

$$(4) \qquad \Diamond(\exists y)(\forall x)(x \in y \leftrightarrow Qx).^{31}$$

However, even with the assumptions needed to obtain a version of the power set axiom we do not obtain greater power than that given by a much more traditional way of saving the comprehension axiom: the simple theory of types. It is clear that without some principle allowing for *transfinite* iteration of something like the above comprehension principle we will not obtain even the possible existence of sets of infinite rank, such as the usual axiom of infinity already requires. For the axiom of infinity, the principle needed is one allowing the conversion of a 'potential' infinity into an 'actual' infinity: we can easily show

$$(5) \qquad \Box(\forall x)\Diamond(\exists y)(y = x \cup \{x\}),$$

but to use (4) to infer that ω possibly exists, we would need to get from (5) to

$$\Diamond(\forall x)(\exists y)(y = x \cup \{x\});$$

in terms of a set-theoretic semantics for the modal language, the possible worlds containing finite segments of ω need to be collected into a single one.

Second, it is clear that there has to be a priority of earlier to later ordinals, whether this is *sui generis* or derivative from the priority of element to set. One could of course assume a well-ordered structure of individuals, within which there would be no ontological priority of earlier to later elements. The axiom of infinity of *Principia* such an assumption. To make it is natural enough, unless we assume a relation to the mind is essential to the natural numbers. Then it seems that smaller numbers are prior to larger ones by virtue of the order of time, as in Brouwer's (and apparently also Kant's) theory of intuition.

For reasons indicated above, no such structure can represent all ordinals. In fact larger ordinals seem conceivable to us only by characteristically set-theoretic means such as assuming that there is already a *set* closed under some operation on ordinals.

Third, it seems to me that the evidence of the axiom of foundation is more a matter of our not being able to understand how non-well-founded sets could be possible rather than in a stricter insight that they are *impossible*. We can understand starting with the immediately actual (individuals) and iterating the 'realization' of higher and higher possibilities. It seems that (at least as long as we hold to the priority of element to set) we do not understand how there could be sets that do not arise in this way. Non-well-founded ∈-structures have been described (simple ones already in Mirimanoff (1917)), but we do not recognize them as structures of *sets* with ∈ as the real membership relation, even when they satisfy the axioms of set theory.[32] We are at liberty to say that the *meaning* of 'set' is, in effect, 'well-founded set', but that does not exclude the possibility that someone might conceive a structure very like a 'real' ∈-structure which violated foundation but which might be thought of as a structure of sets in a new sense closely related to the old.

I shall close with a rather speculative comment. The conception of 'inconsistent multiplicities' as indefinite or ambiguous raises a doubt about whether it is appropriate to talk of *the* cumulative hierarchy as most set theorists do. The definiteness of the power set is maintained even though the hope of deciding such questions as the continuum hypothesis and Souslin's hypothesis by means of convincing new axioms has not been realized. However, in this case the idea of the 'maximality' of the power set gives us some intuitive handle on the plausibility of the hypotheses or of 'axioms' such as $V = L$ that *do* decide them.

Maximality conceptions also contribute to the plausibility of large cardinal axioms. Here it seems conceivable in the abstract that we might see the possibility of a cardinal α with a 'structural property' P and of a cardinal β with such a property Q, where these properties are not 'compossible'; that is, we would see (perhaps even in ZF) that such α and β cannot both exist. That would yield two incompatible possibilities of cumulative hierarchies.

This has not happened with any of the types of large cardinals considered in recent years, where it has generally happened that of two such properties one (say P) implies the other, and indeed $P\alpha$ implies the existence of many smaller β such that $Q\beta$. That this is so has seemed rather remarkable; perhaps it is evidence against the views I have advanced.

However, one reason for thinking that 'incompatible large cardinals' will not arise is that by the Skolem-Löwenheim theorem both would reflect into the countable sets. If our confidence in the uniqueness of $\mathfrak{P}(\omega)$ is so great as to lead us to reject the possibility of incompatible large cardinals, one would still wish for some more direct reason for doing so.[33]

Columbia University

NOTES

[1] Wang (1974, chapter VI). A widely cited writer whose viewpoint I would also describe as genetic is Shoenfield (1967, pp. 238–240).

[2] I shall consider throughout set theories which allow individuals (*Urelemente*); this requires trivial modifications of the most usual axioms, but the choice among possible ways of doing this is of no importance for us. In extensionality and foundation, the main parameters are restricted to sets (or at least nonindividuals, if classes are admitted).

The literature on the foundations of set theory does not sufficiently emphasize that the exclusion of individuals in the standard axiomatizations of set theory is a rather artificial step, taken for the convenience of pure mathematics. An applied set theory would normally have to have individuals. What is more relevant for the present discussion is that some of the intuitions about sets with which set theory starts concern sets of individuals. First-order set theory with individuals is compatible with the assumption that there are no individuals and therefore with the usual individual-free set theory.

[3] From the genetic point of view, this is an artifice: individuals are presumably given prior to any sets, even the empty set, so that if rank directly reflects order of construction, the empty set should have rank 1. The same holds for the alternative viewpoint I present below.

[4] In a set theory with individuals, some usual theorems about ranks, for example that for every ordinal α there is a set R_α of all sets of rank $<\alpha$, require the assumption that there is a set of all individuals. It follows that there cannot be too many individuals; for

example, ordinals cannot all be construed as individuals. The plausibility of this assumption depends on the intended application. It would be a piece of highly dubious metaphysics to assume there is a set of *absolutely* all individuals, if for no other reason because it is not settled once and for all what *is* an individual. Pure mathematics should be independent of this question; for it the individuals can be an arbitrary set, class, or sometimes structure. However, below I shall assume that the individuals constitute a set.
[5] However, in set theories with classes foundation for classes follows from its assumption for sets.
[6] (1932, p. 282). Wang calls this definition "genetic" (1974, p. 188) and speculates that the difference between this and previous ones (in particular, 1932, p. 204) may be due to Cantor's awareness of the Burali-Forti paradox. It should be remarked that a genetic conception of *ordinals* is intimated in (1932, pp. 195–6), a text from 1883.
[7] Formalism apart, one ought not to rule out the possibility of an interpretation of set theory along constructivist lines, particularly in view of the broadening of the intuitionist outlook in recent years. ZF has recently been shown consistent relative to some set theories based on intuitionist logic. See Friedman (1973) and Powell (1975).
[8] Wang may be developing the remark of Gödel (1964, p. 272 n. 40) that the function of the concept of set is "synthesis" in a sense close to Kant's.
[9] Cf. the fact that in intuitionism the classical notion of set splits into those of spread and species.

It is of interest to look at a case, namely the hereditarily finite sets, where something like arithmetical intuition does yield the axiom of separation. Here we can argue by induction on n that there is a w such that

$$(\forall z)(z \in w \leftrightarrow z \in \{x_1 \cdots x_n\} \wedge Fz). \qquad (*)$$

For if $n = 0$, $w = \Lambda$. Suppose w satisfies (*) and consider $\{x_1 \cdots x_{n+1}\}$. If $\neg Fx_{n+1}$ then w satisfies

$$(\forall z)(z \in w \leftrightarrow z \in \{x_1 \cdots x_{n+1}\} \wedge Fz);$$

if Fx_{n+1}, then $w \cup \{x_{n+1}\}$ satisfies the condition. In effect this argument shows us that of the possible subsets of a finite set *one* satisfies the separation condition, without telling us which one.

One might consider interpreting the requirement that a set x can be 'overviewed' as meaning that it can be run through in a well-ordered way and then attempt a transfinite analogue of this inductive argument. It seems that to handle limit cases such an argument requires replacement, and of course separation can be deduced from replacement in a more trivial way without assuming x to be well-orderable.
[10] For example the classic formulation of Bernays (1935, pp. 275–6). Cf. Wang (1974, pp. 183).
[11] This distinction is explicitly made by Steiner (1975, pp. 130–1). However, Steiner regards only *de dicto* mathematical intuition as defensible. I have sketched in previous writing an account of arithmetical intuition on Kantian lines (1965, pp. 201–3; 1969; 1971, pp. 158–62, 166–7). Curiously, Steiner cites me and then says, "No one today, however, upholds hard-core intuition – the direct intuition of mathematical *objects*" (*ibid.*), although he then mentions Gödel as a possible exception. Since I was trying to elucidate the *formal* character of Kantian intuition, perhaps Steiner did not consider

arithmetical intuition on my view to be "direct intuition of mathematical *objects*", particularly in (1971).

Some comment is in order on Gödel's view of mathematical intuition, particularly since he explicitly says, "We do have something like a perception also of the objects of set theory" (1964, p. 270). This seems to commit Gödel to intuition *de re*. His immediately following remark does not give any argument for this; he says only that it "is seen from the fact that the axioms force themselves on us as being true", which implies only intuition *de dicto*.

However, it seems clear that Gödel holds (1964, p. 271 bottom), that our ideas of objects of certain kinds contain "constituents" which are *given* (not "created" by thinking) on the basis of which we "form our ideas" of these objects and postulate theories of them.

> Evidently the "given" underlying mathematics is closely related to the abstract elements contained in our empirical ideas (1964, p. 272).

In the case of set theory, Gödel does not give any indication of wanting to distinguish sets as objects of intuition from other entities (such as "properties of sets" (1964, p. 264 n. 18)) that the axioms might refer to.

Elsewhere Gödel, in contrast to the above passage, contrasts the intuitive with the abstract (1958, p. 281). There he seems to be using 'intuition' in a much narrower and more strictly Kantian sense. Of course there he is writing in German; possibly he would not use 'Anschauung' in the sense in which he uses 'intuition' in (1964).

[12] According to Hintikka, this is Kant's own view. See (1968) and other essays reprinted in (1973) and (1974).

[13] SAR is the statement: if *b* is an operation and b_x is a set for every member *x* of a set *y*, then all these sets b_x form a set (1974, p. 186).

[14] Boolos seems, however, to arrive at his conclusion too easily. He seems to assume that the 'stages' and their ordering have to be given independently of the concept of set, at least for the expression of *the* iterative conception. His actual axioms about stages (and a further possible one he mentions on p. 227) would permit the stages to be ordered by a very simple recursive well-ordering. It seems to me that sets and 'stages' ought to be 'formed' together, so that the formation of certain sets should make possible going on to further stages.

However, the most obvious principle of this kind, that if a well-ordering has been constructed then there is a stage such that the earlier stages are ordered isomorphically to the given well-ordering, is weaker than the axiom of replacement.

[15] Thus Shoenfield (1967, p. 240) deduces replacement from a "cofinality principle": if "we have a set *A*, and $\cdots$ we have assigned a stage S_a to each element *a* of *A*. $\ldots$ There is to be a stage which follows all of the stages S_a" (p. 239). However, he justifies this by saying, "Since we can visualize the collection *A* as a single object (viz., the set *A*), we can also visualize the collection of stages S_a as a single object; so we can visualize a situation in which all these stages are completed" (*ibid.*). Here he is assuming that "visualizability as a single object" is preserved by replacement of *a* by S_a; but that is just the principle of replacement. Wang's picture seems more fundamental than the kind of argument Shoenfield gives.

Wang gives a similar argument (1974, p. 220 n. 4).

The argument does obtain general replacement from the special case where the range of the replacing function consists of stages.

[16] This plausibility is perhaps reinforced by the fact that replacement holds for the hereditarily finite and the hereditarily countable sets.

[17] However, this is not to say that Cantor now conceives sets as having no intrinsic relation to the mind. Hao Wang has pointed out to me that this would be questionable. For example he characterizes a consistent multiplicity as one the totality of whose elements "can be thought of without contradiction as 'being together', so that their being collected together (*Zusammengefasstwerden*) to 'one thing' is possible".

[18] In a sense (   ) ∈ *y is F*, since coextensiveness for concepts is the analogue of identity for objects, but we cannot say that the concepts are *identical*.

[19] (1892). Only it might be a concept after all, since "is a concept" is syntactically such that it takes *object*-names as subjects, and is therefore a predicate of objects.

Of course a voluminous literature has brown up on this question.

[20] (1969, pp. 288–9), a text written in 1924 or 1925. Cf. (1969, pp. 276–7), from 1919. The late evolution of Frege's thought on these matters is discussed in my (1976).

One can question whether the problem of generalizing predicate places is really solved by Frege's approach. Once we have generalizable variables in predicate places, we have new predicates that are not generalized by the variables in question – predicates which in Frege's semantics denote "second-level concepts". Hence the urge to extend the language by nominalization appears in Frege's context in another form. An ultimate Fregean canonical language would have to be a predicate calculus of order ω of which the semantics can no longer be expressed, unless we admit predicates of infinitely many arguments of different types. Surely *we* understand such a language by a means which from this Fregean point of view is a falsification, namely by a recursion in which *in general* variables with argument places of given types range over relations of these arguments, and each type is reached by finite iteration of the ascent from arguments to function. That involves a 'unification of universes' that Frege rejected, and which essentially contains nominalization.

Frege's logic contained bound variables only for objects and first-level functions, and free variables for one type of second-level function. He refers informally in at least one place to a third-level function (1893, p. 41), which would seem to be required by the *semantics* of his system. Formally, he thought higher levels dispensable because second-level functions could be replaced by first-level functions in which the function arguments were replaced by their *Wertverläufe* (1893, p. 42). This was untenable because it depended on the inconsistent axiom V. But of course in a less absolute way to replace functions by sets which are objects is just the procedure of set theory, which then does dispense with 'higher level functions' for most purposes. It is only quite recently, with the discussion of measurable and other very large cardinals, that higher than second-order concepts relative to the universe of sets have had any real application. See especially Reinhardt (1974a) and Wang (1975).

[21] Analogously to the theory of predicatively definable sets of natural numbers, one can explore mathematically the predicative definability of classes relative to the universe of sets. See Moschovakis (1971).

Wang's discussion of the axioms of separation and power set could lead one to think that impredicative reasoning about 'multitudes' is already involved in motivating the axiom of power set. Although this may be psychologically natural, what the power set axiom says is that given a set *x* there is a set of all sub*sets* of *x*, not that there is a set of all 'multitudes' whose elements are elements of *x*. Thus being an arbitrary subset of *x* has to be definite, but the 'multitude' of them is defined without quantifying over arbitrary *multitudes*. The axiom of separation tells us that *any* 'multitude' of elements of x is a subset of x, so that the 'definiteness' of the property of being a subset of x implies that of being a submultitude of x. But we do not need to assume the definiteness of the latter property; indeed if we think of 'multitudes' intensionally (see below), it is only their *extensions* that become a definite totality by this reasoning.

[22] The most natural and elementary application of these principles is in relation to sets of ordinary objects that are the extensions of predicates contingently true of them. I intend to discuss these matters in a paper in preparation; cf. (1974d) and Tharp (1975).

(1)–(3) exactly parallel familiar principles of identity except that identity is usually treated as independent of existence.

(2) implies that set abstracts are not rigid designators. If 'F' is a predicate that holds of an object x, but not necessarily so, then $\Box(E\{z:Fz\} \rightarrow x \in \{z:Fz\})$ is true with the scope of the abstract outside the modal operator but false with the scope within. I assume that in any possible world $\{z:Fz\}$ is the set of existent z such that Fz in that world.

The free variables in (1)–(3) range over all possible objects, although for the present discussion the appropriate modal logic has *bound* variables ranging only over existing objects. If this treatment of free variables is thought to be too Meinongian, then (3) needs to be replaced by a schema

$$\forall x \Box (x \in y \rightarrow Fx) \rightarrow \Box \forall x (x \in y \rightarrow Fx),$$

or, in the second-order case, by the corresponding second-order axiom.

[23] Thus the relation of P and jP does not contradict (2). However, this is due to the special nature of the projected universe: j is an elementary embedding of the (sets and classes of) the actual world into it. (2) is presumably not an appropriate general principle about proper classes.

The explicit application to set theory of a modal conception of mathematical existence and the use of modal quantificational logic to explicate it seem to originate with Putnam (1967). Putnam does not address the questions about 'trans-world identification' of sets that our principles (1)–(3) are meant to answer. However, it appears that his suggested translation of statements containing unrestricted quantifiers (p. 21) requires that a "standard concrete model of Zermelo set theory" should have a structure that is rigid, that is the relations are not changed when considered with respect to an alternative possible world. If this assumption is made, equivalents of (2) and (3) follow from the fact that a standard model is maximal for the ranks it contains (p. 20). On "concreteness", cf. Parsons (1980, footnote 33).

Putnam seems to envisage a first-order formulation, which requires his "models" to be objects. The second-order formulation seems to us more appropriate not only for the set-class distinction but also for explicating the priority of the elements of a set to the set (Section V below).

[24] A reformulation of Reinhardt's ideas in an intensional language would be desirable, in particular in order to eliminate the Meinongian ontology of possible non-actual sets and classes that he uses, especially in (1974b). Montague's intensional logic would not be adequate as it stands for this purpose, since his first-order quantifiers range over all possible objects; however, there is no difficulty in reformulating it to fit an interpretation in which bound variables range over existing objects. If the only alternative possible worlds one wants to consider are those with more ranks, than the version of quantified

modal logic of Schütte (1968) is applicable. This has the additional advantage that free variables also range over existing objects.

Pure modal logic, however, would not suffice to state Reinhardt's schema (S4) (1974a, p. 196), since it expresses a condition on a single 'possible world' for infinitely many formulae.

The question arises how a class P can recur 'in extension' in another possible world such as Reinhardt's projected universe. The answer is that it would be represented by its "rigidification", that is an attribute Q satisfying the condition.

$$\forall x(Px \leftrightarrow Qx) \wedge \square \{\forall x(\square Qx \vee \square \neg Qx) \wedge$$

$$\forall R[\forall x \square (Qx \rightarrow Rx) \rightarrow \square \forall x(Qx \rightarrow Rx)]\}$$

I assume here that bound variables range over existing objects; otherwise Barcan's axiom would hold and the third conjunct would be unnecessary.

Wang (1975) formulates Reinhardt's ideas in the opposite direction, by eliminating the intensional motivation and thinking of V not as the 'actual world' but as a set which is an 'approximation' to the universe. What *mathematical* interest an intensional formulation would have is not clear; perhaps it would suggest 'intuitionistic' approaches to strong reflection axioms.

(*Added in proof.*) The statement that every attribute has a rigidification is just the formulation appropriate to this setting (without the Barcan formula) of the axiom E^{σ} (for $\sigma = (e)$) of Gallin (1975, p. 78). That the second conjunct above is equivalent to Gallin's $\forall x(\square Qx \vee \square \neg Qx)$ follows from Barcan's axiom. However, (1)–(3) (p. 350) and the comprehension axiom of p. 357 are inconsistent with the Barcan formula.

In a paper in preparation, axiomatizations of set theory based on the ideas of this paper will be presented.

[25] Such a conception is hinted at in Zermelo (1930); see especially p. 47.

The conception of quantification over all sets advanced here is close to that of Putnam (1967), except for the addition of the concept of systematic ambiguity.

[26] The existence of an R_{α} such that V is an elementary extension of R_{α} is provable in ZF plus impredicative classes (Bernays-Morse set theory; see Drake (1974, p. 125)). What is essential, however, is not impredicative classes but allowing bound class variables in instances of replacement; one could use the system NB^{+} mentioned in my (1974a, note 15).

[27] Cf. the remarks on the discomfort evinced by use of proper classes in Wang (1975, footnote 8). However, Wang does not make clear whether this discomfort would be removed if we confine ourselves to thinking "of these large classes as extensions or ranges of properties".

The point which I would emphasize is that if the language of set theory with quantifiers read as ranging over 'all sets' has a 'fixed' or 'definite' sense, then it is naturally extended by a satisfaction predicate, and definiteness of sense is preserved. But in the extended language one can of course construe the classes required by the Bernays-Gödel theory. Iteration of the procedure yields more classes.

In Wang's terms, this justification of classes no doubt falls within the conception of them as "extensions or ranges of properties". Still, such an enlargement of the language of set theory seems to be treated with reserve by many set theorists, although the reason could be just that in deductive power it is inferior to stronger axioms of infinity.

Locutions requiring either classes or satisfaction and truth are frequent in writings on set theory (cf. my, 1974c), but the characteristic informal use is very weak and could be captured by a free variable formalism for classes with very elementary operations on them, as George Boolos' comments on (1974c) reminded me.

[28] The 'straightforward reading' involves replacing 'set' by 'class' at certain points in the standard model-theoretic account of (modal) logical validity, just as in the case of ordinary logic in set theory. The same should be the case for set theories with intuitionistic logic, which are suggested by the same considerations as suggest the modal language. It should not be thought that changing to intuitionistic logic will remove the fundamental dilemmas about quantification over all sets.

[29] It is commonly claimed that the axiom of reducibility nullifies Russell's ramification of his hierarchy of types. This claim depends on ignoring, presumably on the grounds that nonextensional features of functions are not significant for mathematics, the fact that Russell thought of propositional functions intensionally.

On the other hand it is hard to see what is left of Russell's no-class theory once the axiom of reducibility is admitted. Russell himself says that the axiom of reducibility accomplishes "what common sense effects by the admission of classes" (1908, p. 167), but he considers the axiom a weaker assumption than the existence of classes. The weakness must consist in the restrictions of the simple theory of types.

[30] In the case of NF, these additional 'multiplicities' would correspond to the proper classes of ML.

If a model of NF is given as a set in the ordinary set-theoretic sense, the domain of the model and the V of the model will of course differ. The membership relation of the model will obviously not be the same as the membership relation 'from outside'.

[31] Thus if in some possible world $(\forall x)(x \in y \leftrightarrow Qx)$ holds, the elements of y are just the objects that *actually* have P. In many cases they will not be just the objects that have P in the world in question.

[32] Perhaps this could be said of trivial variants such as that resulting from identifying individuals with their unit classes. But here of course a slightly modified axiom of foundation holds.

[33] I am indebted to Robert Bunn, William Craig, William C. Powell, Hilary Putnam, and Hao Wang for valuable discussions related to this paper. I regret that time did not permit me to follow up Mr Bunn's remarks on Jourdain's attempt to develop the theory of inconsistent multiplicities.

BIBLIOGRAPHY

Benacerraf, P. and Putnam, H. (eds.): 1964, *Philosophy of Mathematics: Selected Readings*, Prentice-Hall, Englewood Cliffs, N.J.

Bernays, P.: 1935, 'Sur le platonisme dans les mathématiques', *L'enseignement mathématique* **34**, 52–69. Eng. tr. in Benacerraf and Putnam (1964, pp. 274–286). Cited according to translation.

Boolos, G.: 1971, 'The Iterative Conception of Set', *The Journal of Philosophy* **68**, 215–231.

Cantor, G.: 1932, *Gesammelte Abhandlungen mathematischen und philosophischen Inhalts* (ed. by Ernst Zermelo), Springer, Berlin.

Drake, F. R.: 1974, *Set Theory: An Introduction to Large Cardinals*, North Holland, Amsterdam.

Frege, G.: 1892, 'Ueber Begriff und Gegenstand', in *Kleine Schriften* (ed. by I. Angelelli), Olms, Hildesheim, 1967.

Frege, G.: 1893, *Grundgesetze der Arithmetik*, Vol. I, Pohle, Jena.

Frege, G.: 1969, *Nachgelassene Schriften* (ed. by H. Hermes, F. Kambartel, and F. Kaulbach), Meiner, Hamburg.

Friedman, H.: 1973, 'The Consistency of Classical Set Theory Relative to a Set Theory with Intuitionistic Logic', *The Journal of Symbolic Logic* **38**, 315–319.

Gallin, D.: 1975, *Intensional and Higher-Order Modal Logic*, North-Holland, Amsterdam.

Gödel, K.: 1958, 'Ueber eine noch nicht benützte Erweiterung des finiten Standpunktes', *Dialectica* **12**, 280–287.

Gödel, K.: 1964, 'What is Cantor's Continuum Problem?', revised and expanded version in Benacerraf and Putnam (1964, pp. 258–273). Original 1947.

Hintikka, J.: 1968, 'On Kant's Concept of Intuition (*Anschauung*)', in T. Penelhum and J. J. Macintosh (eds.), *The First Critique*, Dickenson, Belmont, Calif.

Hintikka, J.: 1973, *Logic, Language-Games, and Information*, Clarendon Press, Oxford.

Hintikka, J.: 1974, *Knowledge and the Known: Historical Perspectives in Epistemology*, D. Reidel, Dordrecht, Holland.

Lewis, D.: 1968, 'Counterpart Theory and Quantified Modal Logic', *The Journal of Philosophy* **65**, 113–126.

Mirimanoff, D.: 1917, 'Les antinomies de Russell et de Burali-Forti et le problème fondamental de la théorie des ensembles', *L'enseignement mathématique* **19**, 37–52.

Montague, R.: 1970, 'Pragmatics and Intensional Logic', *Dialectica* **24**, 277–302, also *Synthese* **22**, 69–94. Reprinted in R. H. Thomason (ed.), *Formal Philosophy*, Yale, New Haven, 1974.

Moschovakis, Y. N.: 1971, 'Predicative Classes', in *Axiomatic Set Theory* (*Proceedings of Symposia in Pure Mathematics*, Vol. 13), part I, pp. 247–264, American Mathematical Society, Providence, R. I.

Parsons, C.: 1965, 'Frege's Theory of Number', in Max Black (ed.), *Philosophy in America*, Allen and Unwin, London, pp. 180–203.

Parsons, C.: 1969, 'Kant's Philosophy of Arithmetic', in S. Morgenbesser, P. Suppes, and M. White (eds.), *Philosophy, Science, and Method: Essays in honor of Ernest Nagel*, pp. 568–594. St. Martin's Press, New York.

Parsons, C.: 1971, 'Ontology and Mathematics', *Philosophical Review* **80**, 151–176.

Parsons, C.: 1974a, 'Informal Axiomatization, Formalization, and the Concept of Truth', *Synthese* **27**, 27–44.

Parsons, C.: 1974b, 'The Liar Paradox', *Journal of Philosophical Logic* **3**, 381–412.

Parsons, C.: 1974c, 'Sets and Classes', *Noûs* **8**, 1–12.

Parsons, C.: 1974d, 'Sets and Possible Worlds', abstract of a talk at Columbia University, mimeographed.

Parsons, C.: 1977, 'Some Remarks on Frege's Conception of Extension', in M. Schirn and B. Kienzle (eds.), *Studien zu Frege I: Logik und Philosophie der Mathematik*. Fromann, Stuttgart, to appear.

Parsons, C.: 1980, 'Quine on the Philosophy of Mathematics', in P. A. Schilpp (ed.), *The Philosophy of W. V. Quine*, Open Court, La Salle, Ill., to appear.

Powell, W. C.: 1972, *Set Theory with Predication*, Thesis, S.U.N.Y. at Buffalo.

Powell, W. C.: 1975, 'Extending Gödel's Negative Interpretation to ZF', *The Journal of Symbolic Logic* **40**, 221–230.

Putnam, H.: 1967, 'Mathematics without Foundations', *The Journal of Philosophy* **64**, 5-22.

Reinhardt, W. N.: 1974a, 'Remarks on Reflection Principles, Large Cardinals, and Elementary Embeddings', in *Axiomatic Set Theory* (*Proceedings of Symposia in Pure Mathematics*, Vol. 13), part II, pp. 189–206.

Reinhardt, W. N.: 1974b, 'Set Existence Principles of Shoenfield, Ackermann, and Powell', *Fundamenta Mathematicae* **84**, 5–34.

Russell, B.: 1908, 'Mathematical Logic as based on the Theory of Types', reprinted in J. van Heijenoort (ed.), *From Frege to Gödel*, Harvard, Cambridge, Mass. 1967, pp. 150–182.

Schütte, K.: 1968, *Vollständige Systeme modaler und intuitionistischer Logik*, Springer, Heidelberg.

Shoenfield, J. R.: 1967, *Mathematical Logic*, Addison-Wesley, Reading, Mass.

Steiner, M.: 1975, *Mathematical Knowledge*, Cornell, Ithaca.

Takeuti, Gaisi, and Zaring, W. M.: 1971, *An Introduction to Axiomatic Set Theory*, Springer, Berlin-Heidelberg-New York.

Tharp, L. H.: 1975, 'Three Theorems of Metaphysics', unpublished.

Wang, Hao. 1974, *From Mathematics to Philosophy*, Routledge and Kegan Paul, London.

Wang, Hao: 1977, this volume, p. 309.

Zermelo, E.: 1930, 'Ueber Grenzzahlen und Mengenbereiche', *Fundamenta Mathematicae* **16**, 29–47.

VII

PHILOSOPHY OF LOGIC

IAN HACKING

DO-IT-YOURSELF SEMANTICS FOR
CLASSICAL SEQUENT CALCULI,
INCLUDING RAMIFIED TYPE THEORY

1. INTRODUCTION

This paper revives some out-of-date ideas that contribute to the solution of present problems. We philosophical logicians have given up too much of our own past. It is our own fault: we once had the temerity to claim that mathematics is reducible to logic. It isn't. Cowed by the failure of this takeover bid we have neglected a good deal of our heritage which, although irrelevant to the higher mathematics, still matters to our central concerns of truth and meaning, modality and identity, existence and deducibility, syntax and semantics.

The ramified theory of types is a case in point. *Principia Mathematica* was supposed to analyse all mathematics, but the ramified theory on which it is based cannot even make a Dedekind cut without an axiom of reducibility. This frustrated the project because, as the authors well knew, such an extra axiom is no part of logic. Moreover it has turned out that the more powerful simplified theory of types is a poor tool for foundations. Even constructively oriented mathematicians find that the ramifications of *Principia*, for all their predicativity, are unsuitable for most purposes. So *Principia* has become a dusty monument and hardly anyone even remembers how the ramified hierarchy goes.

This outcome might have been foreseen. Russell tried to surmount the set theoretic paradoxes by declaring certain sentences to be meaningless or illformed. This is a special case of what Imre Lakatos (1963) called monster-barring, which is almost always an unsuccessful strategy. There is no reason to suppose that declaring certain sentences to be meaningless will teach us the truth about sets. But ramified theory was also supposed to solve the *other* contradictions which, after Ramsey, have been called semantic. If Ramsey is right, and these paradoxes have to do with grammar, meaning and truth, then a solution like that devised by *Principia* may be more in order. I do not argue that ramification is the only solution: surely it is not. I remark only that *Principia* starts with a grammatical theory, not a mathematical one, and I urge its importance to other parts of philosophical grammar.

It is perhaps Russell's fault that we do not think of ramification as a theory of syntax. At the time he was marrying logic to higher mathematics he was divorcing it from grammar. 'Grammatical form' was castigated as a misleading part of language that leads to philosophical error. What matters is 'logical form'. Now, almost seventy years later, we have before us many proposals for reuniting logic and grammar, including the notion that logic is the 'deep structure' of English. This heretical version of Chomsky's early model of deep and surface structures has attracted some of our colleagues, but it has been plagued by the original sin of nominalism/realism, or extensionalism/intensionalism. Thus Donald Davidson offers a programme for truth, meaning and grammar built on a horizontal extension of first-order logic, while Montague's general grammars have vertical higher-order logics. Since the latter lack the same kind of extensional completeness theorem possessed by first-order logic, they are naturally explained by possible world semantics, anathema to nominalists. The ramified theory is a happy mediator here, for it appears able to handle many of Montague's applications, and yet it has an extensional analysis – a special case of the Do-it-yourself semantics presented in this paper.

My attitude to sematics is a somewhat reactionary hankering after old ideas. The 'logical syntax of language' had a conception of grammar and meaning different from modern model theory. Because of the great wealth of applications of the latter in metamathematics, we have forgotten that the former might still be of use in the philosophy of language. The old idea, I believe, was that logic should bear semantics on its face. That is, there would be a core of atomic sentences whose truth conditions would be given prior to logical syntax. Then it was supposed that a decent logic would make the truth conditions of complex sentences dependent on the truth conditions of classes of simpler ones. The theory of truth would in fact imitate the rules of grammatical formation. This is an old dream going back through Bolzano to Leibniz and beyond. It is one that can be realized for the ramified theory. This is not an isolated *ad hoc* result. In general such semantics can be realized by any logic whose constants are characterized by rules of the sort that Gentzen used for his classical theory of natural deduction or his sequent calculi. Roughly speaking, these are rules for which cut-elimination is finitistically provable, which have the subformula property, and a minor property that I call locality. Prawitz (1965) observed that the ramified but not the simple theory of types could be characterized by rules of this sort.

Indeed within a ramified theory one may also give rules that define directly such constants as identity, the predicate of existence, and Montague's second order predicates of quantification such as *everyman* (under which the predicate *mortal* falls). A theory of quantifying over predicate modifiers–'adverbs'–can also be readily accommodated. The semantics, I repeat, is entirely extensional. This theory is, moreover, not restricted to classical logic, and applies equally to object level non-modal deviant logics such as intuitionist logic and the system of quantum logic based on the orthomodular lattice. Although the latter requires a new semantical idea of compatibility (which is explained in Hacking 197A) the Do-it-yourself semantics and completeness argument of the present paper immediately adapt to the quantum theory. Thus one obtains, for the first time, a complete first-order and ramified 'quantum logic'.

Other philosophical consequences for identity and existence – running even to the identity of indiscernibles and the ontological argument – are described in (Hacking, 197B). Thus once we have satisfied ourselves about the semantics of the ramified theory, we have a powerful tool for philosophical grammar.

Our tools are those developed by Gentzen for his well known sequent calculi, but our approach is in a certain sense the reverse of his. He asked what rules are needed for first-order logic. We ask, what constants can be defined by rules like his? In (Hacking, 197C) I give some '*a priori*' arguments for supposing that this class of rules provides a definite solution to the problem of demarcation for logical constants and logical truth. I do not rely on that in this paper but mention it as providing part of the motivation for these investigations. Logic, on this view, is just exactly that for which a Do-it-yourself semantics exists: logic bears its semantics on its face.

2. PRELOGICAL LANGUAGE

Throughout this paper we adopt what is, nowadays, a slightly eccentric view of logic. We imagine that there is a prelogical language with a definite grammar and such that the truth conditions for each atomic sentence are understood. Then the logical constants are to be planted on top of this prelogical language by means of rules which, in the first place, are rules concerning the inferential relations appropriate to each constant. But we show, by the Do-it-yourself semantics, that these selfsame rules also define the way in which truth conditions for complex sentences are determined by truth conditions for classes of simpler sentences. That is, our syntax generates our semantics. On many points we are following the lead of the *Tractatus*. "Without bothering about sense or meaning, we construct the logical proposition out of others using only *rules that deal with signs*" (6.126). But these rules concerning the 'syntax' of signs will automatically determine the 'meaning' of the propositions constructed.

We make no assumptions about the semantics of the prelogical language, except that the truth conditions be understood. One philosopher may suppose that 'Socrates' in 'Socrates is wise' refers to a character who inhabits many possible worlds; another may insist that it denotes the historical figure of Plato's dialogues. Still another may insist that the truth condition for the sentence is 'Wisdom is possessed by Socrates'. Some will suppose that these different accounts are at the core of philosophical controversy, while others will suppose that there is no issue except one of terminology. We are neutral. At the very most, for metalogical purposes, we suppose that the truth conditions for the prelogical language can be stated in English. We are not concerned here with the theory of reference. Our attitude seems similar to that of Donald Davidson.

The prelogical language will have a grammar which is to be augmented by the introduction of logical constants. We do not need to employ any grammatical theory here, but it is convenient, with the theory of types in view, to suppose that there is a category grammar of the sort invented by Ajdukiewicz for the simplified theory of types (1935). Ramified types are nothing but grammatical categories. In the beginning a category grammar need have only two categories, terms t and sentences s. Sentential functions, or predicates, are of the category

$(t)/s$ for 1-place functions, and $(t, ..., t)/s$ for the various polyadic functions. Note that the introduction of a logical constant may create instances of categories not present in the prelogical grammar. For example, conjunction is $(s, s)/s$; and a language consisting solely of atomic sentences would not possess this category.

The fact that logic augments the scheme of categories is particularly evident for ramified theory. Before explaining, we must note a tedious matter of nomenclature: 'order' means something different in *Principia Mathematica* than in current talk of e.g. second-order logic. The theory of types is the work of Russell, for whom formulae with no quantifiers are 'elementary'. Those with individual quantifiers are of first order. Thus far the old terminology and the new coincide well enough. But now suppose we define a sentential function, whose only free variables are individual variables, by quantifying over first-order functions. For example, in *Principia* identity is the relation $(f)(fx \equiv fy)$, where f is of first order. Identity is said to be a relation of second order. A sentential function with only free individual variables, but defined by quantifying over functions of second order, is said to be of third order, and so on. In modern terminology all the sentential functions so generated are said to be of second order, since they involve quantification over predicates of individuals, while third-order theory involves quantification over predicates of predicates of individuals. Church (1956, Sec. 56) uses the word 'level' where Russell used 'order'. In this paper we do not need to opt for old or new; on the rare occasion when a name is needed, we shall speak of Russell-orders (which are approximately Church-levels).

A prelogical language will contain only elementary predicates. There may be individual constants, predicates of individuals, predicates of predicates of individuals and so forth, but there will be no higher-Russell-order predicates of any type. That is, there will be no predicates that are defined in terms of quantification – a fact implicit in the very word 'prelogical'. As a corollary, all higher-Russell-order predicates are of categories that do not occur in the prelogical language. So we go farther than Wittgenstein quoted above. It is not merely the 'logical propositions' that are constructed by 'rules that deal with signs', but also what we might call the logical predicates of higher order.

Gentzen, in giving his rules characterizing the logical constants, stated their syntax in the initial specification of his language. Since our procedure is the reverse of his, we shall assume a different procedure. We make no *a priori* constraints on the grammar of constants to be defined by rules like Gentzen's, and instead suppose that a Gentzen-style rule has three parts. There is the deduction rule itself. Then there is a statement of the grammatical category of the constant defined. Finally there is a representation rule, which might tell us, for example, whether conjunction is to be represented as in *Principia Mathematica*, in Polish notation, or whatever.

We shall not actually use any grammatical apparatus, but it is important to recognize what is called for in the layout of any specific system of logic to be planted on top of a prelogical language. There is one further feature. As a step towards doing logic, we shall augment our prelogical language with a countable set of variables for each category. Moreover, if the introduction of a logical constant furnishes an instance of a new category, we shall, in planting this constant in the language, add a countable set of variables of the new category. For example, if we move to second-Russell-order quantifiers we shall thereby introduce variables of various new categories. Since the introduction of variables allows us to write down open sentences, we shall in what follows speak of *formulae* as the members of the category s. We would allow, but not demand, that the grammar would have subcategories of s – for example, formulae of different Russell-order – as logic generated entities of higher order.

3. SEQUENT CALCULI

Gentzen's sequent calculi are used here, although his systems of natural deduction would have served as well. The sequent calculus is understood as a theory of deducibility about an object language. The basic form of statement in this theory is $\Gamma \vdash \Delta$. Γ and Δ are sets of sentences of the object language. We prefer this to Gentzen's sequences of sentences in order to obviate rules for permutation and contraction. Γ, A is short for $\Gamma \cup \{A\}$, and the first capital letters of the latin alphabet denote sentences of the object language. The fundamental principles of identity are:

1. The deducibility of identicals. $A \vdash A$.
2. Stability. If $\Gamma \vdash \Delta$, then $\Gamma, A \vdash \Delta$ and $\Gamma \vdash A, \Delta$.
3. Transitivity. If $A \vdash B$ and $B \vdash C$, then $A \vdash C$.

In the presence of (1) and (2), (3) is correctly generalized to Gentzen's rule of *Schnitt*, or cut, namely:

$$\text{If } \Gamma \vdash A, \Delta \text{ and } \Gamma, A \vdash \Delta, \text{ then } \Gamma \vdash \Delta.$$

We regard these as fundamental principles about classical deducibility. The connection with logical consequence is that we intend an interpretation in which if $\Gamma \vdash \Delta$, then some member of Δ is true when all members of Γ are true. However we shall proceed in an entirely syntactic way, and then show that as a byproduct of this syntax, syntactic classical deducibility is indeed equivalent to semantic logical consequence.

Gentzen's sequent calculi consist of rules which introduce each of the familiar logical constants. Apparently he wanted these rules to be thought of as defining the constants. But they cannot all be definitions *ex nihilo* which are intelligible to a person who had no concept of any of our logical constants nor any of their surrogates. For example, the rule for conjunction is explained in terms of something like conjunction, while the rule for the universal quantifier can be used to full effect only by someone who already has mastered a concept akin to 'all'. The latter will be embarrassingly evident in my 'local' versions of Gentzen given in Section 5 below. One should regard Gentzen's rules as characterizations of constants rather than as definitions, although one may literally define many constants *ab initio*, constants of which no one has ever thought before, once one has grasped how the rules work. But to get the rules to work you need to know how to use 'all' and 'not' and 'and' or some stand-ins for these, although it is arguable that all you need to understand are some weak, intuitionistic, or constructive versions of those concepts.

Gentzen gave a set of what he called operational inference rules for each of the familiar constants of first-order logic. A derivation is a finite tree starting with instances of the deducibility of identicals called

initial statements proceeding by applications of thinning, cut, or operational rules to a theorem, a statement of deducibility of the form $\Gamma \vdash \Delta$.
He then proved his *Hauptsatz*, nowadays often called cut-elimination:
Every theorem is derivable without the use of cut. This result is
important for many metamathematical considerations but here we
attach a slightly eccentric significance to it. We regard cut-elimination
as showing that Gentzen's operational rules characterizing logical
constants do so in a way that preserves the transitivity of deducibility.
To put it differently: Gentzen's rules are all stated using some arrow or
turnstile denoting a deducibility relation. But once the rules are stated
(and nothing else is done) is the relation a deducibility relation at all?
Must we add some further rule constraining the turnstile to be a
deducibility relation? No. The *Hauptsatz* shows us that we do not need
to supplement the operational rules with an additional rule, *Schnitt*,
that insists that the turnstile is transitive. Gentzen's operational rules
have the peculiar feature that transitivity is guaranteed. From my
perspective, of planting logic on top of a prelogical language, I prefer
to say that transitivity is preserved. Not all rules will do this: that is the
lesson of Prior's 'tonk' (1960). Evidently transitivity – cut-elimination –
is not on the face of it sufficient to show that the turnstile occurring in
Gentzen's rules is a deducibility relation. We need to connect it up
with logical consequence, which is the aim of the present paper. But
the *Hauptsatz* is an essential first syntactic step in that direction.

Gentzen's papers take for granted the class of logical constants to be
characterized. We reverse his procedure and ask: what constants can
be defined by rules like his? Such a question demands a definition of
what it is to be 'like' his rules, and there are several possible explica-
tions. I do not claim that the present paper gives the only account, but
only that it is a productive one. A first question to settle is about the
Hauptsatz. Any rules 'like' his ought to have such a result. Gentzen's
own proof was finitistic. There are fairly constructive but non-finitistic
proofs for stronger systems, and many workers have now investigated
the kinds of constructivity that these involve. Here we shall consider
only systems for which cut-elimination is finitistically provable. My
(197C) mentions one *a priori* justification for this: 'Logic' is supposed
to be our standard of consistency, and anything that is to count as pure
logic ought to be able to prove its own consistency.

Another important property of Gentzen's own rules is that they are all in a sense 'building rules'. They take a set of premises of the form.

$$\{\Gamma, \Pi_i \vdash \Sigma_i, \Delta\}$$

and have a conclusion of one or the other of the following forms:

$$\Gamma, A \vdash \Delta \qquad \Gamma \vdash A, \Delta.$$

The *principal formula* A is in a fairly literal sense built up out of (parts of) the *components*, namely Π_i, Σ_i, and a new logical constant, such as the universal quantifier or conjunction, which I henceforth call the *new constant* provided by the rule. In what follows I assume that the new constant always occurs in A, and also that the class of components is non-empty. A rule whose principal formula occurs on the left of the turnstile is called an *elimination* rule (since we are as it were deducing something out of it) while one that occurs on the right is an *introduction* rule, for here we are deducing a new sentence. In Gentzen's calculi each constant has at least one introduction and at least one elimination rule. The collection of all introduction and elimination rules for a constant will be called *the* rule for that constant. Since the principal formula is always built up out of the components and the new constant, Gentzen's rules are said to have the *subformula property*, which will now be defined.

4. THE SUBFORMULA PROPERTY

Subformulae are defined for example in Prawitz (1965) for specific first-order and ramified calculi. A rule is said to satisfy the subformula property if all the components are subformulae of the principal formula. Many mathematical logicians would say this property is of little interest but the standpoint of grammar is different. There is a familiar argument to the effect that people learn natural languages from scratch, yet their languages have no upper bound on the number of sentences. Hence there must be a finite base and a set of rules determining the meaning of complexes in terms of simples – or else the language could never be acquired. Gentzen rules having the subformula property conform to this point of view. The rules provide for more and more complex sentences, but the derivations are always from subformulae; if we understand the subformulae and the rule, then the use of the complex formula is determined. Unfortunately it is not obvious when one formula is more 'complex' than another, and since, unlike Prawitz, we are concerned with all possible Gentzen rules, we cannot merely define 'subformula' for a designated set of rules. The basic requirement is a *complexity ordering* of sentences: a partial ordering of the domain of formulae such that if A is a proper part of B, then B is more complex than A. We shall choose orderings with the further feature that if A is a proper part of B, and X is a variable free in A but not in B, then all substitution instances of A are less complex than B. The relationship between Ft and $(x)Fx$ is a case in point; the former is commonly said to be a subformula of the latter, yet the term t may be as long as you please. Hence long formulae may be less complex than short ones. In general we take quantified formulae to be more complex than their instances. In what follows we suppose that 'proper part', 'free', 'bound', 'alphabetic change of bound variable', 'substitution instance' and the like are defined in terms of the underlying category grammar.

A is a *subformula* of B if (1) it is a proper part of B, or (2) there is a subformula of B say C, in which the variable X occurs free, and such that those occurrence of X free in C are not free in B, and such that A is a substitution instance of $C(X)$, or (3) A is a subformula of a formula C which under alphabetic change of bound variable either is B or is a subformula of B.

A sequent calculus has the *subformula property* under a complexity ordering if (1) no operational rule lacks components (2) every component in every operation rule is a subformula of the principal formula, and (3) under the complexity ordering every component is less complex than the principal formula.

The third clause is not vacuous. Suppose for example that, as suggested in Section 2, identity were taken to be a second order relation and that the definite description operator were defined, as in Russell, in terms of this relation. One might expect the following to be a correct application of the rule of $(\exists x)$-introduction:

$$\frac{F((\imath y)(\exists x)(Fxy)), y) \vdash F((\imath y)(\exists x)(Fxy)), y)}{F((\imath y)(\exists x)(Fxy)), y) \vdash (\exists x)(Fxy)} \ .$$

Here we introduce $(\exists x)(Fxy)$ on the basis of Fty, and the conclusion below the line is certainly true. Yet the principal formula is a proper part of the component, so no rule with the subformula property can license this step. The problem arises, of course, only in a ramified calculus where identity is not primitive, and in that case the rule of $(\exists x)$-introduction, for which cut-elimination is provable, will require that the term t in Ft must be of first order if $(\exists x)Fx$ is to be introduced. The true conclusion of the above invalid derivation is, of course derivable, but not in such a direct way. Instead one uses the rule for $\imath$-elimination to obtain the formula on the left containing the definite description.

Evidently ramification and the subformula property are closely connected. If an unramified rule introducing quantification over predicates were stated as on the left below, then the inference on the right would, apparently, be permitted. That would violate the subformula property. But in a ramified calculus such as that of Prawitz (1965), the rule on the left requires that X and P be of the same Russell-order, while $(\exists X)(F(X))$ is of higher order. This is of course precisely the way that *Principia Mathematica* tried to avoid paradoxes that use steps like that on the right.

$$\frac{\Gamma \vdash F(P), \Delta}{\Gamma \vdash (\exists X)(F(X)), \Delta} \qquad \frac{\Gamma \vdash F((\exists X)(F(X, Y), Y), \Delta}{\Gamma \vdash (\exists X)(F(X, Y)), \Delta} \ .$$

5. Locality

A *replication* occurs in a derivation if a formula is both parameter and principal formula at some step. For example:

1. $Fc, Fa \vdash Fa, Fb.$

2. $\quad Fc \vdash Fa, Fa \supset Fb.$

3. $\quad Fc \vdash Fa, (y)(Fa \supset Fy).$

4. $\quad \vdash Fc \supset Fa, (y)(Fa \supset Fy).$

5. $\quad \vdash Fc \supset Fa, (\exists x)(y)(Fx \supset Fy).$

6. $\quad \vdash (y)(Fc \supset Fy), (\exists x)(y)(Fx \supset Fy).$

7. $\quad \vdash (\exists x)(y)(Fx \supset Fy).$

The principal formula in (7) is also a parameter. One must first obtain this formula at (5) and again at (7), and there is no way of permuting steps so that the rule used at (6) precedes that used at (5). A replication is said to be *condensed* when, between the first replication of A and the last, there occur only steps in which A is principal formula. In the above derivation the replication is not condensed because step (6) intervenes between the first introduction of A and the second. In Gentzen's sentential systems, all replications can be condensed, but this is not true of first-order logic. (This explains why his calculi do not provide first-order decision procedures.)

Restrictions on side formulae are what prevent us from condensing replications. For example, the rule for introducing the universal quantifier says that $\Gamma \vdash (x)Fx, \Delta$ may be inferred from $\Gamma \vdash Fa, \Delta$, where x captures all and only occurrences of a in F, *so long as a does not occur free in* Γ, Δ. A rule is called *local* if it places no restrictions on side formulae.

Gentzen's calculi are *formal* in the sense that one can actually write down any derivation in the object language. Proofs involve finitely many sentences. One may exchange formality for locality by allowing countably many premises. However, so long as cut-elimination is finitistically provable, and the subformula property obtains, there will be an upper bound on the length of each branch in a local derivation. This is in contrast to, for example, Schütte's (1951) systems for arithmetic.

Locality is easily achieved in first-order logic. The standard proof of cut-elimination uses the fact that whenever $\Gamma \vdash Fa, \Delta$ is derivable, and a does not occur free in Γ, Δ, then there exists a derivation of equal length of $\Gamma \vdash Ft, \Delta$, for any closed term or variable t. Hence whenever we might apply Gentzen's rules for (x) we could equally apply:

$$\frac{\Gamma, [Ft_i]_i \vdash \Delta}{\Gamma, (x)Fx \vdash \Delta} \qquad \frac{\{\Gamma \vdash Ft_i, \Delta\}_i}{\Gamma \vdash (x)Fx, \Delta} \; .$$

The subscripted *curly* brackets mean that all statements of a certain form indexed by i are to count as premises. The subscripted *square* brackets mean that in the single premise, one or more occurrences of sentences of a certain form indexed by i are to be found. Local rules are completely permutable, i.e. if rule R is applied at a step directly above rule S, and the principal formula in R is a side formula in S, then S may be applied before R. The rules just stated are equivalent in theorems to Gentzen's. Whenever his introduction rule can be applied, each of the premises for the local version can be obtained. Conversely let a be the first variable not occurring free in the conclusion. Then $\Gamma \vdash Fa, \Delta$ will be among the premises and may be used as a premise for applying Gentzen's rule. Note that since the derivation of that statement may be mimicked with a replaced by any closed term, or variable not in the conclusion, there is an upper bound on the length of branches required to derive any of the premises for the (x)-introduction rule. As for the elimination rule, note that a premise with countably many sentences of form Ft_i is never needed so long as we permute rule applications. Suppose then that all premises are of finite length; then the local (x)-elimination rule is equivalent to finite number of applications of Gentzen's rule.

The local rules for the universal quantifier make plain to the point of absurdity that you have to understand 'all' or some surrogate in order to use the introduction rule, and 'some' or somesuch to employ the elimination rule. This makes explicit the feature of Gentzen's rules discussed in Section 3. We cannot regard his calculi as 'defining' all of the logical constants for a person who understood none of them. At best they characterize and mark out the constants called logical, and may be used to introduce some logical constants with which one was not acquainted before.

Since rules in local calculi may be permuted, the replications which occur in Gentzen's calculi may always be condensed. To take the example that began this section, we obtain:

$$\left\{ \begin{array}{l} \{\vdash Fc \supset Ft_i, \, Ft_i \supset Ft_j\}_j \\ \vdash Fc \supset Ft_i, \, (y)(Ft_i \supset Fy) \end{array} \right\}_i$$

$$\vdash (y)(Fc \supset Fy), \, [(y)(Ft_i \supset Fy)]_i$$

$$\vdash (\exists x)(y)(Fx \supset Fy).$$

Since replications can always be condensed, it is possible to consider the set of possible ways in which A might have been derived on the left or the right of the turnstile, using only steps in which A is principal formula, and starting with premises in which A does not occur. This leads us to the idea of 'protopremises'.

An *introduction protopremise* for a formula A is a class of statements $(\Pi_i \vdash \Sigma_i)$ such that (1) A does not occur in (Π_i, Σ_i); (2) $\vdash A$ may be derived from it by successive applications of operational rules in which A is principal formula; and (3) for no i can any formula be deleted from Π_i, Σ_i without making such a derivation of $\vdash A$ impossible. A *weak* introduction protopremise is one for which (1) and (2) hold, but not (3). Likewise for elimination protopremises and weak elimination protopremises. Note that a protopremise is a class of premises, not a premise.

It will be observed that if $\Gamma \vdash A, \Delta$ is derivable, then there is a derivation of it whose bottom part consists of a weak introduction protopremise, each member of which is flanked by Γ on the left and Δ on the right, and concluding with $\Gamma \vdash A, \Delta$ using a sequence of replications of A. Likewise for $\Gamma, A \vdash \Delta$.

The introduction and elimination rules of Gentzen's calculi match each other in the sense that if, in an introduction rule, a certain component occurs on one side of the turnstile, there will be an elimination rule with that component on the other side of the turnstile. This may be stated generally for local systems. Let $(\Pi_i \vdash \Sigma_i)$ be an introduction protopremise for A. A *matching set* is a set of sets of premises $(\Pi'_j \vdash \Sigma'_j)_k$ with the feature that

$$\bigcup_i \Pi_i = \bigcup_{jk} \Sigma'_{jk} \qquad \bigcup_i \Sigma_i = \bigcup_{jk} \Pi'_{jk}$$

$$\bigcup_i \Pi_i \cap \bigcup_{jk} \Pi'_{jk} \subseteq \bigcap_i \Pi_i \cup \Sigma_i \qquad \bigcup_i \Sigma_i \cap \bigcup_{jk} \Pi'_{jk} \subseteq \bigcap_i \Pi_i \cup \Sigma_i$$

Local systems for which cut elimination is provable, and which have the subformula property, have the feature that for each introduction protopremise, there is a matching set of elimination protopremises. We call this the lemma on matching protopremises; the parallel result holds when we interchange 'introduction' and 'elimination'.

6. DEFINABILITY OF CONSTANTS

A *G-calculus* is a classical sequent calculus for which cut-elimination is provable, which has the subformula property, and which is equivalent to a calculus all of whose rules are local. In this and the following section when we speak of *G*-calculi we shall be referring to a local version, and rules in this version will be called *G*-rules. Among the philosophically useful features of *G*-calculi is a result about *definability*. Suppose there is a logical constant c which is defined in terms of other logical constants $c_1, ..., c_k$ by what Russell used to call implicit or contextual definition. That is, given any formula in which c occurs, there is a scheme for unpacking this formula into a scheme in which the only logical constants are $c_1, ..., c_k$. The definite description operator is the paradigm example. Our definability result states that if there are *G*-rules for $c_1, ..., c_k$, then there is a *G*-rule for c which does not involve those constants. In short, *an implicit definition can always be replaced by an explicit G-definition.*

An implicit definition is of the following form. Let there be $A_1, ..., A_r$ variables of categories $g_1, ..., g_r$ and let c be of category $(g_1, ..., g_r)/s$. Let the representation of c applied to the A's be Z, in which $A_1, ..., A_r$ occur as free variables except for those that are bound by c. Then Z is defined as Y, where Y is a representation of the result of some successive applications of the constants $c_1, ..., c_k$ to the A's. This definition is a schema for all definitions that result from replacing other expressions of the same categories for the free A's. We now construct a *G*-rule for c, assuming that there is at least one variable in the implicit definition. We merely consider protopremises for $\vdash Y$, then protopremises for sentences occurring in those protopremises and so on, until atomic variables are obtained. The resulting set of 'hyper-protopremises' for Y constitute the introduction protopremises for Z; likewise for the elimination protopremises. For example, let identity be defined as in Principia, with f a first-order quantifier, in Russell's sense of order: $x = y =_{df} (f)(fx = fy)$. The rules for identity are precisely those that would give the definiens.

$$\frac{\Gamma \vdash Xa, \Delta \quad \Gamma, Xb \vdash \Delta}{\Gamma, a = b \vdash \Delta} \qquad \frac{\Gamma \vdash Xb, \Delta \quad \Gamma, Xa \vdash \Delta}{\Gamma, a = b \vdash \Delta} \qquad \frac{\Gamma, Xa \vdash Xb, \Delta \quad \Gamma, Xb \vdash Xa, \Delta}{\Gamma \vdash a = b, \Delta}$$

or, in the local versions of the first elimination and the introduction rules:

$$\frac{\Gamma \vdash [X_i a], \Delta \; \Gamma, [X_i b] \vdash \Delta}{\Gamma, a = b \vdash \Delta} \qquad \frac{\{\Gamma, X_i a \vdash X_i b, \Delta\}_i \; \{\Gamma, X_i b \vdash X_i a, \Delta\}}{\Gamma \vdash a = b, \Delta}_i.$$

Throughout X is of first order and identity is of second order. In the 'Gentzen' version of the introduction rule, X must not occur free in the conclusion. The above procedure does not work if no free variable occurs in the definition, as occurs when the constant absurd proposition is defined as $F =_{\text{df}} (p) \, (p \& \sim p)$, and indeed rules often given for F do not satisfy our version of the subformula property.

There are many definable constants for which there do not exist Gentzen-style rules, but for which there are local rules, for example $(Qxy)(Fxy) =_{\text{df}} (x)(\exists y)Fxy$. Moreover using second-Russell-order ramified logic one may construct some branching quantifiers which, as is well known, cannot be done in first-order logic at all. Since the branching quantifiers do not seem to use any 'logical intuitions' significantly different from those required for other logical enterprises. We take this to support our contention that constants definable by G-rules constitute *the* class of logical constants.

At any rate we see that the class of logical constants characterized by G-rules has no second-class citizens. There is not a class of basic constants, in terms of which others are defined by implicit or contextual definition. Anything definable by contextual definition in terms of constants characterized by G-rules, can be characterized by its own G-rule.

7. Do-it-yourself semantics

Let there be an assignment of truth values to sentences of the prelogical language. Then truth values are assigned to complex formulae in a routine way based on the protopremises for the G-rules. Elimination and introduction protopremises for A will be represented by,

$$(\Pi_i' \vdash \Sigma_i') \qquad (\Pi_i \vdash \Sigma_i).$$

That is to say, $A \vdash$ can be derived from the former, and $\vdash A$ from the latter.

A is assigned the value t if and only if there is an introduction protopremise for A such that for each i, either some formula in Π_i is assigned f or some formula in Σ_i is assigned t.

By the lemma on matching protopremises, it follows that A is assigned f if and only if there is some elimination protopremise such that for every i, some Π_i' is assigned f or some Σ_i' is assigned t.

A statement $\Gamma \vdash \Delta$ is t under an assignment of values if some formula in Γ is f or some formula in Δ is t.

For example, consider the following rules for a 'new' connective denoted '?'.

$$\frac{\vdash A,B}{A?B\vdash} \qquad \frac{A\vdash}{\vdash A?B} \qquad \frac{B\vdash}{\vdash A?B}.$$

What does this new connective 'mean'? It is a connective that takes t if and only if one or the other of A or B is false, that is, it is Sheffer's connective. In this way the 'meaning' of logical constants can be read off-immediately from the G-rules that define them. For brevity I shall spare the reader the rule for the Russellian definite description operator, which has over a dozen premises. If I were simply to state the rules, without saying what they define, few readers would immediately understand what is defined–and yet by applying the Do-it-yourself semantics, the 'meaning' can be immediately worked out.

Soundness

If each member of a set of premises takes the value t, then any statement that can be derived from these will also take the value t under the same assignment of values. Certainly initial statements all take t, for the same formula will occur on each side of the turnstile. Now consider any step in a derivation from t premises, and suppose, by induction, that all the premises for the last step take t. If one of the side formulae Γ takes f, or one of the side formulae Δ takes t, then the conclusion of this step will take t. Otherwise consider the components (Π_i, Σ_i). If all the Γ are t and all the Δ are f, then for each i, one of Σ_i must be t or one of Π_i must be f. But since (Π_i, Σ_i) constitutes a protopremise for A, then if this is an introduction rule, A takes the value t, or, if an elimination rule, A takes the value f.

Some Definitions

$\Gamma \vdash \Delta$ is *valid* if it takes the value t under every assignment of truth values to sentences of the prelogical language. A set of statements is said to be *inconsistent* if there is a formula C such that there are two derivations, using as premises statements in the given set, and terminating in $C\vdash$ and $\vdash C$. Otherwise the set of statements is said to be *consistent*. Let $(\Pi_i \vdash \Sigma_i)$ be a protopremise for A. Then $(\Gamma, \Pi_i \vdash \Sigma_i, \Delta)$ is a *predecessor* of $\Gamma \vdash A$, Δ or Γ, $A \vdash \Delta$, according as we have an introduction or elimination protopremise. Evidently if a set of statements is consistent, so is a set of predecessors of that statement. Finally we extend the complexity ordering associated with the subformula property, so that it applies to sets of statements. A set T is *more complex* than a set S if there a homomorphism from S to T such that each member of S is less complex than the corresponding member of T. Evidently the predecessor of a statement is less complex than the statement itself.

LEMMA.   *If a set of statements is consistent, there is an assignment of values in which every member of the set takes t.*

This is shown by induction on the complexity of the set. The lemma holds for sentences of the prelogical language, so we suppose that complex formulae occur in statements in the set. Consider any set of predecessors for statements in the set with complex formulae, together with all non-complex statements in the set. Then this is a consistent set and by induction will take t under some interpretation. But if the statements that make up a predecessor of a given statement all take t, then the statement itself must take t.

LEMMA.   *Suppose that $A_1, ..., A_k \vdash B_1, ..., B_m$ is not derivable. Then the set consisting of $B_i \vdash$, $\vdash A_j$, for $i \leq k$ and $j \leq m$, is a consistent set.*

For suppose that we could derive both $C\vdash$ and $\vdash C$ from members of this set. Then mimic the two derivations, starting $B_1 \vdash B_1, ..., A_k \vdash A_k$ as premises, and derive, using thinning where necessary, both $A_1, ..., A_k$, $C \vdash B_1, ..., B_m$ and $A_1, ..., A_k \vdash C$, $B_1, ..., B_m$. Then, by applying cut, we contradict the assumption of non-derivability.

Completeness

If $A_1, ..., A_k \vdash B_1, ..., B_m$ is valid, it is derivable. For if it is not derivable, then the set of statements $(B_i \vdash, \vdash A_i)$, $i \leq k$, $j \leq m$, is consistent. Hence there is an assignment of values in which each B_i takes f and each A_j takes t, contradicting the assumption of validity.

Since the primitive constants and variables of each category constitute a countable domain, our first lemma amounts to a generalized Löwenheim-Skolem theorem for each category. That is, a consistent theory in a G-calculus has a model in which the domain of each category is countable.

This is not the occasion to develop ramified types in detail, but we note that the rules for ramified quantifiers as given in Prawitz (1965) are readily made local. Hence our results automatically provide a semantics for the ramified hierarchy. Some time ago David Kaplan proved an as yet unpublished completeness theorem for his ramified theory. My emphasis here is not on the existence of such a theorem, but on the fact that it is a direct corollary of the correct proof theoretic formulation of the theory. At first glance it may seem puzzling that ramified theory has a model in which there are only countably many attributes at every type, but this is to be expected. The ramified theory was to be what Russell called a 'no-class' theory, in which attributes did all the work of classes. If a Löwenheim-Skolem theorem holds for standard set theory, it must certainly hold for the ramified theory of types at every type.

Finitistic cut-elimination, the subformula property, and the existence of local variants define the class of calculi 'like' those invented by Gentzen. G-rules carry their own semantics with them. This confirms Gentzen's suggestion that his operational rules give the meaning of the logical constants: not just the inferential properties, but also the *meaning*, in any viable sense of that contentious word. As the *Tractatus* says at 6.126: "Without bothering about sense or meaning, we construct the logical proposition out of others using only *rules that deal with signs*". We have shown that the rules that deal with signs, and which construct *logical* predicates and propositions, are of a very special sort. Although they are syntactic, and deal strictly with signs, they absolutely determine the meanings of those signs.

Stanford University

BIBLIOGRAPHY

Ajdukiewicz, K.: 1935, 'Die syntaktische Konnexität', *Studia Philosophia* **I**, 1–27; translated in *Polish Logic* (ed. by S. McCall), Oxford 1967.

Church, A.: 1956, *An Introduction to Mathematical Logic*, Princeton.

Hacking, I.: 197A, 'Why Orthomodular Quantum Logic is Logic.'

Hacking, I.: 197B, 'On the Reality of Existence and Identity'.

Hacking, I.: 197C, 'What is Logic'?

Lakatos, I.: 1963, 'Proofs and Refutations', *British Journal for the Philosophy of Science* **XIV**, 1–25, 120–39, 221–45, **XV**, 296–342.

Prawitz, D.: 1965, *Natural Deduction*, Stockholm.

Prior, A.: 1960, 'The Runabout Inference Ticket', *Analysis* **XXI**, 38–9.

Schütte, K.: 1951, 'Beweistheoretische Erfassung der unendlichen Induktion in der Zahlentheorie', *Mathematische Annalen* **CXXII**, 369–89.

V. V. TSELISHCHEV

SOME PHILOSOPHICAL PROBLEMS OF HINTIKKA'S POSSIBLE WORLDS SEMANTICS

A paradoxical situation has taken place of late in discussion of the philosophical problems of quantified modal logic. On the one hand, the works by S. Kripke, J. Hintikka, and S. Kanger[1] have shown the possibility of constructing a satisfactory semantics for modal systems, the so-called possible worlds semantics. On the other hand, this formal satisfactoriness has left a lot of philosophical problems unsolved, the most significant of which have been suggested by W. V. Quine[2]. A number of recently proposed concepts are calculated to refute the philosophical objections of W. V. Quine, proceeding from semantic methods: the concept of individuating function, introduced by Hintikka, is the most important of them[3].

In the present paper some philosophical problems related to the notion of individuating function are put under consideration. Materials supplied to the author by the Institute of Philosophy, University of Helsinki, and by the Academy of Finland, as well as results of author's interviews with Prof. J. Hintikka have been used in writing the paper[4].

Quantification into modal contexts is possible only if semantic considerations include methods for individuating objects, or, in other words, methods for tracing individuals through several possible worlds. A meaningful discussion of the same individual as a member of different possible worlds is a necessary condition for such quantification. The concept of individuating function is closely related to the concept of world line, which connects the different 'roles' of the same individual in different possible worlds.

Individuals are picked out by functions from possible worlds to individuals. An individuating function indicates the same individual in different possible worlds in some acceptable sense of the identity of individuals.

Hintikka's semantics for modal logic is philosophically based in a sense. Because individuation is a very complicated and difficult procedure, it demands a philosophical explanation. The status of possible

Butts and Hintikka (eds.), Logic, Foundations of Mathematics and Computability Theory, 391–403.

worlds in epistemic logic is a very important component part of such an explanation.

A possible world is not to be understood metaphysically, as something different from the real one. It is simply another way to describe situations in the real world. Without possible descriptions it is impossible to grasp the workings of a conceptual scheme in the process of knowing the real world. Understanding language includes of necessity the admission of alternatives in the description of the world and of its counterfactual situations. In particular understanding a singular term implies not only knowing what object the term refers to, but also knowing what objects the term would refer to in different situations.

Thus, a possible world is conceived in modal logic not as an alternative to the real world, but as an alternative to some description of the real world.

Further, the notion of possible world is an ultimate and irreducible one. This circumstance results from an inevitable counter-factuality involved in the description of the real world: possible worlds semantics reflects in a sufficient degree this state of affairs by means of its technical formulation of the notions of possibility and necessity.

The transition from models to possible worlds semantics brings about an important change in the understanding of the relationship of a singular term and the object referred to. Quine insists on a separation of the theory of reference from the theory of meaning, as he thought that the understanding of a singular term includes knowing the object which is referred by the term as well as knowing the empirical procedures related to reference. Any meaning of a term surpassing this knowledge Quine considers a 'myth'. Possible worlds semantics eliminates the difference between reference and meaning, connecting both notions into one whole by means of the notion of individuating function. Understanding a singular term includes not only knowing the object referred to, but also knowing the way it is referred to and the rules governing their reference. These 'ways' or 'rules' are understood as functions from possible worlds to individuals. The fact is that these 'ways of reference' involve different situations, i.e. a singular term indicates different unrealized individuals in different possible worlds. The function determining this transition from a term to an individual is the meaning of the term.

Thus, understanding a singular term b is related to a function f_b which picks out individuals, indicated by the term, in different possible worlds.

The arguments of this function are determined by the whole of the possible world from which the individual in question is picked out. The specification of the function's arguments can, of course, be narrowed by an analysis of the relevant situation, but in principle the argument cannot include anything beyond the treatment of the situations in which the corresponding singular term can be used.

As an individuating function is intensional by its very nature, it can be identified with the *sense* of a term, in Frege's sense, or with an individual concept of R. Carnap. The explication of intensional concepts in terms of individuating functions, presented in Hintikka's works, proceeds from the assumption that because intensional entities are abstract ones, they are to be presented logically as individuals. This very abstractness makes one to look for ways of referring rather than merely for an individual referred to, when having to understand the term, because such abstractness implies the need to consider more than one situation concerning the real world.

Quantification over a set of individuating functions gives rise to a number of serious problems in the light of Quine's criterion of ontological commitment. The concept of individuating function is a rather peculiar one. On the one hand, it is an intensional entity; on the other hand, it picks out ordinary individuals referred to by a singular term. The double nature of the individuating function is closely related to the problem of quantification into modal contexts. Quantification makes sense only if the same individual can be traced through several possible worlds, i.e. if the world line of an individual can be extended through possible worlds. It is precisely by means of individuating functions that an identification procedure of the appropriate kind is ensured. The availability of a function or a world line permits us to quantify into modal contexts by presenting to us an authentic individual.

A very important question arises in this connection. What is the value of a quantified variable? On the one hand, individuating functions are the very entities which constitute the range of quantifiers. Because these functions are intensional entities, the possibility of quantifying in must be guaranteed through commitment to an intensional ontology. On the other hand, the functions being ordinary objects, traced through possible worlds, one has to agree that the ordinary objects are what is being quantified over, and, hence, an ontology of ordinary objects should be accepted. Accepting an individuating function as a part of an ontology is, according to the criterion, at variance with the intended interpretation of Hintikka. Admitting the objectivity of individuating functions, he first considered them not as ontological entities, but as a part of an ideology, i.e. as a part of a conceptual apparatus, intended for the treatment of courses of events different from the real one[5]. But later on Hintikka gave up understanding individuating functions as a part of an ideology and presented no definite account of their nature.

One can propose two more accurate versions, I think, which comply with some statements by Hintikka himself. One of them is to a considerable degree determined by a wish to preserve the significance of quantification into modal contexts.

Individuating functions serve to answer the question as to how the quantification of a variable within the scope of a modal operator which ranges over certain entities is possible at all. Thus, values of a bound variable are to be thought of as individuating functions, when we are speaking about the concept of quantification, about its possibility. A condition for quantification, conceived of in the usual sense, is the functional character of the reference by a term to an individual.

The question 'how' obviously does not require ontological specification, as it is in fact related to ways of describing the real world. However, the notion of individuating function is necessary for the explication of quantification (this latter notion being a primitive one) and employed for constructing a conceptual scheme. This makes it clear why it makes no sense to relate functions to conceptual contents, which are assumed to be inferred in a sense. The primacy of individuating functions is an essential feature of this situation. Whether the primary notions of epistemological analysis form a part of conceptual contents, is partly a matter of terminology. But in most cases the discussion of a conceptual scheme can be reduced to a discussion of possible worlds. The possibility of such a discussion lies in the definite use of language for describing the world. It is clear that questions about the existence of objects are secondary ones when one is investigating the ways of describing the world. In this case ontological problems turn out to be secondary insofar as ontology is affected by the totality of possible worlds, which are relativized to an individual person.

Another clarification is a more important one. Hintikka blames the paradoxes of modal logic on quantificational rules and not on the concept of identity. Further, the concept of existential quantifier comprises two seemingly different ideas: existence in a particular world and identity in several possible worlds. Hence, the 'ontology-ideology' dichotomy should be replaced by the dichotomy 'ontological vs. functional' in the use of a quantifier. To be more exact, the interpretation of quantifiers determine, in our opinion, the use of the concept of existence or the use of the concept of identity in different possible worlds. We require an ontological language when "we insist on quantifying over normal, down-to-earth sorts of individuals in the normal 'objectual' sense of quantification"[6]. The choice of objectual vs substitutional interpretation of quantification is of great importance for clarification of the same paradoxes of quantification in the modal contexts. The difference between ontological and functional interpretation of quantifiers does not coincide with the difference between objectual and substitutional quantification, but it is closely related to it. It promises interesting prospects for further investigations.

There are also other ontological considerations, to the effect that individuating functions do not affect one's ontology as determined by Quine's criterion.

Hintikka has also results on the differences between two cross-identification methods. Two kinds of quantification can be defined on the same universe of discourse, the so-called descriptive quantification and the perceptual quantification, respectively. These two kinds of quantifiers differ not in the individuals they run over in any particular world, but in relying on different methods of cross-identification of individuals through possible worlds.

Descriptive identification makes use of cross-identification by means of the similarity in the properties and relations that an individual has in different possible worlds, i.e. by means of the sameness of the roles the individual in question is to play in a totality of possible worlds. Perceptual quantification employs cross-identification of individuals by means of perceptual data (sense-data). We do not intend to produce here arguments to the effect that in counter-factual situations the usual descriptive and the perceptual way of reference to individuals result in different individuals in different possible worlds.

Hintikka's application of the notions of cross-identification and world-lines to the analysis of perceptual conceptions gives rise to a number of problems. Let us consider, in particular, the reconstruction of the notion of sense-datum by means of perceptual cross-identification. Hintikka's presupposition is that sense-data are hypostatizations of lines of perceptual cross-identification; and there are two reasons, in his opinion, why they are not ontological entities. Firstly, quantification over world lines does not mean commitment to an ontology of intensional entities, for the corresponding individuating functions do not contribute to ontology, but to ideology. Secondly, perceptual quantification differs from a descriptive one not with respect to ontology, but by the way of cross-identification. Both reasons entail consequences of some difficulty.

Consider the argument against the hypostatization of sensedata. It should be noted here that the thesis that quantification over world lines does not commit us to an intensional ontology concerns both perceptual and descriptive world lines, while the sense-datum problem is a matter of perceptual world lines only. The importance of the difference can be seen from the fact that the values of variables in an ontological sense for descriptive lines are ordinary objects considered as members of possible worlds, i.e. considered as points on a world line. The recognition of points on a similar perceptual line results, in Hintikka's view, in a hypostatization of sense-data as ontological entities. But, if the argument against sense-data proceeds from the first reason, then we have to admit points on the corresponding world lines as values of variables in the case of perceptual quantification.

Two logically admissible circumstances determine the specification of points on perceptual world lines. On the one hand, they must be ordinary individuals, their only distinctive feature consisting in a perceptual world line extended through them. On the other hand, the specification of these points can be determined by veritable supplementary conditions in addition to the requirements that they be ordinary individuals. The semantical nature of these supplementary conditions does not matter here.

Let us consider the problem of how to bring an ontology to light by picking out points on a world line, assuming that the specification of points on perceptual world lines is determined by the supplementary conditions just mentioned. Let this point be ontologically an ordinary individual. The specification of the supplementary conditions needed for picking out the points in question is not related to its ontological status. This assumption means admitting hypothetical cases in which an ordinary individual is a cross-point of two lines, a perceptual one and a descriptive one. We shall pronounce on the meaning of the consequence later.

Let the ontological status of a point on a perceptual line be determined to some degree by the supplementary conditions. In this case we have to abandon the simple ontology of ordinary individuals in favour of a mixed one, with sense-data added. Hence one cannot imagine a cross-point of the two kinds of world lines. But this variant cannot be acceptable to Hintikka, for it takes us back to the hypostatization of sense-data. Thus, if the values of perceptual quantification variables are correlated with the ontology in the same way as in the case of descriptive quantification, the rejection of sense-data is obtained at the cost of an important consequence: an ordinary individual must be assumed to be cross-points of the two types of world lines.

But it is difficult to see conditions for recognizing such a cross-point. Indeed, what kind of individual is a cross-point of two lines? Individuation itself is surely a cross-indentification, hence knowing 'this individual' presupposes knowing a corresponding line. Even if an individual happens to be a cross-point, the primacy of world lines prevents us from seeing the sense of this event.

It is for this very reason that the problem of sense-data reconstruction should be considered only as a special problem concerning perceptual quantification.

The perceptual world lines intended to represent sense-data should be considered in this case as primary entities in relation to the corresponding individuals.

In other words, sense-data are rejected because they are hypostatizations of lines of perceptual quantification over ordinary individuals;

the whole range of ontological problems is then determined by the primacy of perceptual lines. In this case we do have a complication of our ontology even if we do not consider quantification over corresponding individuating functions as contributing to ontology and assume only an ontology of ordinary objects. Indeed, Hintikka observes more than once that the extension of a perceptual line through possible worlds does not permit us to trace it back to the real world (at least, it is a difficult problem, for the relationship of the real world and possible one is of causal nature here). That is why an individual of the real world, related causally to perceptually cross-identified individuals, differs ontologically from these individuals. At least such an interpretation is to be found in Hintikka's explication of *Dinge-an-sich*.

The main contribution of R. Montague and J. Hintikka is that a rule for the transition from a term to an object is by its very nature functional. The point is that the rule is expressed by a function, which picks out an object depending on the circumstances. In order to understand a singular term, for instance b, we have to understand a function f_b picking out objects in different possible worlds.

Thus, an individuating function is associated with a singular term. But there is a difficulty in understanding the relationship of the function and term. From the standpoint of possible world semantics, one of the reasons why the rule for substitution of identicals in modal contexts is violated is that an object can have different names in different possible worlds. But if a function 'tracing' the world line of an individual is rigidly associated with a term, for instance b, then any object has the same name in all possible worlds. The very situation which allows for violations of the rule for substitution of identicals becomes inconceivable.

To get out of the difficulty, we have to assume one of the following possibilities:

An individuating function is an example of what Kripke called 'rigid designator', i.e. the individual has the same name in all possible worlds.

Individuating functions do not specify well-defined individuals in most cases.

Both these possibilities make sense only if a singular term associated with the function is a proper name.

Another possibility consists in the refusal to present a singular term as a proper name and in replacing it by a description.

Finally, the difficulty can be avoided if a function is taken to be associated with an individual rather than with a term. This variant will be discussed later on, while at present we shall concentrate on the three earlier ones.

Let us now consider the first variant, using the notion of rigid designator. We do not undertake here a detailed discussion of whether this notion can be used as a permissible one in explaining the nature of modalities. Our point is in the present case that the idea underlying rigid designators is not appropriate for associating it with an individuating function. This is well argued by Hintikka himself in 'Quine on Quantifying In'. Let us, therefore, consider another variant which is related to the notion of a well-defined individual.

Not every singular term, for instance a, picks out in possible worlds manifestations of the same individual. Such terms do not specify well-defined individuals and so are not liable to existential generalization.

Proceeding from an arbitrary correlation between possible worlds one cannot imagine any sufficient number of well-defined definite individuals. In fact, it is not very plausible that in different possible worlds the same individual should have the same name, as the very notion of possibility proposes for discussion different names for the same individual. The situation does not change radically if a transition takes place from arbitrary relations between worlds (as is the case with alethic modalities) to very natural relations between individuals in different worlds in the case of epistemic modalities. Even if one gets rid of the purely occasional character of alethic modalities, a host of epistemic inadequacies comes to replace it (delusion, false belief, false knowledge, etc.).

Hence, in that sense, well-defined individuals are a rarity. This reminds us of an interesting parallel with the genuine proper names of Bertrand Russell, who reduced the class of proper names to the names of egocentric particulars. One is right in saying that the class of well-defined individuals is reducible to such entities as can deserve only proper names.

But the resemblance with Russell is incomplete. Genuine proper names were reserved by Russell for sense-data, known by means of knowledge by acquaintance. From the point of view of possible world semantics, however, knowledge by acquaintance is knowledge in a totality of epistemically possible worlds, in which the tracing of individuals is based on perceptual individuation. But insofar as difference between sense-data, as they are understood by Russell, and sense-data required by possible world semantics consists not in a difference between different kinds of entities, but in a difference in our ways of tracing individuals, insofar the problems of naming are of no concern here. At any rate, the presentation of a singular term as a proper name does not help us to understand why violation of substitutivity for identicals occurs, if the term is provided with an individuating function. There is only one way out, and that is to admit that a singular term can only be presented by a description.

Let us now consider questions related to an introductory function's mechanism of operation. The first remark concerning the mechanism of such a function's operation is this. The functional character of reference means the primacy of function in the process of reference. To specify the rule, according to which we 'come' from a term to the object referred, is to specify the function which picks for us an object depending on the circumstances. The essence of possible worlds semantics consists exactly in the insights that some arguments of individuating functions are necessarily counter-factual, i.e. are among unrealized states of affairs. These arguments are simply not to be found in the real world.

The process of seeing, for example, takes place in the real world only. Therefore, we cannot begin with arguments which may happen not to be found in the real world, and then look for objects using a suitable function. We begin, on the contrary, with the function and consider it to be primary, not derivative. It is precisely in this that the objectivity of possible worlds and individuating functions consists.

Turning our attention to the mechanism of the operation of a function we can describe the following process. The arguments of these functions are in a totality of possible worlds. The function 'looks through' possible worlds with a view of founding a suitable argument. Our attention is diverted in this case from the function itself to the domain of its definition.

The second remark on the mechanism of the function's operation is concerned with a particular kind of function. This kind depends on many factors, but basically it depends on the possible world as a whole in which the corresponding arguments are situated. Of course, a detailed analysis permits us to simplify in each case the kind of dependence a function exhibits, but taken as a whole this dependence is very complicated. Here our attention proceeds from a possible world to a function.

Thus, we cannot hope for a simple expression for an individuating function, because the domain of its definition and the particular kind of function it is are closely interrelated, as is seen from the two remarks above on the mechanism of a function's operation. Though we begin with a function, we do not know its particular kind, while the arguments of the function, discovered by a particular function, determine this particular kind through their particular kind.

The third remark on the mechanism of a function's operation has to do with whether a particular kind of function is an algorithm. It is natural enough to assume that the function provides for a connection of different individual's names in different possible worlds, and that it is an algorithm for the searching of these names.

Violations of the rule for substitutivity of indenticals and existential generalization when quantifying into modal contexts are caused by violations of the correlation between individuals and names that takes place in different possible worlds. A general consideration of this correlation's failure can result in a false conception of an individuating function's role. Indeed, the violation can arise when the same individual has different names in different possible worlds as well as when different objects are picked out by the same term in different possible worlds.

If the first situation is taken as a basis for the introduction of individuating functions, the impression may arise that an individuating function is an algorithm, providing for the correlation of an individual with different names in different possible worlds.

Such an interpretation is wrong for the following reasons. The idea of reference to an individual by means of a term together with the related idea of alternativity of worlds (and also the sense of a term as a way of referring) should be replaced in a sense by a reverse correlation.

In the same process the same individual acquires different names in different possible worlds. Thus, the individuating function is related not to the reference of an individual by means of a term, but to an association of an individual with names, or else the function is related not to a term which is to be understood, but with an object which exists in a metaphysical sense. Should this be the case, the individuating function would play the role of postulating a certain metaphysical universe of possible worlds. Furthermore, this would complicate the interpretation of the possible world semantics, for we have to do with terms referring to objects, and not the other way round. The asymmetry meant here is of considerable importance. The point is that the two interpretations of an individuating function are closely related to the acknowledgement of the status of a possible world as a metaphysical one or else as an epistemic one. One can easily imagine empirical situations where the same object can acquire different names. Although these problems deserve attention, they are not relevant here. It is the epistemic situation that is under discussion.

We also have to comment on certain shortcomings in the semantic treatment of individuating functions. A treatment of functions as abstract mathematical entities does not contribute to our understanding of individuating functions as meanings of language terms. The use of language (situations of reference is what is under consideration here) means the presence of some algorithm for discovering the objects which are indicated by these terms. A general discussion of these functions has shown that they cannot be such algorithms, for this particular kind of function depends on the whole possible world.

However, not the whole world, but only its fragments are taken into considerations in practice. Only future investigation would allow to determine the 'algorithmicity' degree of the functions in this case.

U.S.S.R. Academy of Sciences, Novosibirsk

NOTES

[i] S. Kripke, 'Semantical Considerations on Modal Logics', *Acta Philosophica Fennica* **16** (1963) 83–94; S. Kanger, *Provability in Logic*, Stockholm Studies in Philosophy, vol. I, Stockholm, 1957; J. Hintikka, 'Modality as Referential Multiplicity' *Ajatus* **20** (1957) 49–64.

[2] W. V. O. Quine, 'Three Grades of Modal Involvement' *The Ways of Paradox and Other Essays*, N.Y., 1966.

[3] J. Hintikka, 'The Semantics of Modal Notions and the Indeterminacy of Ontology' *Synthese* **21** (1970), 708–727.

[4] I am very grateful to Prof. Hintikka for his permission to use results unpublished at the time, especially 'Quine on Quantifying In' and 'Carnap's Semantics in Retrospect'. Later these papers appeared as Chapters 5 and 6 of Hintikka's *Intentions of Intentionality and Other New Models for Modalities*, D. Reidel, Dordrecht-Holland, 1975. I am also grateful for Prof. Hintikka's valuable suggestions during my visit to the Institute of Philosophy, University of Helsinki, summer 1974.

[5] J. Hintikka, 'On the Logic of Perception' *Models for Modalities*, D. Reidel, Dordrecht-Holland, 1969, p. 179.

[6] J. Hintikka, 'Carnap's Heritage in Logical Semantics' p. 91 (note 4 above).

INDEX OF NAMES*

---

* This index does not list names included in bibliographies following papers.

THE UNIVERSITY OF WESTERN ONTARIO SERIES IN PHILOSOPHY OF SCIENCE

A Series of Books on Philosophy of Science, Methodology, and Epistemology
published in connection with
the University of Western Ontario Philosophy of Science Programme

Managing Editor:

J. J. LEACH

Editorial Board:

1. J. Leach, R. Butts, and G. Pearce (eds.), *Science, Decision and Value.* Proceedings of the Fifth University of Western Ontario Philosophy Colloquium, 1969. 1973, vii + 213 pp.
2. C. A. Hooker (ed.), *Contemporary Research in the Foundations and Philosophy of Quantum Theory.* Proceedings of a Conference held at the University of Western Ontario, London, Canada, 1973, xx + 385 pp.
3. J. Bub, *The Interpretation of Quantum Mechanics.* 1974, ix + 155 pp.
4. D. Hockney, W. Harper, and B. Freed (eds.), *Contemporary Research in Philosophical Logic and Linguistic Semantics.* Proceedings of a Conference held at the University of Western Ontario, London, Canada. 1975, vii + 332 pp.
5. C. A. Hooker (ed.), *The Logico-Algebraic Approach to Quantum Mechanics.* 1975, xv + 607 pp.
6. W. L. Harper and C. A. Hooker (eds.), *Foundations of Probability Theory, Statistical Inference, and Statistical Theories of Science.* 1976, three volumes: xii + 308 pp.; xii + 455 pp.; xii + 241 pp.